Remote Sensing and Image Interpretation

Remote Sensing and Image Interpretation

Edited by Catherine Williamson

SYRAWOOD
PUBLISHING HOUSE
New York

Published by Syrawood Publishing House,
750 Third Avenue, 9th Floor,
New York, NY 10017, USA
www.syrawoodpublishinghouse.com

Remote Sensing and Image Interpretation
Edited by Catherine Williamson

International Standard Book Number: 978-1-64740-403-1 (Hardback)

Cataloging-in-publication Data

Remote sensing and image interpretation / edited by Catherine Williamson.
p. cm.
Includes bibliographical references and index.
ISBN 978-1-64740-403-1
1. Remote sensing. 2. Remote sensing--Analysis. 3. Remote-sensing images. I. Williamson, Catherine.
G70.4 .R46 2023
621.367 8--dc23

TABLE OF CONTENTS

PREFACE

Over the recent decade, advancements and applications have progressed exponentially. This has led to the increased interest in this field and projects are being conducted to enhance knowledge. The main objective of this book is to present some of the critical challenges and provide insights into possible solutions. This book will answer the varied questions that arise in the field and also provide an increased scope for furthering studies.

Remote sensing is a type of geospatial technology that is used to acquire information about an object or phenomenon without making any physical contact. Some major advantages of remote sensing include observation and monitoring of surface area and objects, and collecting information over large spatial areas. It is used to characterize physical objects or natural features on the ground. Data collected through remote sensing is used with other available data, which helps to facilitate decision-making. Remote sensing imagery must be interpreted in order to derive meaningful information. Image interpretation involves identifying and measuring various targets (features or objects) in an image. Before image analysis, the images can also be processed digitally. The results of image interpretation can be used for various applications such as classifying and categorizing vegetation and water management. This book unravels the recent studies on remote sensing and image interpretation. It will serve as a valuable source of reference for graduate and postgraduate students.

I hope that this book, with its visionary approach, will be a valuable addition and will promote interest among readers. Each of the authors has provided their extraordinary competence in their specific fields by providing different perspectives as they come from diverse nations and regions. I thank them for their contributions.

Editor

1

Multi-Scale Semantic Segmentation and Spatial Relationship Recognition of Remote Sensing Images Based on an Attention Model

Wei Cui *, Fei Wang, Xin He, Dongyou Zhang, Xuxiang Xu, Meng Yao, Ziwei Wang and Jiejun Huang

School of Resources and Environmental Engineering, Wuhan University of Technology, Wuhan 430070, China; flyking@whut.edu.cn (F.W.); 2962575697@whut.edu.cn (X.H.); gis@whut.edu.cn (D.Z.); xxx1049731704209@whut.edu.cn (X.X.); yaomeng@whut.edu.cn (M.Y.); zwei@whut.edu.cn (Z.W.); hjj@whut.edu.cn (J.H.)

* Correspondence: cuiwei@whut.edu.cn

Abstract: A comprehensive interpretation of remote sensing images involves not only remote sensing object recognition but also the recognition of spatial relations between objects. Especially in the case of different objects with the same spectrum, the spatial relationship can help interpret remote sensing objects more accurately. Compared with traditional remote sensing object recognition methods, deep learning has the advantages of high accuracy and strong generalizability regarding scene classification and semantic segmentation. However, it is difficult to simultaneously recognize remote sensing objects and their spatial relationship from end-to-end only relying on present deep learning networks. To address this problem, we propose a multi-scale remote sensing image interpretation network, called the MSRIN. The architecture of the MSRIN is a parallel deep neural network based on a fully convolutional network (FCN), a U-Net, and a long short-term memory network (LSTM). The MSRIN recognizes remote sensing objects and their spatial relationship through three processes. First, the MSRIN defines a multi-scale remote sensing image caption strategy and simultaneously segments the same image using the FCN and U-Net on different spatial scales so that a two-scale hierarchy is formed. The output of the FCN and U-Net are masked to obtain the location and boundaries of remote sensing objects. Second, using an attention-based LSTM, the remote sensing image captions include the remote sensing objects (nouns) and their spatial relationships described with natural language. Finally, we designed a remote sensing object recognition and correction mechanism to build the relationship between nouns in captions and object mask graphs using an attention weight matrix to transfer the spatial relationship from captions to objects mask graphs. In other words, the MSRIN simultaneously realizes the semantic segmentation of the remote sensing objects and their spatial relationship identification end-to-end. Experimental results demonstrated that the matching rate between samples and the mask graph increased by 67.37 percentage points, and the matching rate between nouns and the mask graph increased by 41.78 percentage points compared to before correction. The proposed MSRIN has achieved remarkable results.

Keywords: multi-scale; semantic segmentation; image caption; remote sensing; LSTM; U-Net; upscaling; downscaling

1. Introduction

Deep neural networks [1,2] are gradually being applied to high-resolution remote sensing image analysis [3], especially in scene classification [4–9], semantic segmentation [10], or single-scale remote sensing object recognition [11,12], and they all have achieved good results. Unfortunately, most of the existing studies do not address the interpretation of spatial relationships between remote sensing objects, which limits the understanding of remote sensing objects, especially when the phenomenon of different objects with the same spectrum in remote sensing appears.

The phenomenon of different objects with the same spectrum in remote sensing is quite common. It is difficult to identify objects only by their own textures, spectra, and shape information. Object identification requires multi-scale semantic information and spatially adjacent objects to assist in decision-making. The spatial relationship between remote sensing objects is of great significance to the recognition of remote sensing objects when different objects have the same spectrum, for example, many different types of buildings with similar shapes and spectral features, such as commercial buildings and workshops. The traditional object recognition methods [13–15] can only identify the object by its spectral, texture, and shape features without considering its adjacent objects. Therefore, it is impossible to accurately distinguish the different objects with the same spectrum without additional information. However, commercial buildings are often adjacent to green spaces and squares, and workshops are more adjacent to other factories and warehouses. In this way, it is possible to effectively identify commercial buildings and workshops through adjacent object categories.

According to existing research, scene classification describes the entire patch of the sample but does not involve remote sensing objects. Although semantic segmentation can identify the location and boundaries of remote sensing objects, it does not include the interpretation of complex spatial relationships between remote sensing objects, which leads to a certain degree of incomplete semantic understanding of remote sensing images. How to carry out a comprehensive semantic description of remote sensing objects and their spatial relationships is an issue that still needs further study.

The prosperity of image captions based on recurrent neural networks (RNNs) [16], especially the attention-based LSTM [17], can provide not only image description but also the attention location corresponding to the currently generated word at different time steps, which provides a new way to address the problems above. Chen et al. proposed a novel group-based image captioning scheme (termed GroupCap) [18], which jointly models the structured relevance and diversity among group images towards an optimal collaborative captioning. Previous works only used the global or local image feature. A model with 3-Gated model [19] was proposed to fuse the global and local image features together for the task of image captioning. In recent years, more studies have focused on the relationship between generated words and corresponding regions in the image. An attribute-driven attention model [20] was proposed to focus on training a good attribute-inference model via the RNN for image captioning. The uniqueness of the model lied in the usage of an RNN with the visual attention mechanism to observe the images before generating captions. Khademi et al. presented a novel context-aware, attention-based deep architecture [21] that employed a bidirectional grid LSTM for image captioning. The bidirectional grid LSTM took visual features of an image as the input and learned complex spatial patterns based on two-dimensional context.

In recent years, the application of reinforcement learning [20–23] in image caption has also been a hot topic, which adjusts the generation strategies using the change of the reward functions in the caption generation process to dynamic vocabulary generation.

However, most of the current studies focus on the scene semantic description of ordinary digital images [24,25]. To use the deep RNN or LSTM to execute the semantic analysis [26–30] of remote sensing objects, the following problems must be solved:

Location ambiguity: At different time steps, the attention mechanism is based on 14×14-sized image features and corresponds to 196 spatial locations in remote sensing images. There are some deviations [31], however, that limit the application in remote sensing object recognition.

Boundary ambiguity: the nouns (label of objects) in captions cannot accurately segment the boundaries of remote sensing objects in an image; thus, it is impossible to identify the spatial relationship between the objects.

Spatial scale ambiguity: Everything is related to everything else, but near things are more related to each other [32]. The surroundings of objects are various, which makes it difficult to detect remote sensing objects using a uniform scale model. Sometimes we need a large scale to contain the neighboring and context information to identify remote sensing objects accurately.

To solve the above problems, we present the MSRIN, which is based on an FCN, a U-net, and an attention-based LSTM. The MSRIN can generate remote sensing image descriptions at multi-spatial scales, segment objects in images, and recognize their spatial relationships from end-to-end. First, a remote sensing image is semantically segmented through an FCN and a U-net on two spatial scales such that each pixel in the original image is labelled with two semantic labels; therefore, a hierarchical relationship of a multi-scale remote sensing object can be formed. Second, the features of the same image obtained using a pre-trained Visual Geometry Group 19 (VGG-19) network are input for the attention-based LSTM, which outputs the captions that describe the two-scale remote sensing objects and their spatial relationships. Finally, the relationship between the nouns in the caption and the object mask graphs is established through the attention weight matrix. In this way, the remote sensing objects from the U-Net get their spatial relationship from captions. To overcome the spatial deviations of the attention weight matrix from the LSTM, the MSRIN designs an attention-based multi-scale remote sensing object identification and correction mechanism. Our method produces a complete semantic interpretation of remote sensing images.

In summary, the main contributions of this paper are as follows:

1. A multi-scale semantic caption strategy is proposed. Based on this strategy, a parallel network (the MSRIN) is designed to completely interpret the semantic information of remote sensing images.
2. We discuss the remote sensing object recognition and correction mechanism based on the attention weight matrix and multi-scale semantic segmentation using the FCN and the U-Net, simultaneously realizing the instance segmentation of the remote sensing images and the spatial relationship identification from end-to-end.

The remainder of this paper is organized as follows: Section 2 discusses related work. Multi-scale semantic segmentation and spatial relationship recognition of remote sensing images based on an attention model is presented in Section 3. Experiments and analysis are listed in Section 4. Discussion is presented in Section 5. Finally, the conclusion is presented in Section 6.

2. Related Work

2.1. Scene Classification

Because of the GPU memory limitations, high-resolution imagery must be segmented into patches for Convolutional Neural Networks (CNN) models, and the label is always attached to a remote sensing image sample [8,10,33–35] in scene classifications. To manage and retrieve patches easily, the previous research studies modified the structure of traditional deep convolution neural networks into two different forms, i.e., cascade and parallel models, according to the characteristics of remote sensing images [36]. In the cascade model, the corresponding layer structure in a traditional CNN is transformed to reduce the total parameters. For example, a global average pooling layer is used to replace the fully connected network as the classifier [10] or to insert a region-based cascade pooling (RBCP) method between the last normal down-sampling layer and the classifier to aggregate convolutional features from both the pre-trained and the fine-tuned convolutional neural networks [36]. Parallel models try to extract more abundant features for scene classification by designing parallel network structures [35]. All of the above methods achieved satisfactory results. However, the interpretation of remote sensing images more than meets the needs of spatial operations of geographical objects; therefore, it is not enough to obtain only the labels of remote sensing image patches.

2.2. Semantic Segmentation

Semantic segmentation algorithms assign a label to every pixel in an image and are the basis of instance segmentation. The research includes two cases: CNN series and RNN series.

2.2.1. CNN Series

A CNN is used not only to label the samples but also to classify the pixels to achieve semantic segmentation [37]. Meanwhile, semantic segmentation has been applied to the remote sensing recognition of buildings and other objects [38,39]. High-resolution imagery must be segmented into patches for CNNs due to Graphics Processing Unit (GPU) memory limitations, thus in a limited area, to make full use of the output features of different convolution layers to achieve a better semantic segmentation effect, the researchers often use a multi-depth network model [40] or design a multiple-feature reuse network in which each layer is connected to all the subsequent layers of the same size, enabling the direct use of the hierarchical features in each layer [41]. Emerging new networks, such as U-Net [42] and DenseNet [43], have also been applied in remote sensing image semantic segmentation [44]. The application scenario also extends from surface geographic objects to continuous phenomena such as highly dynamic clouds [45]. Some studies introduce the attention mechanism [46,47] to achieve an ideal segmentation effect by suppressing low-level features and noise through high-level features.

In general, semantic segmentation of remote sensing images based on a CNN has been developed from a simple transplanting network structure to the design of a creative network structure [48] according to the characteristics of remote sensing and has achieved good results. Application scopes are expanded from building extracting [49], built-up area extracting [48], and mapping impervious surfaces [50] to oil palm tree detection [51].

2.2.2. RNN Series

RNNs [52] are an important branch of the deep learning family, which are widely used for sequence analysis. In hyperspectral remote sensing images, there are tens of hundreds of spectral bands, which can be regarded as a related spectral sequence. Therefore, an RNN is proposed for hyperspectral image classification and achieves excellent classification performance [53,54]. With further research, the RNN model, which takes both spatial and spectral features into account [55], has also been applied to hyperspectral remote sensing semantic segmentation. In this way, the comprehensive utilization of spectral-spatial information is realized and good results are achieved. Another method is to input bands information into an LSTM network as an l-length sequence [56], which also achieves good results. The ability of an RNN to process time series data can also be used to process synthetic aperture radar (SAR) images [57]. The fine structure of SAR images can be retained as much as possible by filtering noise from multi-temporal radar images.

Most of researchers regard the spectral features of each individual pixel as one sequential feature for the RNN input layer. Recently, a novel strategy for constructing sequential features is proposed [58], and similar pixels collected from the entire image are used to construct the respective sequential features and the strategy achieves significant improvements in classification.

2.3. Remote Sensing Image Captioning

Remote sensing image captioning aims to generate comprehensive captions that summarize the image content at a semantic level [26]. Relevant research originated from natural language descriptions of images [16,24] in the field of computing. More recently, attention-based LSTMs [17] have emerged to describe not only the semantic information of images but also the image region corresponding to the words generated at the current moment through a 14×14 weight matrix. Because the image captions and image features are input into the RNN at the same time, there is a debate about whether it focuses on captions or on images at each time step (i.e., at each time step the model decides whether to attend to the image or to the visual sentinel) [25]. Although image captioning of ordinary digital images has achieved good results, it has encountered many difficulties in remote sensing fields. Compared with ordinary digital images, the remote sensing images from satellites or aircraft have a unique "view of God," which makes the remote sensing images have no directional distinction and lack a focused object or centre. All of these factors increase the difficulty of obtaining natural language descriptions of remote sensing images. Despite these difficulties, some researchers have made useful advances. Qu et al. [26] used an RNN to describe remote sensing images using a natural language. Shi et al. [27] proposed a remote sensing image captioning framework that leverages the techniques of a convolutional neural network (CNN). Both methods used a CNN to represent the image and to generate the corresponding captions from recurrent neural networks or pre-defined templates. To better describe the remote sensing images, and after a comprehensive analysis of scale ambiguity, category ambiguity and rotation ambiguity, a large-scale benchmark dataset of remote sensing images is presented to advance the task of remote sensing image captioning [28]. Wang et al. [29] used semantic embedding to measure the image representation and the caption representation. The captioning performance is based on CNNs, and the authors regarded caption generation task as a latent semantic embedding task, which can be solved via matrix learning. Zang et al. [30] presented a new model with an attribute attention mechanism for the description generation of remote sensing images by introducing the attributes from the fully connected layer of CNN, where the attention mechanism perceives the whole image while knowing the correspondence between regions and words, and the proposed framework achieves robust performance. Then, various image representations and caption generation methods were tested and evaluated. This work made a great step forward in the research on remote sensing image captions.

Although research on remote sensing image captioning has recently made some achievements, there are still some limitations, such as words in image captions that cannot correspond to remote sensing objects one by one, and relatively weak descriptions of the spatial relationships. In particular, the image region corresponding to the attention weight matrix often does not match the remote sensing object corresponding to the word at the same time step [31]. To better understand the semantic information of remote sensing images, further research is still needed.

3. Methodology

The MSRIN was defined first as a multi-scale remote sensing image caption strategy. A parallel network structure was designed to identify multi-scale remote sensing objects and spatial relationships based on attention.

3.1. *Strategy of Multi-Scale Caption Design*

According to Tobler's first law of geography, everything is related to everything else, but near things are more related to each other [32]. We propose a strategy for multi-scale captioning:

1. Each caption consists of small-scale remote sensing objects and their spatial relationships, which implicitly constitute a large-scale object, as shown in Figure 1 and Table 1.
2. Usually, one object is selected as the main object in a small-scale image caption, while other objects are subordinate to it through spatial relationships. In this way, each class of small-scale objects will not repeat within one large-scale object.
3. If there are two or more large-scale objects, the corresponding number of captions are joined by the word "with."

Table 1 shows the multi-scale classification system in our experiment. Large-scale and small-scale categories are encoded in 1X and 2x respectively, where 1 and 2 represent large-scale and small-scale information respectively, X represents a large-scale category number, and x represents a small-scale category number. In our system, there are 9 large-scale categories and 10 small-scale categories. Each large-scale category has a one-to-many relationship with the small-scale categories.

Following the caption strategy, the multi-scale image caption output from the LSTM looks like this:

$$\text{noun}_1\ R_{12}\ \text{noun}_2, \ldots, \text{with noun}_i\ R_{ij}\ \text{noun}_j, \ldots, \text{noun}_n$$

In the example above, two large-scale objects, which are composed of small-scale objects and their spatial relationship, are on both sides of "with." Thus, the caption contains two scales of spatial semantic information. The noun$_i$ describes a remote sensing object O_i, and R_{ij} describes the spatial relationship between the remote sensing objects O_i and O_j (e.g., "road cross residence with road next to service"). Noun$_i$ and noun$_j$ are always different in one clause.

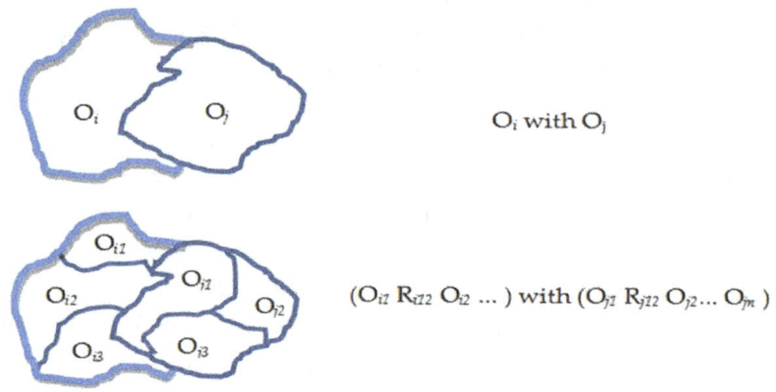

Figure 1. Multi-spatial scale semantic segmentation and image caption. It shows a two-scale hierarchy of one image. An image contains many large-scale objects, each large-scale object contains many small-scale objects, and there are spatial relationships between objects of the same scale. Our strategy of captioning is to describe both information of scale and spatial relationship contained in an image as completely as possible.

Table 1. Classification of multiscale remote sensing objects.

Value	10	11	12	13	14
large-scale	residence region	industry region	service region	village region	forest region
value	15	16	17	18	
large-scale	uncompleted region	road region	other region	Green-space region	
value	20	21	22	23	24
small-scale	residence	industry	service	village	forest
value	25	26	27	28	29
small-scale	uncompleted	road	other	Green-space	waterbody

Our strategy is still valid when generating captions for more complex scenes, as shown in Figure 2. We use small-scale objects and spatial relationships between them to describe the large-scale objects, and then connect each clause with "with." When there are more large-scale objects in an image, we use this method to iterate to form a complete caption containing multi-scale semantic information.

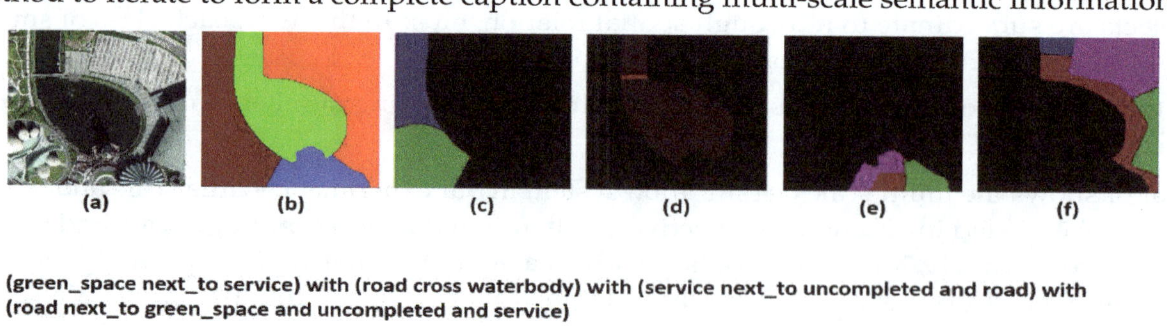

Figure 2. Sample with complex scenes. It shows an image with more complex scenes that contain four large-scale objects. (**a**) is the input image; (**b**) is the large-scale segmentation map of (**a**); (**c–f**) are the small-scale objects contained in each large-scale object. (**c**) corresponds to the clause "green_space next_to service," (**d**) to the clause "road cross waterbody," (**e**) to the clause "service next_to uncompleted and road," and (**f**) to the clause "road next_to green_space and uncompleted and service.".

The advantages of a multi-scale semantic caption strategy are as follows:

Hierarchically describing the spatial relationship between objects according to scale effect can simplify the type of spatial relationship (including "next_to", "near", "cross", "surround" and "surround_by"). Due to the scope limitation of sample patch, only the spatial neighbourhood relationship between large-scale objects is considered and described using "with" such that the network training is facilitated.

3.2. *Multi-Scale Network Structure*

Corresponding to the multi-scale semantics caption strategy, the MSRIN consists of three different deep neural networks: an FCN and a U-Net for multi-scale semantic segmentation, and an attention-based LSTM for image caption, both the FCN and the U-Net are used to semantically segment the same remote sensing image at two different spatial scales. The output of FCN and U-Net are masked to obtain the location and boundaries of remote sensing objects.

Meanwhile, the same sample is input into one attention-based LSTM network, with the output captions following the principles of multi-scale remote sensing captioning. To match $noun_t$ with the remote sensing object via attention and to overcome the location deviation of attention, we designed a multi-scale remote sensing object recognition and correction mechanism. The structure of the MSRIN is shown in Figure 3.

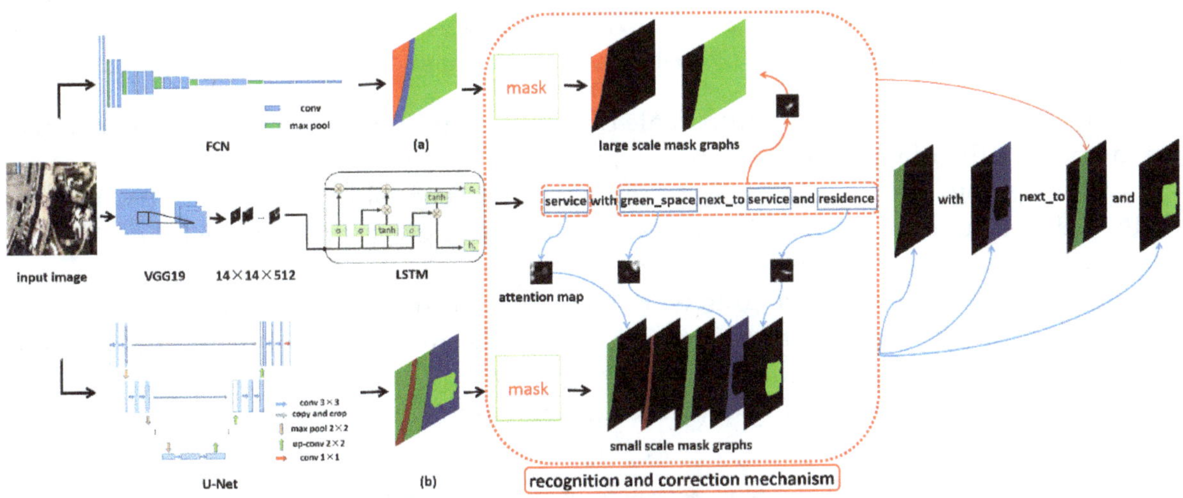

Figure 3. Network structure. It shows the overall network structure of the MSRIN. In our network, one remote sensing image is input into three branch networks. (**a**) is the large-scale segmentation map of the FCN output, (**b**) is the small-scale segmentation map of the U-Net output, and they are masked to obtain the location and boundaries of remote sensing objects. The LSTM outputs image captions and attention areas. The process of identification and correction is given in Section 3.3. The multi-scale objects recognition and correction mechanism attaches the object (the mask graphs from U-Net) to nount through the weight matrix at time step t.

The FCN [37] can achieve pixel-to-pixel classification by using full convolution, up-sampling, and jump structure. The U-Net [42] follows the idea of FCN for image semantic segmentation and combines the features of coding–decoding structures and jumping networks. From the encoder to the decoder, there is usually a direct information connection to help the decoder recover the target details better. Considering the features of small-scale objects are more complex and large-scale objects are more abstract, we used a U-Net to segment small-scale objects and a FCN to segment large-scale objects, and the segmentation effect is shown in Figure 4.

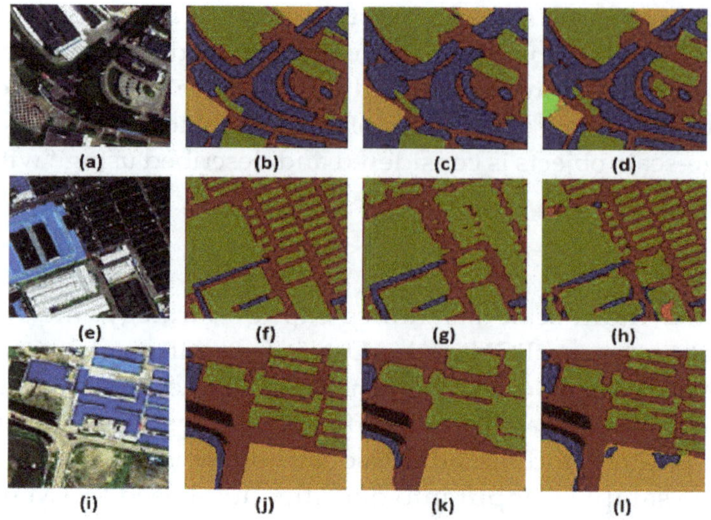

Figure 4. Segmentation effect of FCN and U-Net. It shows three examples of FCN and U-Net segmentation effect comparison. (**a,e,i**) are the input images. (**b,f,j**) are corresponding ground truth of the images. (**c,g,k**) are segmentation maps of FCN. (**d,h,l**) are segmentation maps of U-Net. It was found that FCN performed better at segmenting large objects, while smaller objects were easier to aggregate into blocks, so FCN was more suitable for large-scale segmentation. U-Net worked well when smaller objects were segmented but tended to misclassify some small fragments of the large-scale objects when it was being segmented, so U-Net was more suitable for small-scale segmentation.

The core of the networks is the multi-scale objects recognition and correction mechanism, which attaches the object (the mask graphs from U-Net) to $noun_t$ through the weight matrix at time step t. In this way, the remote sensing objects get their spatial relationship from captions. In other words, the MSRIN will output not only a series of the remote sensing objects (the mask graphs) but also the spatial relationships between them from image captions.

Unfortunately, the attention weights were computed from a 14×14 size feature map. Thus, the spatial location accuracy was relatively low, leading to a mismatch between $noun_t$ in the caption and objects in the image at some time step t, as shown in Figure 5. To solve this problem, our paper proposes a multi-scale remote sensing object recognition and correction mechanism.

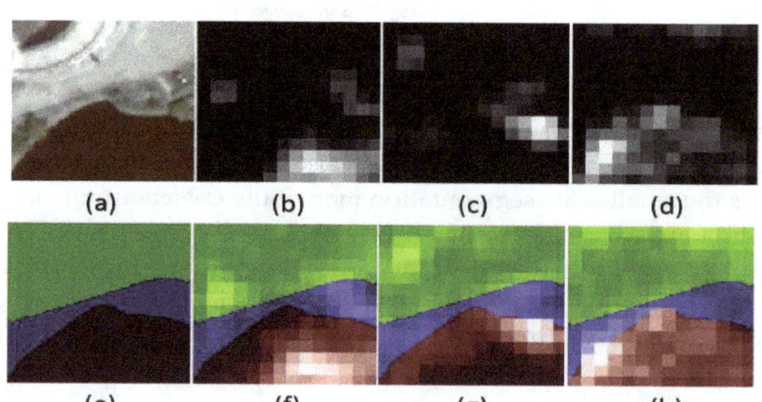

Figure 5. Attention weight matrix error. It shows mismatches between nouns in the caption and objects in the image. (**a**) is the input image; (**b–d**) are the attention maps for generating "green_space," "service," and "waterbody," respectively; (**e**) is a small-scale segmentation map of (**a**); (**f–h**) are overlaid maps of (**b–d**), respectively, with (**e**). As shown in the figure, the attention area of the first generated noun "green_space" corresponds to the object "waterbody" in the image, which resulted in mismatch. Of the three nouns contained in the generated image caption, only the third noun matched the right object.

3.3. *Remote Sensing Objects Recognition and Correction Mechanism*

Attention-based LSTM provides a 14 × 14 weight matrix at different time steps, which is the basis for implementing the remote sensing object recognition and correction.

3.3.1. Remote Sensing Object Recognition

The remote sensing object recognition is based on the attention weight matrix and U-Net mask graphs. First, we resample the 14 × 14 attention weight matrix to a 210 × 210 size. Then, we denote the attention map (weight matrix) at location $(i, j) \in L \times L (L = 210)$ at time step t as a^t_{ij}, and the U-Net mask graphs at location $(i, j) \in L \times L$ as m_{ij}. In the mask graphs, the area where the object is located has a pixel value of the class label C and the rest is 0:

$$m = \begin{cases} C, (i,j) \in object \\ 0, (i,j) \notin object \end{cases} \tag{1}$$

The values of intersect areas can be computed using:

$$v_{ij} = \frac{1}{C} \alpha^t_{ij} . m_{ij} \tag{2}$$

where C is the normalization factor such that v_{ij} sums to 1. The mean value of the intersect areas (weight mean value) can be computed using:

$$v_{mean} = \frac{1}{n} \sum_{ij} v_{ij} \tag{3}$$

where n is the total number of pixels of the remote sensing object. Then, the mask graph with the largest mean value will be selected. If the class label of the selected mask graph (object) is consistent with the $noun_t$ in the caption at current time step t, it means the mask graph represents the $noun_t$ of current time t, and the location and boundary of the remote sensing object will be identified using the selected mask graph.

However, at time step t, the label of the selected mask graph often does not match $noun_t$ in the captions, as shown in Figure 5. To solve this problem, we propose a multi-scale remote sensing object correction algorithm.

3.3.2. Remote Sensing Correction Algorithm

If the mismatch happens, the correction algorithm needs to upscale and search for the large-scale object, which the current weight matrix pays attention to first. The detailed method is shown below:

The MSRIN first scales up to the large-scale objects region that are large-scale mask graphs output from the FCN, then calculates the weights mean value in each large-scale object and takes the maximum one as a candidate object. In the candidate object, the MSRIN downscales to small-scale mask graphs and selects the remote sensing object whose class label corresponds to $noun_t$ using a one-to-one relationship, thus completing the correction.

The key to the above process is that the strategy of multi-scale captions made of small-scale objects of each class will not be repeated within each large-scale object such that in the large-scale object region, the small-scale object that matches to $noun_t$ can be selected. The Algorithm 1 is shown as follows:

Algorithm 1. For Multi-Scale Remote Sensing Objects Recognition and Correction

Input: noun, weight matrix at time step t.
Small scale object set $o = \{o_i\}, i \in 1, n]$, n is the number of mask graphs from U-Net in one sample patch,
Large scale object set $O = \{O_j\}, j \in [1, m]$, m is the number of mask graphs from FCN in the same sample patch with the U-Net.
weight_graph is a visual graph of the weight matrix generated at the current moment.
Output: $\mathbf{o_{selected}}$.

For $i = 1$ to n; step = 1; do //search the small-scale object that the current weights graph pay attention to
{ weight_graph intersect with o_i; //determine the area of attention on a small-scale object
Calculate mean value of intersect area; //basis for selecting small-scale candidate
Update o_i to small_candidate when weights mean value is the current maximum mean; //update candidate based on the mean value
}
If the class label of the small_candidate is equal to noun$_t$; //the generated noun matches the object
Then $o_{selected}$ = small_candidate; //the object was recognized.
Else //there is a mismatch between noun$_t$ and the candidate, so a correction process will start
{//upscale, search the candidate large-scale object that the current weights graph pay attention to
For $j = 1$ to m; step = 1; do //search the large-scale object that the current weights graph pay attention to
{weight_graph intersect with O_j; //determine the area of attention on a large-scale object
Calculate mean value of intersect area; //basis for selecting large-scale candidate
Update O_j to large_candidate when weights mean value is the current maximum mean; //update candidate based on the mean value
}
Downscaling in large_candidate; //downscale, determine small-scale object based on the large-scale candidate
search the small-scale object oi which class label is corresponded to noun$_t$ in large_candidate; //the target small scale object in the large-scale candidate
$o_{selected} = o_i$; //thus the object was recognized and corrected.
}

3.3.3. Case Analysis

The following example analyzes the process of multi-scale remote sensing object recognition and correction, as shown in Figures 6–8. The generated caption is "service with green_space next_to service and surround residence." The reference caption is "service with green_space next_to service and surround residence." The mean value of remote sensing objects at each time is shown in Table 2.

Table 2. Mean value of remote sensing objects.

	The Mean Value of Weight at Every Object When Generating the First "Service"	The Mean Value of Weight at Every Object When Generating the Second "Service"
service_0	0.000061702 (correct)	0.000016412
service_1	0.000015618	0.000019851
road_0 (with)	0.000029001	0.000027496
green_space _0	0.000015094	0.000023904
residence_0	0.000021753	0.000046018 (incorrect)
service_region	0.000061702	0.000016288
residence_region	0.0000164134	0.0000265598

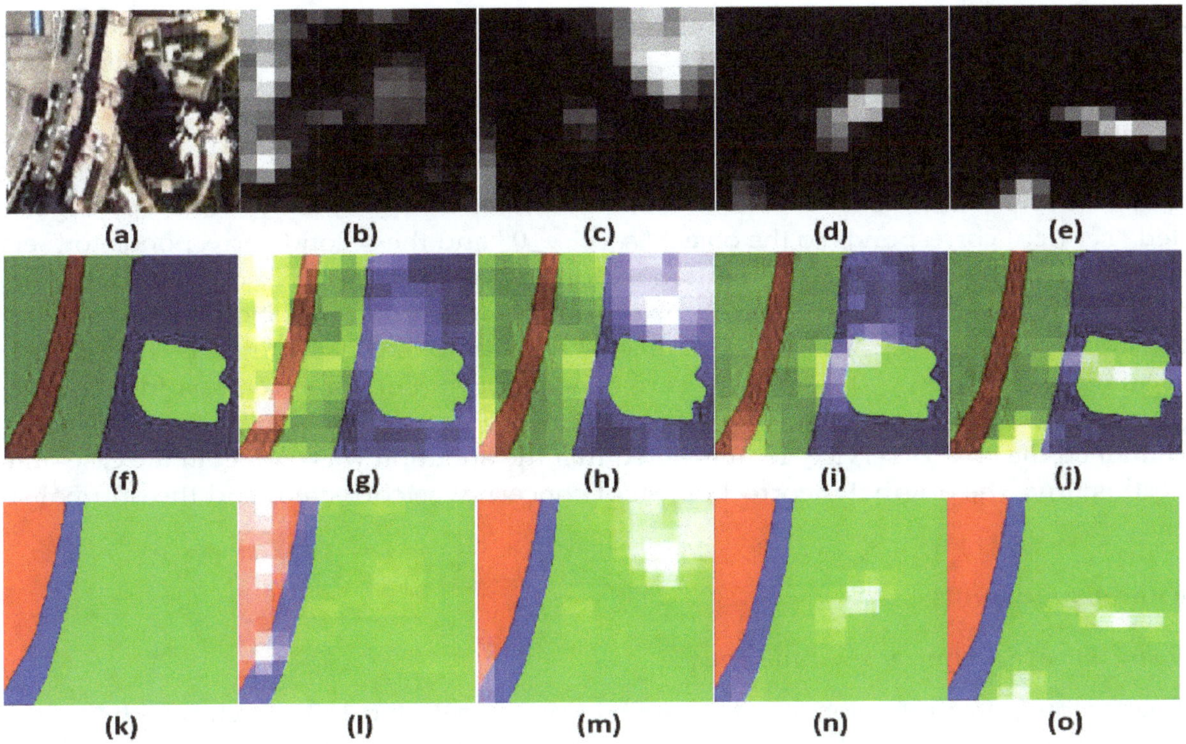

Figure 6. Remote sensing object recognition and correction. It shows the process of multi-scale remote sensing object recognition. (**a**) is the input image; (**b–e**) are the attention maps for generating "service," "green_space," "service," and "residence," respectively; (**f**) is a small-scale segmentation map of (**a**); (**g–j**) are overlaid maps of (**b–e**), respectively, with (**f**); (**k**) is a large-scale segmentation map of (**a**); (**l–o**) are overlaid maps of (**b–e**), respectively, with (**k**). As shown in (**I,n**), when generating the second "service," the spatial location of attention weights is incorrect at the small scale, but it is correct at the large scale.

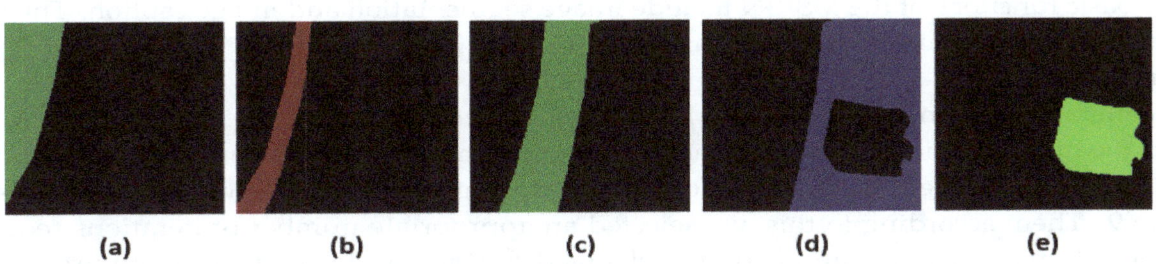

Figure 7. Small-scale objects. It shows the small-scale objects of the image. (**a**) is service_0 (in order to distinguish between different objects of the same class, we number each object); (**b**) is road_0; (**c**) is service_1; (**d**) is green_space _0; and (**e**) is residence_0.

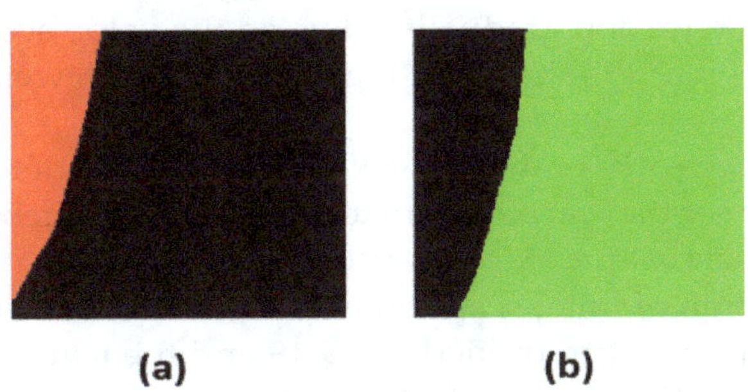

Figure 8. Large-scale objects. It shows the large-scale objects of the image. We divided the image into two large-scale objects by the road. (**a**) is service_region, which contains small-scale object service_0; (**b**) is residence_region, which contains small-scale object service_1, green_space_0, and residence_0.

Table 2. The mean value of weight at every object when two "service" words are generated. The object of concern on the small scale was correct when generating the first "service" and it was incorrect when generating the second "service".

The caption "service with green space next to next_to service and surround residence" was divided into two parts using "with" (representing a road in the image). "Service" can describe both remote sensing objects "service_0" and "service_1." The optimal case is that the attention object of the first generated "service" corresponds to the object "service_0" and the second corresponds to "service_1." The sub-optimal case is that the two attention regions of the "service" are both aimed at "service_1." By comparing the mean value of small-scale objects, it was found that the object of concern was correct when the first generated "service" aimed at object "service_0" and it was incorrect when the second "service" aimed at "residence_0." Thus, the correction algorithm upscaled, comparing the mean value of large-scale objects; the attention region of the second generated "service" was "residence_region," and it contained the object "service_1." It is noted that the attention was correct in the case of the large scale. At this time, the original incorrect object of concern was corrected, and the optimal situation was achieved.

4. Data and Experiments

4.1. Introduction of Experiment Area and Sample

To better integrate professional and research ideas, we selected 1835 patches with the longitudes ranging from 114°23′50″ *E* to 114°25′7″ *E*, the latitudes ranging from 30°27′50″ *N* to 30°30′37″ *N*, and a total area of 9.06 km^2 of remote sensing images in Guanggu in 2009. To make the number of samples in the verification set and training set sufficient and the results reasonable, we allocated 1167 samples to the training set and 668 samples to the verification set. For each sample image, we gave three captions that were as different as possible.

4.2. Network Parameters and Experiment

The basic functions of the MSRIN include image segmentation and image caption. The function of semantic segmentation of the original images is obtained based on a pre-trained FCN and a pre-trained U-Net.

In general, when fine-tuning network parameters, in order to reduce the learning cost, we first adjusted the number of iterations of the network. For example, in FCN, we first set a larger number of iterations and observed the loss function change during the iterations, where the trend is shown in Figure 9. Then, according to this, we selected an appropriate number of iterations required for the network to reach stability and kept it unchanged during the subsequent tuning process. When adjusting other parameters, we followed the principle of single factor experiments: fine-tune a certain parameter while keeping other parameters unchanged until the parameter is optimal, and then adjust other parameters one by one. For example, when adjusting the batch size of LSTM, we set the initial value to 25 according to experience and gradually adjusted the value according to the trend of Bleu_1, where the change process of Bleu_1 is shown in Figure 10. Finally, we selected the batch size when Bleu_1 was the highest.

After adjusting the sub-networks of the MSRIN one by one, we determined the basic parameters of each network. In FCN, we set the learning rate to 1×10^{-5}, the batch size to 1, and the number of iterations to 60,000. In U-Net, we set the learning rate to 1×10^{-4}, the batch size to 20, and the number of iterations to 120. The function of image caption was obtained based on an attention-based LSTM. The original image was input into a pre-trained VGG-19 and the features of conv5_3 were extracted. The size of the feature map was $14 \times 14 \times 512$, which was used as a part of the input of the LSTM. In the LSTM, we set the hidden layers to 1024, the embedding dimension of the word vector to 512, the learning rate to 0.001, the batch size to 20, the number of iterations to 120, and we used the softmax function as the nonlinear activation function.

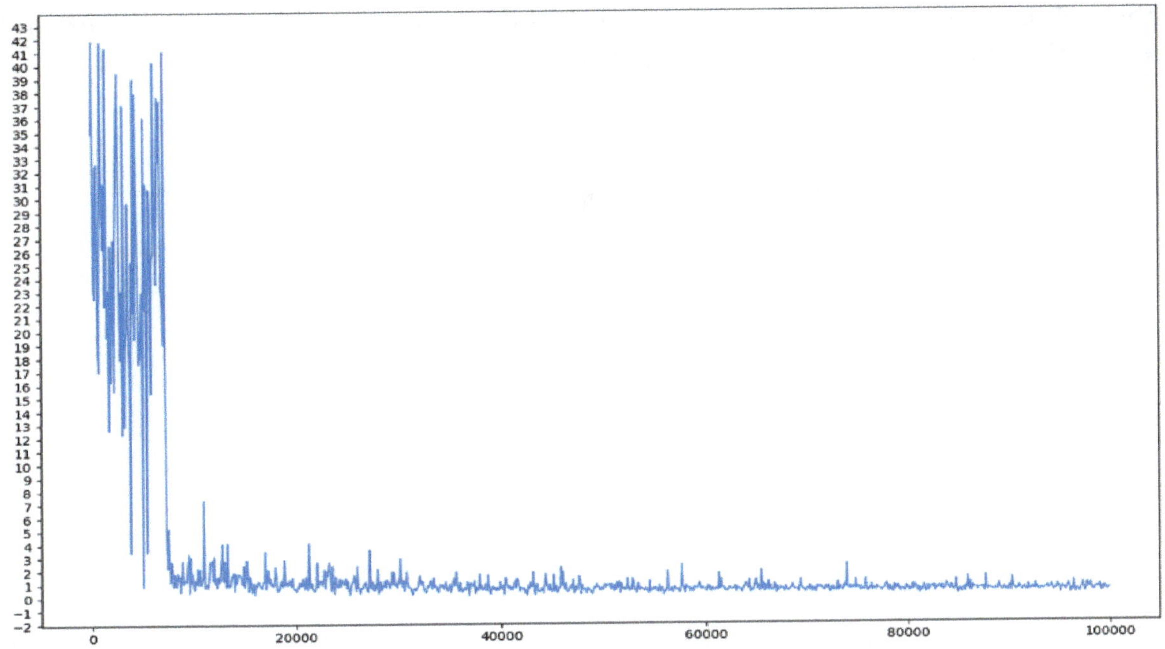

Figure 9. The loss value of FCN during training. It shows the trend of loss values during training. From the figure, we can see that in the early period of the iteration (about before 5000 times), the loss value violently oscillated and then dropped sharply. In the medium term (around 5000–50,000), the loss value decreased slightly and tended to be stable. In order to ensure that the network has stabilized, we chose 60,000 as the number of iterations.

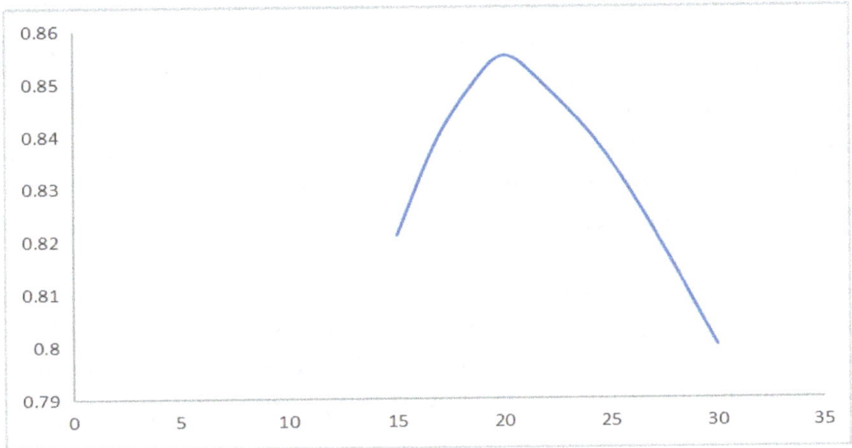

Figure 10. Bleu_1 of different batch sizes. It shows the trend of Bleu_1 when the other parameters were constant and only the batch size was changed. As the batch size increased, Bleu_1 increased first and then decreased, and the effect of batch size on Bleu_1 was obvious, so a suitable batch size was necessary.

4.3. Experiment Evaluation

We randomly allocated the total samples to the training set and validation set according to the sample set sizes in Section 4.1. After an image caption experiment, we obtained a set of satisfactory experimental results, in which Bleu_1 was 0.893, Bleu_2 was 0.744, Bleu_3 was 0.655, and Bleu_4 was 0.587. Then, we kept the number of training sets and validation sets unchanged, randomly allocated samples, and performed nine independent Monte Carlo runs. We compared the Bleu_1, Bleu_2, Bleu_3, and Bleu_4 of those 10 experiments, where the trend of Bleu is shown in Figure 11. In these ten experiments, the mean values of Bleu_1, Bleu_2, Bleu_3, and Bleu_4 was 0.8982, 0.7466, 0.6521, and 0.5866, respectively, and the standard deviations were 0.004, 0.009, 0.011, and 0.012, respectively, which proved the stability and reliability of the experimental results. We selected the results of the first experiment as the basis for the subsequent experiments and analyses.

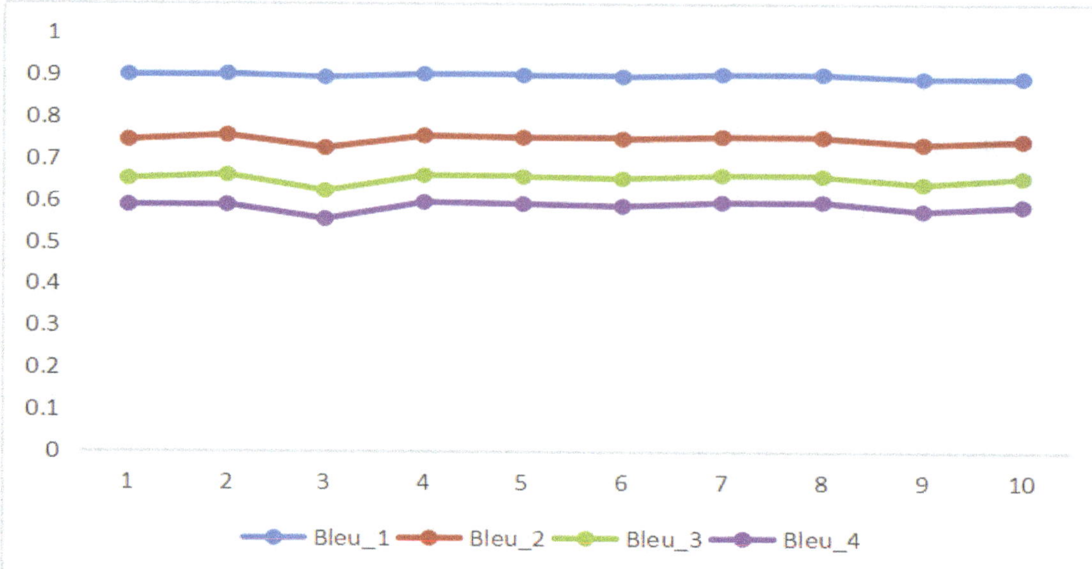

Figure 11. Bleu trend of ten experiments. It shows the trend of Bleu. From the figure, we can see that in ten experiments, the variation amplitudes of Bleu_1, Bleu_2, Bleu_3, and Bleu_4 are small, which can prove the randomness of data distribution and the robustness of the algorithm.

Next, we selected 200 samples from the remote sensing image captioning data set (RSICD) [28] as a validation set to test our experimental model. As shown in our test result, compared with using the VGG-19+LSTM model from the original paper [28], our model outperformed in all the evaluation metrics, where the comparison result of metrics is shown in Table 3. Bilingual Evaluation Understudy of n gram (Bleu_n) calculates the matching degree between n-dimensional phrases and reference captions (GT) [59]; Metric for Evaluation of Translation with Explicit Ordering (METEOR) adds synonym matching on the basis of Bleu to make it more strongly correlated with manual discrimination [60]; Recall-Oriented Understudy for Gisting Evaluation of Longest Common Subsequence (ROUGE_L) and Consensus-based Image Description Evaluation (CIDEr) evaluate the similarity between the generated caption and the GT using the recall rate and cosine similarity, respectively [61,62]. In general, these metrics calculate the matching degree between the generated caption and the GT in different ways, and the larger the metrics values are, the better the generated caption is.

Table 3. Results comparison using VGG-19+LSTM.

	Bleu_1	Bleu_2	Bleu_3	Bleu_4	METEOR	ROUGE_L	CIDEr
Ours	0.774	0.535	0.406	0.314	0.359	0.663	2.745
RSICD	0.583	0.423	0.331	0.270	0.261	0.519	2.033

Table 3 shows the comparison of the results of our model and the model (VGG-19+LSTM) from Reference [28]. Our metrics are higher than that from Reference [28].

There are two reasons for our network performing better on the RSICD:

1. The multi-scale caption design strategy makes the spatial relationship more concise such that the vocabulary size is relatively small, which makes it easier for the network training.
2. The RSICD is a shared test dataset such that the purpose of their experiment is to verify the credibility of the dataset. Therefore, in their test experiment, accuracy is not the main indicator.

We made statistical comparisons between the evaluation metrics and similar work in existing studies, and the results are shown in Table 4:

Table 4. Comparing evaluation metrics.

	Bleu_1	Bleu_2	Bleu_3	Bleu_4	METEOR	ROUGE_L	CIDEr
Mean	0.670	0.509	0.399	0.310	0.235	0.560	0.978
Ours	0.893	0.744	0.655	0.587	0.455	0.779	5.044

Table 4 shows the comparison of the evaluation metrics between the experiment and the mean values from 18 experiments in several related papers. Our evaluation metric values are higher than the mean values.

Comparing our experimental results with the mean values of 18 experimental evaluation metrics from related papers [17,25,26,28,29,63], it is obvious that all of our evaluation metrics scores were better than the mean values. Moreover, in our experience, the pixel accuracy of FCN and U-net for semantic segmentation were 0.89 and 0.93, respectively, which also reached a good level. Therefore, our experimental results are credible and can support subsequent recognition and correction experiments.

We analyzed the reasons for the better experimental results. By comparing the generated captions with the GT, we found that among 668 samples in the validation set, 300 samples contained the word "with" in GT. A total of 256 samples of generated captions containing the word "with," accounting for 85%. The analysis shows that the multi-scale labeling strategy we proposed is feasible and can be used in experiments. This multi-scale spatial relationship description strategy not only greatly reduces the sum of spatial relationship vocabulary in reference captions and the difficulty of network learning but also accurately describes the complex spatial relationship between remote sensing objects. Both reasons are important for increasing the evaluation metric values of the experiment.

Combining words in all sample generation captions, we analyzed the reliability of caption descriptions. The analysis results are shown in Table 5.

Table 5. Reliability analysis for generated captions.

Samples	Total Number of Samples: 668		
	Correct	Incorrect	Total
Words	3417	368	3785
Proportion	90.28%	9.72%	100%
S.R. Word	1396	170	1566
Proportion	89.14%	10.86%	100%
Nouns	2021	198	2219
Proportion	91.08%	8.92%	100%

Table 5 shows the results of our analysis of the generated captions of 668 samples in the validation set. In the table, "word" is the sum of "S.R. word" and "nouns" in the captions, "S.R. word" was the spatial relationship word, and "nouns" were the category nouns. "Correct" means that the word existed in the reference captions (GT), and "incorrect" means that it did not exist.

From Table 5, it is obvious that a total of 3785 words existed in the 668 generated captions, of which 3417 were correct, accounting for 90.28%, and 368 were incorrect, accounting for 9.72%. The Bleu_1 value in our experiment was slightly lower than 0.9028. There are two possible reasons: (1) the influence of other words (such as "and" and "with"), and (2) the introduction of a penalty mechanism in the calculation of the Bleu scores. We divided the 3785 words into nouns and relatives and calculated the statistics. Among them, there were 2219 nouns in total and 2021 were correct, accounting for 91.08%. Moreover, there were 1566 relative words, and 1396 were correct, accounting for 89.14%. Because of the high accuracy of nouns and relative words, it was possible to recognize and correct image objects.

After the recognition and correction of objects, the effect is shown in Table 6.

Table 6. Number of matched nouns before and after correction.

Samples	Total Number of Samples: 668			
	Pre-Correction Matched	Post-Correction Matched	Unmatched	Total
Nouns	929	1856	363	2219
Proportion	41.87%	83.64%	16.36%	100%

Table 6 shows the matching of the nouns and the object mask graphs before and after recognition and correction. "Pre-corrected matching" means that the attention area of the noun was matched with the object mask graphs before correction. "Post-correction matched" means that the attention area of the noun was matched after correction. "Unmatched" means that the attention area of the noun was unmatched with the object mask graphs before and after correction.

5. Discussion

The object recognition and correction mechanism proposed in this paper performs object recognition based on the $noun_t$ generated at time step t and the corresponding attention weight matrix, and multi-scale correction for the mismatched object concerned by the attention weight matrix. Therefore, in order to better analyze the object recognition and correction effects, we divided the 668 samples into two subsets: Sample Set 1 and Sample Set 2. The nouns contained in the generated captions of the samples in Sample Set 1 were all correct, and the captions generated by the samples in Sample Set 2 contained error nouns. In this way, Sample Set 1 could fully realize object recognition and correction, and Sample Set 2 could only identify and correct the remote objects corresponding to the correct nouns. We further analyzed the two sample sets. The overall results before and after correction are shown in Table 7.

Table 7. The results of the overall analysis of the subsets.

Samples	Total Number of Samples: 668							
	All Nouns are Correct: 477				Not All Nouns are Correct: 191			
	Correct	Incorrect	Total		Correct	Incorrect	Total	
S.R. Word	1013	76	1089		383	94	477	
Proportion	93.02%	6.98%	100%		80.29%	19.71%	100%	
	Pre-Correction Matched	Post-Correction Matched	Unmatched	Total	Pre-Correction Matched	Post-Correction Matched	Unmatched	Total
Nouns	781	1541	26	1567	148	315	337	652
Proportion	49.84%	98.34%	1.66%	100%	22.70%	48.31%	51.69%	100%

Table 7 shows the overall analysis of Sample Set 1 and Sample Set 2. From the table, we can see the words contained in the captions generated using the two sample sets and the comparison results of the number of nouns matched with the object before and after correction.

The 477 samples in Sample Set 1 generated a total of 2656 words, of which 2580 words are correct, and 76 words were incorrect. In the generated 2656 words, there were 1567 object nouns, and all of them were correct; 1089 words were relational words, of which 1013 words were correct, and 76 words were incorrect. There were a total of 1129 words contained in the 191 generated captions of Sample Set 2, of which 837 words were correct, and 292 words were incorrect. Among the 1129 words, there were 652 object nouns, of which 454 words were correct, 198 words were incorrect; 477 words were relational words, of which 383 words were correct, and 94 words were incorrect.

In Sample Set 1, we first performed an analysis based on the samples, and the results are shown in Table 8. There were 477 samples in the validation set (i.e., the sample set to be analyzed). Among them, 87 samples, accounting for 18.24%, did not need to be corrected because all nouns were matched with objects. There were 337 samples with incorrect recognition objects before correction, but nouns were all matched with objects after correction, accounting for 70.65%. Generally, after the implementation of our correction method, there were a total of 424 samples in which each noun in the generated captions were matched with the corresponding objects, accounting for 88.89%, an increase of 70.65 percentage points, which was a remarkable effect. In addition, there were 37 partially corrected samples, accounting for 7.76%; only 16 samples were not corrected, accounting for 3.35%, basically achieving the purpose of recognizing remote sensing objects and the spatial relationship between them.

Table 8. Sample-based analysis of Sample Set 1 before and after correction.

Sample Class	Completely Corrected	Partial Correction	No Corrective Effect	No Need for Correction	Total
Number	337	37	16	87	477
Proportion	70.65%	7.76%	3.35%	18.24%	100%

Table 8 shows the sample-based analysis of Sample Set 1. As shown in the table, more than 78% of the samples were completely or partially corrected.

Next, we conducted an analysis based on category nouns included in the 477 generated captions of Sample Set 1. There were 1567 nouns in all generated captions from the 477 samples. There were 781 nouns whose attention areas were matched with objects before the correction, the proportion was 49.84%, and the matched ones increased to 1541 after correction, accounting for 98.3%, an accuracy increase of 48.50 percentage points. The effect was greatly improved, which means that the method we proposed can meet the demand of remote sensing interpretations.

We performed a similar analysis of 191 samples in Sample Set 2. The results of the sample-based analysis are shown in Table 9. There were multiple objects in one or more classes of some samples, and in the generated captions of these samples, nouns of these classes appeared more than once, and a many-to-many relationship could therefore be constructed. However, the matched objects' judgement of the generated correct nouns in the image will be affected by the incorrect nouns. These samples were classified into having no corrective effect on the statistics, totaling 55 samples. One sample whose words in the generated captions were all incorrect and was classified into no corrective effect.

Table 9. Sample-based analysis of Sample Set 2 before and after correction.

Sample Class	Completely Corrected	Partial Correction	No Corrective Effect	No Need for Correction	Total
Number	113	0	56	22	191
Proportion	59.16%	0.00%	29.32%	11.52%	100%

Table 9 shows the sample-based analysis of Sample Set 2. As shown in the table, only approximately 59% of the samples were corrected. The correction effect was worse than that of Sample Set 1.

The nouns-based analysis of Sample Set 2 showed that 148 (22.70%) of the nouns matched with the mask graphs before the correction, and it increased to 315 after correction, accounting for 48.31%. The proportion increased by 25.61 percentage points, which was less of an effect than for Sample Set 1. However, 337 nouns still could not be corrected, accounting for 51.69%, which indicated that the correction algorithm could only solve the mismatching problem between the attention weight matrix and the object but could not correct the incorrect words generated by LSTM. In addition, 315 of 341 nouns could not be corrected in Sample Set 2, indicating that for a sample, the higher the Bleu scores, the better the recognition and correction mechanism performs.

The sample-based and nouns-based overall correction effect analysis of Sample Set 1 and Sample Set 2 are shown in Figure 12. We conducted a comprehensive analysis of the corrective effect with a combined Sample Set 1 and Sample Set 2. Before the correction, the number of each noun in the captions generated by the samples matching with the mask graph was 109. When the correction finished, the number rose to 559, equivalent to a proportion increase of 67.37 percentage points. Before and after correction, the number of nouns matching with the mask graph increased from 929 to 1856, a proportion increase of 41.78 percentage points.

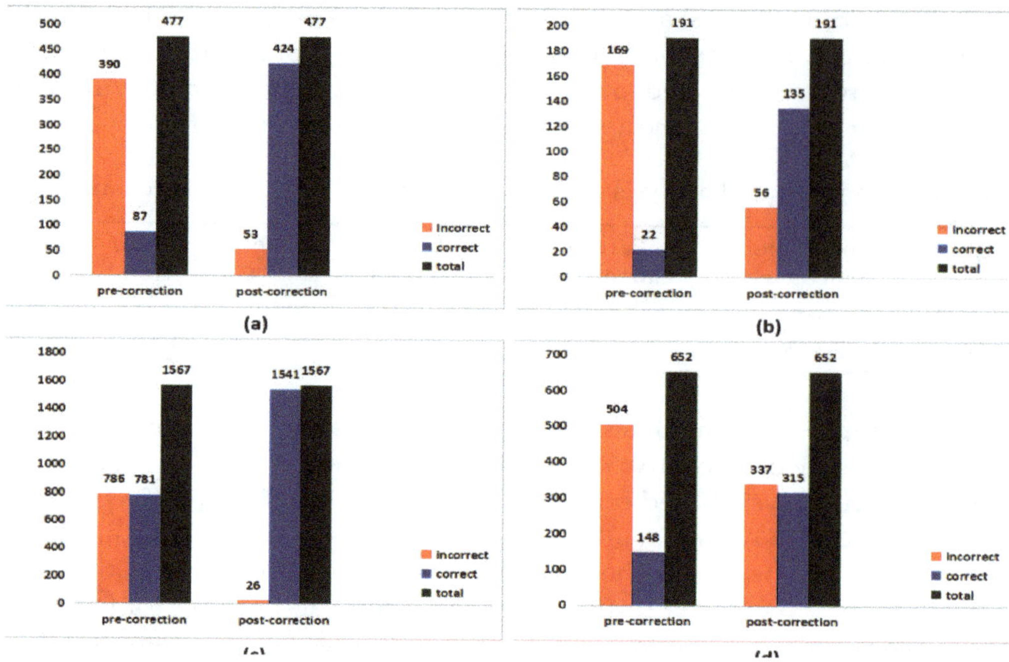

Figure 12. Correction effect analysis. It shows the corrective effect of Sample Set 1 and Sample Set 2. (**a**,**b**) are the sample-based overall correction effect for Sample Set 1 and Sample Set 2, respectively; (**c**,**d**) are the noun-based overall correction effect for Sample Set 1 and Sample Set 2, respectively. As shown in the figure, whether from the perspective of samples or nouns, the correction algorithm proposed in this paper achieved good results. The correction effect of Sample Set 1 was better than that of Sample Set 2.

From the above analysis, the following conclusions can be drawn:

- When the noun in the generated caption was correct and the spatial relationship was incorrect, the remote sensing object could still be recognized, but the spatial relationship could not be corrected.
- When both the nouns and spatial relationship were incorrect, the proposed method was ineffective. This requires further research.

6. Conclusions

In this paper, a multi-scale remote sensing image interpretation network (the MSRIN) was proposed for identifying remote sensing objects and their spatial relationships from end-to-end. First, a remote sensing image was semantically segmented through an FCN and a U-net on two spatial scales such that each pixel in the original image was labelled with two semantic labels; therefore, a hierarchical relationship of a multi-scale remote sensing object could be formed. Second, the features of the same image obtained using a pre-trained VGG-19 network were input for the attention-based LSTM, which outputted the captions that described the two-scale remote sensing objects and their spatial relationships. Finally, the relationship between the nouns in the caption and the object mask graphs was established through the attention weight matrix. In this way, the remote sensing objects from the U-Net got their spatial relationship from the caption. To overcome the spatial deviations of the attention weight matrix from the LSTM, the MSRIN designed an attention-based, multi-scale remote sensing object identification and correction mechanism. Our method produced a complete semantic interpretation of remote sensing images.

Identifying remote sensing objects and their spatial relations was based on the attention weight matrix. In the future, we will improve the attention weight calculation method to achieve more accurate positioning.

The objectivity of the evaluation method based on Bleu depends on the completeness of the captions, which greatly aggravates the burden of sample preparation. Therefore, we will concentrate on exploring the establishment of evaluation metrics and methods that are suitable for remote sensing image captioning.""

The experimental data in this paper was mainly optical remote sensing images. In the future, we will further verify the recognition and correction mechanism proposed in this paper by using other types of data, such as SAR.

Author Contributions: W.C. contributed toward creating the original ideas of the paper. W.C. conceived and designed the experiments. F.W. and X.H. prepared the original data, performed the experiments and analyzed the experimental data with the help of D.Z., X.X., and M.Y. and W.C. wrote and edited the manuscript. F.W., X.H., and Z.W carefully revised the manuscript. J.H. contributed constructive suggestions on modifying the manuscript.

References

1. Lecun, Y.; Bengio, Y.; Hinton, G. Deep learning. *Nature* **2015**, *521*, 436. [CrossRef]
2. Schmidhuber, J. Deep learning in neural networks: An overview. *Neural Netw.* **2015**, *61*, 85–117. [CrossRef]
3. Lecun, Y.; Bottou, L.; Bengio, Y.; Haffner, P. Gradient-based learning applied to document recognition. *Proc. IEEE* **1998**, *86*, 2278–2324. [CrossRef]
4. Cheng, G.; Han, J.; Guo, L.; Liu, Z.; Bu, S.; Ren, J. Effective and efficient midlevel visual elements-oriented land-use classification using vhr remote sensing images. *IEEE Trans. Geosci. Remote Sens.* **2015**, *53*, 4238–4249. [CrossRef]
5. Cheng, G.; Zhou, P.; Han, J. Learning rotation-invariant convolutional neural networks for object detection in vhr optical remote sensing images. *IEEE Trans. Geosci. Remote Sens.* **2016**, *54*, 7405–7415. [CrossRef]
6. Han, J.; Zhang, D.; Cheng, G.; Guo, L.; Ren, J. Object detection in optical remote sensing images based on weakly supervised learning and high-level feature learning. *IEEE Trans. Geosci. Remote Sens.* **2015**, *53*, 3325–3337. [CrossRef]
7. Han, X.; Zhong, Y.; Zhao, B.; Zhang, L. Scene classification based on a hierarchical convolutional sparse auto-encoder for high spatial resolution imagery. *Int. J. Remote Sens.* **2017**, *38*, 514–536. [CrossRef]
8. Hu, F.; Xia, G.-S.; Hu, J.; Zhang, L. Transferring deep convolutional neural networks for the scene classification of high-resolution remote sensing imagery. *Remote Sens.* **2015**, *7*, 14680–14707. [CrossRef]
9. Hu, W.; Huang, Y.; Wei, L.; Zhang, F.; Li, H. Deep convolutional neural networks for hyperspectral image classification. *J. Sens.* **2015**, *2015*, 1–12. [CrossRef]
10. Zhong, Y.; Fei, F.; Zhang, L. Large patch convolutional neural networks for the scene classification of high spatial resolution imagery. *J. Appl. Remote Sens.* **2016**, *10*, 025006. [CrossRef]
11. Cui, W.; Zheng, Z.; Zhou, Q.; Huang, J.; Yuan, Y. Application of a parallel spectral–spatial convolution neural network in object-oriented remote sensing land use classification. *Remote Sens. Lett.* **2018**, *9*, 334–342. [CrossRef]
12. Cui, W.; Zhou, Q.; Zheng, Z. Application of a hybrid model based on a convolutional auto-encoder and convolutional neural network in object-oriented remote sensing classification. *Algorithms* **2018**, *11*, 9. [CrossRef]
13. Cannon, R.; Dave, J.; Bezdek, J.; Trivedi, M. Segmentation of a thematic mapper image using the fuzzy c-means clusterng algorthm. *IEEE Trans. Geosci. Remote Sens.* **1986**, *GE-24*, 400–408. [CrossRef]
14. Jeon, B.; Landgrebe, D.A. Classification with spatio-temporal interpixel class dependency contexts. *IEEE Trans. Geosci. Remote Sens.* **1992**, *30*, 663–672. [CrossRef]
15. Baatz, M.; Schape, A.; Multiresolution segmentation: An optimization approach for high quality multi-scale image segmentation. Angew. *Geogr. Inf. 2000, XII, 12–23.* Available online: https://pdfs.semanticscholar.org/364c/c1ff514a2e11d21a101dc072575e5487d17e.pdf?_ga=2.55340014.416308819.1554177081-320853791.1554177081 (accessed on 2 April 2019).
16. Vinyals, O.; Toshev, A.; Bengio, S.; Erhan, D. Show and tell: A neural image caption generator. In Proceedings of the 2015 IEEE Conference on Computer Vision and Pattern Recognition (CVPR), Boston, MA, USA, 7–12 June 2015; pp. 3156–3164. [CrossRef]
17. Xu, K.; Ba, J.; Kiros, R.; Cho, K.; Courville, A.; Salakhutdinov, R.; Zemel, R.; Bengio, Y. Show, attend and tell:

Neural image caption generation with visual attention. In Proceedings of the International Conference on Machine Learning, Lille, France, 6–11 July 2015; pp. 2048–2057.
18. Chen, F.; Ji, R.; Sun, X.; Wu, Y.; Su, J. GroupCap: Group-based image captioning with structured relevance and diversity constraints. In Proceedings of the 2018 IEEE Conference on Computer Vision and Pattern Recognition (CVPR), Salt Lake City, UT, USA, 18–22 June 2018; pp. 1345–1353. [CrossRef]
19. Yuan, A.; Li, X.; Lu, X. 3G structure for image caption generation. *Neurocomputing* **2019**, *330*, 17–28. [CrossRef]
20. Chen, H.; Ding, G.; Lin, Z.; Zhao, S.; Han, J. Show, observe and tell: Attribute-driven attention model for image captioning. In Proceedings of the Twenty-Seventh International Joint Conference on Artificial Intelligence, Stockholm, Sweden, 13–19 July 2018; pp. 606–612. [CrossRef]
21. Khademi, M.; Schulte, O. Image caption generation with hierarchical contextual visual spatial attention. In Proceedings of the 2018 IEEE/CVF Conference on Computer Vision and Pattern Recognition Workshops (CVPRW), Salt Lake City, UT, USA, 18–22 June 2018; pp. 2024–2028. [CrossRef]
22. Ranzato, M.; Chopra, S.; Auli, M.; Zaremba, W. Sequence Level Training with Recurrent Neural Networks. Available online: https://arxiv.org/abs/1511.06732 (accessed on 2 April 2019).
23. Shi, H.; Li, P.; Wang, B.; Wang, Z. Image captioning based on deep reinforcement learning. In Proceedings of the 10th International Conference on Internet Multimedia Computing and Service(ICIMCS), Nanjing, China, 17–19 August 2018; pp. 45–49. Available online: https://arxiv.org/10.1145/3240876.3240900 (accessed on 2 April 2019).
24. Karpathy, A.; Li, F.-F. Deep visual-semantic alignments for generating image descriptions. In Proceedings of the 2015 IEEE Conference on Computer Vision and Pattern Recognition (CVPR), Boston, MA, USA, 7–12 June 2015; pp. 3128–3137. [CrossRef]
25. Lu, J.; Xiong, C.; Parikh, D.; Socher, R. Knowing when to look: adaptive attention via a visual sentinel for image captioning. In Proceedings of the 2017 IEEE Conference on Computer Vision and Pattern Recognition (CVPR), Honolulu, HI, USA, 21–26 July 2017; pp. 3242–3250. [CrossRef]
26. Qu, B.; Li, X.; Tao, D.; Lu, X. Deep semantic understanding of high resolution remote sensing image. In Proceedings of the 2016 International Conference on Computer, Information and Telecommunication Systems (CITS), Kunming, China, 6–8 July 2016; pp. 1–5. [CrossRef]
27. Shi, Z.; Zou, Z. Can a machine generate humanlike language descriptions for a remote sensing image? *IEEE Trans. Geosci. Remote Sens.* **2017**, *55*, 3623–3634. [CrossRef]
28. Lu, X.; Wang, B.; Zheng, X.; Li, X. Exploring models and data for remote sensing image caption generation. *IEEE Trans. Geosci. Remote Sens.* **2018**, *56*, 2183–2195. [CrossRef]
29. Wang, B.; Lu, X.; Zheng, X.; Liu, W. Semantic descriptions of high-resolution remote sensing images. *IEEE Geosci. Remote Sens. Lett.* **2019**, *99*, 1–5. [CrossRef]
30. Zhang, X.; Wang, X.; Tang, X.; Zhou, H.; Li, C. Description generation for remote sensing images using attribute attention mechanism. *Remote Sens.* **2019**, *11*, 612. [CrossRef]
31. Wang, Y.; Lin, Z.; Shen, X.; Cohen, S.; Cottrell, G.W. Skeleton key: Image captioning by skeleton-attribute decomposition. In Proceedings of the 2017 IEEE Conference on Computer Vision and Pattern Recognition (CVPR), Honolulu, HI, USA, 21–26 July 2017; pp. 7378–7387. [CrossRef]
32. Tobler, W.R. A computer movie simulating urban growth in the Detroit region. *Econ. Geogr.* **1970**, *46*, 234. [CrossRef]
33. Nogueira, K.; Penatti, O.A.B.; dos Santos, J.A. Towards better exploiting convolutional neural networks for remote sensing scene classification. *Pattern Recognit.* **2017**, *61*, 539–556. [CrossRef]
34. Zhang, F.; Du, B.; Zhang, L. Scene classification via a gradient boosting random convolutional network framework. *IEEE Trans. Geosci. Remote Sens.* **2016**, *54*, 1793–1802. [CrossRef]
35. Li, E.; Xia, J.; Du, P.; Lin, C.; Samat, A. Integrating multilayer features of convolutional neural networks for remote sensing scene classification. *IEEE Trans. Geosci. Remote Sens.* **2017**, *55*, 5653–5665. [CrossRef]
36. Ge, Y.; Tang, Y.; Jiang, S.; Leng, L.; Xu, S.; Ye, F. Region-based cascade pooling of convolutional features for HRRS image retrieval. *Remote Sens. Lett.* **2018**, *9*, 1002–1010. [CrossRef]
37. Shelhamer, E.; Long, J.; Darrell, T. Fully convolutional networks for semantic segmentation. *IEEE Trans. Pattern Anal. Mach. Intell.* **2017**, *39*, 640–651. [CrossRef]
38. Huang, Z.; Cheng, G.; Wang, H.; Li, H.; Shi, L.; Pan, C. Building extraction from multi-source remote sensing images via deep deconvolution neural networks. In Proceedings of the 2016 IEEE International Geoscience and Remote Sensing Symposium (IGARSS), Beijing, China, 10–15 July 2016; pp. 1835–1838. [CrossRef]

39. Cui, W.; Li, R.; Yao, Z.; Chen, S.; Tang, S.; Li, Q. Study on the optimal segmentation scale based on fractual dimension of remote sensing images. *J. Wuhan Univ. Technol.* **2011**, *33*, 463–467. [CrossRef]
40. Xia, J.; Yang, X.; Jia, L. A multi-depth convolutional neural network for SAR image classification. *Remote Sens. Lett.* **2018**, *9*, 1138–1147. [CrossRef]
41. Li, L.; Liang, J.; Weng, M.; Zhu, H. A multiple-feature reuse network to extract buildings from remote sensing imagery. *Remote Sens.* **2018**, *10*, 1350. [CrossRef]
42. Ronneberger, O.; Fischer, P.; Brox, T. U-Net: Convolutional networks for biomedical image segmentation. In *Medical Image Computing and Computer-Assisted Intervention—MICCAI 2015*; Navab, N., Hornegger, J., Wells, W.M., Frangi, A.F., Eds.; Springer International Publishing: Cham, Switzerland, 2015; Volume 9351, pp. 234–241, ISBN 978-3-319-24573-7.
43. Huang, G.; Liu, Z.; van der Maaten, L.; Weinberger, K.Q. Densely connected convolutional networks. In Proceedings of the 2017 IEEE Conference on Computer Vision and Pattern Recognition (CVPR), Honolulu, HI, USA, 21–26 July 2017; pp. 2261–2269. [CrossRef]
44. Tao, Y.; Xu, M.; Lu, Z.; Zhong, Y. DenseNet-based depth-width double reinforced deep learning neural network for high-resolution remote sensing image per-pixel classification. *Remote Sens.* **2018**, *10*, 779. [CrossRef]
45. Drönner, J.; Korfhage, N.; Egli, S.; Mühling, M.; Thies, B.; Bendix, J.; Freisleben, B.; Seeger, B. Fast cloud segmentation using convolutional neural networks. *Remote Sens.* **2018**, *10*, 1782. [CrossRef]
46. Yang, H.; Wu, P.; Yao, X.; Wu, Y.; Wang, B.; Xu, Y. Building extraction in very high resolution imagery by dense-attention networks. *Remote Sens.* **2018**, *10*, 1768. [CrossRef]
47. Panboonyuen, T.; Jitkajornwanich, K.; Lawawirojwong, S.; Srestasathiern, P.; Vateekul, P. Semantic segmentation on remotely sensed images using an enhanced global convolutional network with channel attention and domain specific transfer learning. *Remote Sens.* **2019**, *11*, 83. [CrossRef]
48. Zhang, T.; Tang, H. A comprehensive evaluation of approaches for built-up area extraction from landsat oli images using massive samples. *Remote Sens.* **2019**, *11*, 2. [CrossRef]
49. Sun, G.; Huang, H.; Zhang, A.; Li, F.; Zhao, H.; Fu, H. Fusion of multiscale convolutional neural networks for building extraction in very high-resolution images. *Remote Sens.* **2019**, *11*, 227. [CrossRef]
50. Fu, Y.; Liu, K.; Shen, Z.; Deng, J.; Gan, M.; Liu, X.; Lu, D.; Wang, K. Mapping impervious surfaces in town–rural transition belts using China's GF-2 imagery and object-based deep CNNs. *Remote Sens.* **2019**, *11*, 280. [CrossRef]
51. Li, W.; Dong, R.; Fu, H.; Yu, L. Large-scale oil palm tree detection from high-resolution satellite images using two-stage convolutional neural networks. *Remote Sens.* **2019**, *11*, 11. [CrossRef]
52. Hochreiter, S.; Schmidhuber, J. Long short-term memory. *Neural Comput.* **1997**, *9*, 1735–1780. [CrossRef] [PubMed]
53. Mou, L.; Ghamisi, P.; Zhu, X.X. Deep recurrent neural networks for hyperspectral image classification. *IEEE Trans. Geosci. Remote Sens.* **2017**, *55*, 3639–3655. [CrossRef]
54. Wu, H.; Prasad, S. Convolutional recurrent neural networks forhyperspectral data classification. *Remote Sens.* **2017**, *9*, 298. [CrossRef]
55. Ndikumana, E.; Minh, D.H.T.; Baghdadi, N.; Courault, D.; Hossard, L. Deep recurrent neural network for agricultural classification using multitemporal sar sentinel-1 for Camargue, France. *Remote Sens.* **2018**, *10*, 1217. [CrossRef]
56. Liu, B.; Yu, X.; Yu, A.; Zhang, P.; Wan, G. Spectral-spatial classification of hyperspectral imagery based on recurrent neural networks. *Remote Sens. Lett.* **2018**, *9*, 1118–1127. [CrossRef]
57. Liu, Q.; Zhou, F.; Hang, R.; Yuan, X. Bidirectional-convolutional LSTM based spectral-spatial feature learning for hyperspectral image classification. *Remote Sens.* **2017**, *9*, 1330. [CrossRef]
58. Ma, A.; Filippi, A.M.; Wang, Z.; Yin, Z. Hyperspectral image classification using similarity measurements-based deep recurrent neural networks. *Remote Sens.* **2019**, *11*, 194. [CrossRef]
59. Papineni, K.; Roukos, S.; Ward, T.; Zhu, W.-J. BLEU: A method for automatic evaluation of machine translation. In Proceedings of the 40th Annual Meeting on Association for Computational Linguistics—ACL'02, Philadelphia, PA, USA, 7–12 July 2002; pp. 311–318. [CrossRef]
60. Lavie, A.; Agarwal, A. Meteor: An automatic metric for MT evaluation with high levels of correlation with human judgments. In Proceedings of the Second Workshop on Statistical Machine Translation—StatMT'07, Prague, Czech Republic, 23 June 2007; pp. 228–231. [CrossRef]

61. Lin, C. ROUGE: A package for automatic evaluation of summaries. In Proceedings of the Workshop on Text Summarization Branches Out—ACL'05, Barcelona, Spain, 25–26 July 2004; pp. 74–81. Available online: https://www.aclweb.org/anthology/W04-1013 (accessed on 2 April 2019).
62. Vedantam, R.; Zitnick, C.L.; Parikh, D. CIDEr: Consensus-based image description evaluation. In Proceedings of the 2015 IEEE Conference on Computer Vision and Pattern Recognition (CVPR), Boston, MA, USA, 7–12 June 2015; pp. 4566–4575. [CrossRef]
63. You, Q.; Jin, H.; Wang, Z.; Fang, C.; Luo, J. Image captioning with semantic attention. In Proceedings of the 2016 IEEE Conference on Computer Vision and Pattern Recognition (CVPR), Las Vegas, NV, USA, 26 June–1 July 2016; pp. 4651–4659. [CrossRef]

2

A Fast and Effective Method for Unsupervised Segmentation Evaluation of Remote Sensing Images

Maofan Zhao [1,2,3], Qingyan Meng [1,3,*], Linlin Zhang [1,3], Die Hu [1,2,3], Ying Zhang [1,2,3] and Mona Allam [1,3,4]

1 Aerospace Information Research Institute, Chinese Academy of Sciences, Beijing 100101, China; zhaomaofan19@mails.ucas.ac.cn (M.Z.); zhangll@radi.ac.cn (L.Z.); hudie@radi.ac.cn (D.H.); zhangying2016@radi.ac.cn (Y.Z.); monaecri@yahoo.com (M.A.)
2 University of Chinese Academy of Sciences, Beijing 101400, China
3 Sanya Institute of Remote Sensing, Sanya 572029, China
4 Environment & Climate Changes Research Institute, National Water Research Center, El Qanater El khairiya 13621/5, Egypt
* Correspondence: mengqy@radi.ac.cn

Abstract: The segmentation of remote sensing images with high spatial resolution is important and fundamental in geographic object-based image analysis (GEOBIA), so evaluating segmentation results without prior knowledge is an essential part in segmentation algorithms comparison, segmentation parameters selection, and optimization. In this study, we proposed a fast and effective unsupervised evaluation (UE) method using the area-weighted variance (WV) as intra-segment homogeneity and the difference to neighbor pixels (DTNP) as inter-segment heterogeneity. Then these two measures were combined into a fast-global score (FGS) to evaluate the segmentation. The effectiveness of DTNP and FGS was demonstrated by visual interpretation as qualitative analysis and supervised evaluation (SE) as quantitative analysis. For this experiment, the "Multi-resolution Segmentation" algorithm in eCognition was adopted in the segmentation and four typical study areas of GF-2 images were used as test data. The effectiveness analysis of DTNP shows that it can keep stability and remain sensitive to both over-segmentation and under-segmentation compared to two existing inter-segment heterogeneity measures. The effectiveness and computational cost analysis of FGS compared with two existing UE methods revealed that FGS can effectively evaluate segmentation results with the lowest computational cost.

Keywords: unsupervised evaluation; image segmentation; geographic object-based image analysis; area-weighted variance; difference to neighbor pixels; remote sensing

1. Introduction

With the rapid development of remote sensing technology, high spatial resolution remote sensing images can be more easily obtained and widely used in a variety of applications [1–4]. Compared with the low- and medium-resolution remote sensing image, the high-resolution remote sensing image contains more detailed spatial information, but the spectral resolution is lower [5]. If the pixel-based analysis method which only uses the spectral information of the image is applied to the high-resolution image, its rich spatial information will be ignored and more noise will be produced [6,7]. Therefore, geographic object-based image analysis (GEOBIA) has begun to emerge and can achieve better accuracy in the high-resolution image [8–10].

The purpose of GEOBIA is to effectively utilize the spatial and texture information of high-resolution images [11,12]. The first step of GEOBIA is to segment the image into a series of meaningful objects. Therefore, the segmentation effect has an important impact on the subsequent feature extraction, classification, etc. [13]. A large number of segmentation algorithms have been applied in GEOBIA. For example, the deep learning method [14,15], multi-resolution segmentation [16], watershed segmentation [17,18], and hybrid segmentation [19,20] are widely used in GEOBIA. In most of the segmentation algorithms, there is a parameter called "scale" to control the size of the object, which greatly affects the final segmentation results and the effect of GEOBIA [21–24]. Hence, evaluating image segmentation is critical to select optimal scale and obtain better segmentation results for subsequent analysis.

Direct evaluation methods can be classified into three variants: analytical evaluation (AE), supervised evaluation (SE), and unsupervised evaluation (UE) [25]. AE determinates the optimal scales by comparing a series of segmentation results by human visual inspection and it is widely used for simplicity [26]. However, it is time-consuming and greatly influenced by human subjectivity. SE evaluates segmentation quantitatively using discrepancy criteria between the segmentation results and the reference segmentation result [11,27,28]. SE overcomes human error and produces a more reliable, objective, and quantitative evaluation. Therefore, SE needs to build a reference dataset manually. However, building the whole reference dataset for high-resolution remote sensing image is tedious and time-consuming and involves some subjectivity. Moreover, the computation process and the matching algorithm between objects are very complex. The above disadvantages make it difficult to be applied in specific practical applications. UE needs no reference segmentation result and measures segmentation quality by defining evaluation criteria, which is based on human interpretation of the ideal segmentation result [24,29–32]. In the absence of prior knowledge of segmented images, UE can objectively evaluate the segmentation results without being affected by human subjectivity. UE can not only be used to evaluate the segmentation results but also be used for the selection of the segmentation algorithm and the setting of the segmentation parameters. Furthermore, since no prior knowledge is required, it is also used for segmentation parameters optimization which is not available in other segmentation evaluation methods [29,33,34]. Therefore, compared with AE and SE, UE is more applicable and more practical.

The widely accepted ideal state of segmentation results is as follows: for the features that can express the characteristics of objects, the objects should be consistent and homogeneous; the adjacent objects should have significant heterogeneity [35]. In the UE, the homogeneity goodness index is used to measure the homogeneity within the objects, and the heterogeneity goodness index is used to measure the heterogeneity between the objects, and then the two indexes are combined to evaluate the segmentation quality. There are texture-based and spectrum-based methods for homogeneity [36,37]. However, because computing the spectral information is simple and most of the segmentation methods are based on the spectral information, the research of the goodness based on the spectral information is far greater than the goodness based on the texture information. Compared with homogeneity, only spectrum-based methods are available for heterogeneity.

The area-weighted variance (WV) and global Moran's I (MI) are often used to compute the homogeneity within objects and the heterogeneity between objects, respectively [29,30,38,39]. Higher WV values indicate the higher dispersion of spectral values and the lower homogeneity within the object. Higher MI values indicate the higher the correlation and the lower the heterogeneity between objects. Wang et al. [22] proposed a UE method using WV and Jeffries–Matusita (JM) distance to evaluate segmentation quality. JM computes the spectral separability distance between objects to measure the spectral heterogeneity. Wang et al. [24] used between-segment heterogeneity (BSH) instead of MI to compute the heterogeneity between objects. The main difference between BSH and MI is that BSH uses the local statistics of the target object and its adjacent objects, while MI uses the global statistics of the whole image. The spectral value of a pixel can be represented as a vector, so Yang et al. [31]

used the spectral angle to measure the spectral homogeneity between objects. Although the existing methods play a certain role in the quality evaluation of image segmentation, most of the heterogeneity evaluation methods need to compute the differences between adjacent objects. In other words, we need to construct the region adjacency graph (RAG) to get the adjacency between objects. However, the process of RAG construction is very complex and time-consuming, especially when the image size is large and the segmentation scale value is low. As the size of the image increases, the number of segmented objects is increasing, and the construction time of RAG is also increasing geometrically, which makes the existing methods difficult to apply in practice. Up to now, far too little attention has been paid to computational cost. Therefore, a fast and effective UE method is needed.

In this paper, we present a fast and effective UE method for remote sensing images using the WV and difference to neighbor pixels (DTNP). Different from most existing heterogeneity evaluation measures, the difference with neighboring pixels was calculated in DTNP which avoids the time-consuming RAG construction and computation with neighboring objects. The WV and DTNP were used to compute the intra-segment homogeneity and inter-segment heterogeneity, respectively. The WV and DTNP were then combined into a fast-global score (FGS) to evaluate segmentation. In order to demonstrate the effectiveness and computational cost of DTNP and FGS, the other two existing methods were compared in the qualitative and quantitative analysis.

The main contributions of this paper are as follows: (1) Most of the existing inter-segment heterogeneity measures are based on RAG to compute the difference with neighboring objects, but the proposed DTNP computes the difference with neighboring pixels and is sensitive to both under-segmentation and over-segmentation. (2) The proposed FGS performs comparable to JM and WM, but the computational cost is much lower than JM and WM, which can be used for optimal segmentation parameter selection and segmentation parameter optimization. (3) Evaluating image segmentation by combining intra-segment homogeneity and inter-segment heterogeneity is more effective than evaluating image segmentation only by inter-segment heterogeneity, indicated by comparison of Sections 3.1 and 3.2

The rest of this paper is arranged as follows. In Section 2, we elaborate on the specific details of the proposed method as well as the experimental areas and data. Section 3 gives the experimental results. Sections 4 and 5 describe the discussion and conclusions, respectively.

2. Materials and Methods

2.1. Overview

The schematic of segmentation evaluation through FGS which combines WV and DTNP is shown in Figure 1. The FGS was computed by the intra-segment homogeneity and the inter-segment heterogeneity under a series of different segmentation scales. First, we represented intra-segment homogeneity by computing the WV. Second, we represented inter-segment heterogeneity by computing the DTNP. Third, WV and DTNP were combined into FGS, and the curve of FGS with scale was generated for segmentation evaluation. Note that the WV and DTNP were computed in eCognition Developer 9.0 [40], FGS was computed in Python 3.6, equipped in a computer with 64-bit Windows 10, Intel Core i5-8265U CPU at 1.8 GHz, and 8 GB RAM. The algorithm has been open source (https://github.com/mfzhao1998/FGS).

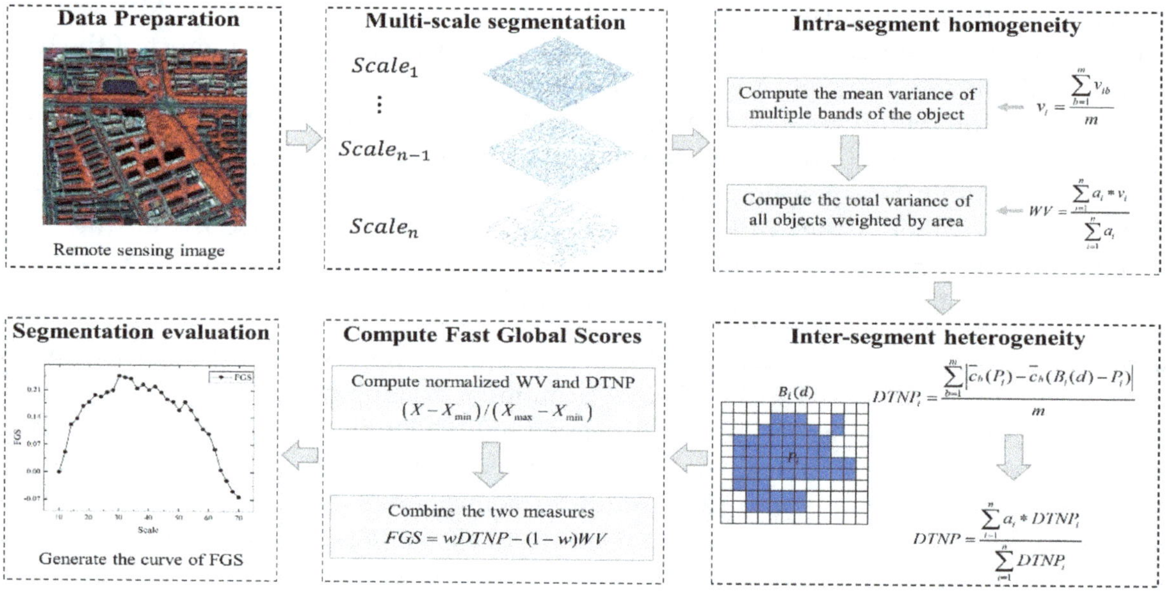

Figure 1. Schematic of the proposed method for segmentation evaluation.

2.2. Study Area and Data

A Gaofen-2 (GF-2) scene in Tongzhou District, Beijing, China, which was acquired on 9 September 2018, was applied to evaluate the performance of the proposed method. The GF-2 images contain four multi-spectral bands (blue, green, red, and near-infrared) with a resolution of 3.2 m and a panchromatic band with a resolution of 0.8 m. The NNDiffuse Pan Sharpening [41] which performs best when the combination of all multispectral bands covers the spectral range of the panchromatic raster in ENVI software [42] was used to fuse panchromatic data with multispectral data to obtain multispectral data with a resolution of 0.8 m. Four subsets of a residential area, an industrial area, a farmland area, and a mixed area in the GF-2 scene were selected for testing the proposed method (Figure 2).

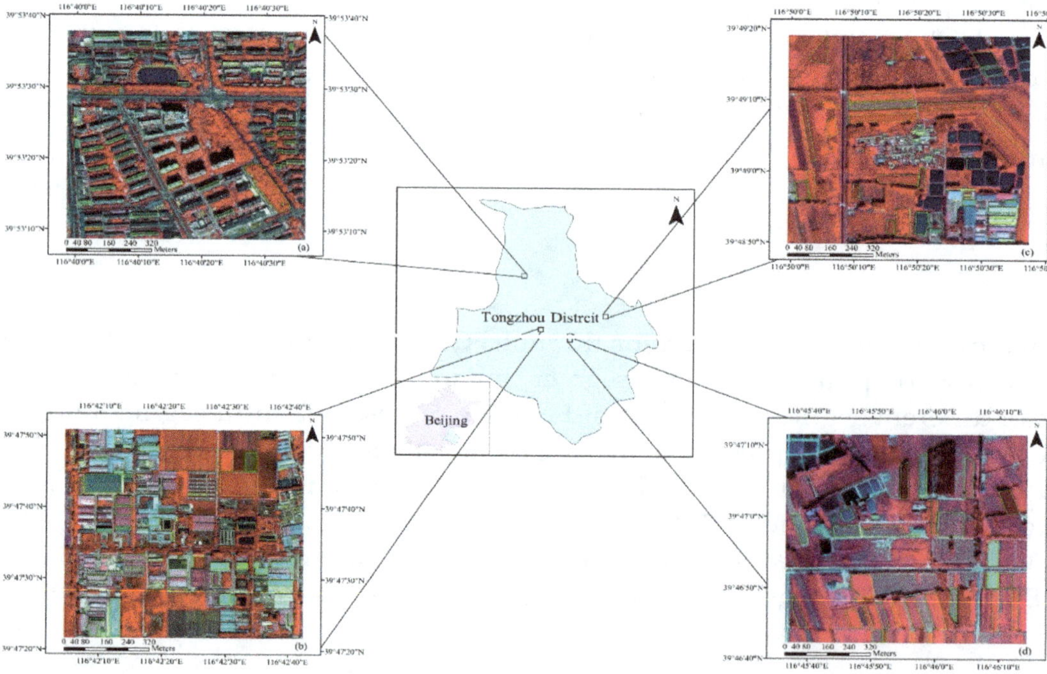

Figure 2. Four typical study areas of the GF-2 image: (**a**) T1, residential area; (**b**) T2, industrial area; (**c**) T3, mixed areas of residential areas, industrial areas, and farmland; and (**d**) T4, farmland area. Green polygons represent the manually digitized reference.

2.3. Image Segmentation

Image segmentation is a technique that segments the image into different image regions with high homogeneity and mutual connection, corresponding to the objects or spatial structure features of interest. The "Multi-resolution Segmentation" algorithm in eCognition Developer 9.0 was used to perform image segmentation. The details of the algorithm can refer to Benz et al. [43]. The algorithm starts from a single pixel and uses the "bottom-up" region merging method to form polygon objects with different sizes (scales). Two adjacent image regions merge and grow. The judgment of merging depends on the heterogeneity of adjacent objects. Several small objects can be merged into a large object through several steps. The merging of two objects must ensure that the heterogeneity of them is less than the given threshold. The final object size is decided by this threshold. Therefore, this threshold can be regarded as a scale parameter. Other segmentation parameters include color factor and shape factor, and the weights between them. The color factor is the spectral characteristics of the image—the weight of each band. The shape factor is composed of two parameters, smoothness and compactness, which can prevent the image object from being broken too much, so that the segmented object has the shape characteristics of the actual object.

In this paper, the color factor to shape factor ratio was set to 9:1. All spectral bands were used and given the same weight. Smoothness and compactness were given the same weight. A series of segmentation produced by the Multi-resolution Segmentation was carried out with different scale parameters at intervals of 2, ranging from 10 to 70, so as to adapt to different scenes such as cities and suburbs.

2.4. Unsupervised Evaluation Using Fast-Global Score

Most evaluation methods mainly consider intra-segment homogeneity and inter-segment heterogeneity, excluding internal continuity and boundary complexity that are difficult to apply to remote sensing images [25,30]. Appropriate segmentation parameters were defined that minimized intra-segment homogeneity and maximized inter-segment heterogeneity [29]. In other words, the difference within an object is the lowest, and the difference between the objects is the highest. First, the variance was used as global intra-segment homogeneity, weighted by each segment's area. It is defined as follows:

$$v_i = \frac{\sum_{b=1}^{m} v_{ib}}{m} \tag{1}$$

where m is the band number of the image, v_{ib} is the variance of object i in band m, and v_i is the mean variance of object i averaged by all bands.

$$WV = \frac{\sum_{i=1}^{n} a_i * v_i}{\sum_{i=1}^{n} a_i} \tag{2}$$

where n is the total object number and a_i is the area of object i. WV gives the same weight to each band, and we give v_i different weight according to the size of the area, avoiding the instability caused by small segments. Third, the DTNP of each object was computed as follows:

$$DTNP_i = \frac{\sum_{b=1}^{m} \left|\bar{c}_b(P_i) - \bar{c}_b(B_i(d) - P_i)\right|}{m} \tag{3}$$

$B_{i(d)}$ is the extended bounding box of object i with distance d (Figure 3),which is equal to $\left\{(\mathrm{x},\mathrm{y}) : \mathrm{x}_{\min}(\mathrm{i}) - \mathrm{d} \leq \mathrm{x} \leq \mathrm{x}_{\max}(\mathrm{i}) + \mathrm{d}, \mathrm{y}_{\min}(\mathrm{i}) - \mathrm{d} \leq \mathrm{x} \leq \mathrm{y}_{\max}(\mathrm{i}) + \mathrm{d}\right\}$, P_i is the set of pixel of object i, $\bar{c}_b$ is the mean of band b. $DTNP_i$ is the DTNP of object i. A global DTNP is formed by area weighting $DTNP_i$ of each segment:

$$DTNP = \frac{\sum_{i=1}^{n} a_i * DTNP_i}{\sum_{i=1}^{n} DTNP_i} \tag{4}$$

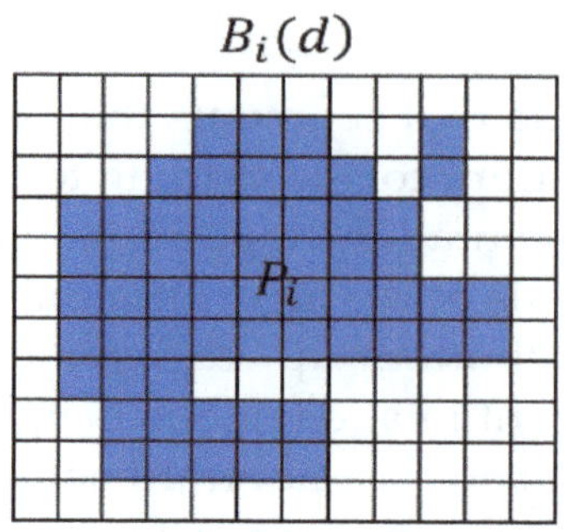

Figure 3. The surrounding area of object *i* defined by its bounding box with the distance of one pixel.

This approach takes local spectral difference into full consideration and make the result more reasonable by area weighting. Compared with other methods [24,30], the computation of DTNP is more convenient because it does not need to construct RAG and compute heterogeneity with multiple neighboring objects. Finally, WV and DTNP are combined into FGS to evaluate the quality of segmentation considering both the intra-segment homogeneity and inter-segment heterogeneity. In order to consider intra-segment homogeneity and inter-segment heterogeneity equally, WV and DTNP were normalized to a 0–1 scale using the formula in Equation (5):

$$(X - X_{min})/(X_{max} - X_{min}) \tag{5}$$

where X is one of WV or DTNP obtained by a set of segmentation parameters, X_{min} and X_{max} are the minimum and maximum value of WV or DTNP. Note that Low WV and high DTNP values represent better intra-segment homogeneity and inter-segment heterogeneity, respectively. To assign an overall FGS to each segmentation result, WV and DTNP were combined using the formula in Equation (6):

$$FGS = wDTNP - (1 - w)WV \tag{6}$$

where weight w can determine the relative weights of the intra-segment homogeneity and inter-segment heterogeneity. In this paper, the same weight was given to WV and DTNP. Note that the higher FGS value indicates better segmentation quality.

2.5. Accuracy Assessment Measures for the Proposed Method

The supervised quantitative evaluation method is often used to evaluate the accuracy of UE methods and has been proposed in quantities [25]. The index of quality rate (QR), over-segmentation (OS), under-segmentation (US), and D which is composed of US and OS were used to validate the effectiveness of the proposed method [26]. QR represents the similarity of the reference object and the corresponding object, and ranges from 0 to 1. Zero means that the similarity between them is maximum and the segmentation result is the best. OS and US evaluate the degree of over-segmentation and under-segmentation by calculating the ratio of over-segmented area to reference object and the ratio of under-segmented area to the corresponding object, respectively. Both OS and US have a range of (0, 1). Zero means neither over-segmentation nor under-segmentation. D combines OS and US and considers both over-segmentation and under-segmentation. A lower D value reflects higher segmentation quality.

In this experiment, manual visual interpretation in ArcGIS 10.6 was used to generate reference objects for four typical study areas, and the above four accuracy assessment measures were computed to verify the effectiveness of DTNP and FGS. In this paper, we generated 40 reference objects for T1, T2, and 30 reference objects for T3, T4 (a total of 140 reference objects), as shown in Figure 2 (Green polygon).

2.6. Comparison with Other UE Methods and Inter-Segment Heterogeneity Measures

To demonstrate the effectiveness of the proposed method, the two methods proposed by Johnson et al. [30] and Wang et al. [24] were compared to FGS. The three methods use the same intra-segment homogeneity measure, but the inter-segment heterogeneity measure is different. Johnson's method used MI to measure the correlation between all segments, and its effectiveness has been proven in many studies [22,24,29,30,39]. The BSH was used by Wang's method to measure inter-segment heterogeneity, focusing on enhancing the objectivity of heterogeneity with local spatial statistics compared to MI. Different from MI and BSH, DTNP computes the difference to neighbor pixels.

To individually evaluate the effectiveness of DTNP, the UE method proposed by Yang et al. [31], which only uses intra-segment homogeneity and inter-segment heterogeneity was used. As the scale parameter increases, the inter-segment heterogeneity decreases while DTNP increases. When representative objects are matched, since inter-segment heterogeneity almost remains after reaching the optimal scale parameters, the growth trend of inter-segment heterogeneity will suddenly weaken or stop. The index of $\dot{H}$ which measures changes in inter-segment heterogeneity between scales is defined by the expression given in Equation (7):

$$\dot{H} = \frac{H(l) - H(l - \Delta l)}{\Delta l} \tag{7}$$

where $H(l)$ is the inter-segment heterogeneity value when the scale is l and Δl is the interval of scale parameter. Based on the above analysis, when the segmentation result with the segmentation scale parameter $l - \Delta l$ is close to the appropriate segmentation result, at the scale parameter l, $\dot{H}(l)$ is usually higher or lower than $\dot{H}(l - \Delta l)$ and $\dot{H}(l + \Delta l)$. Higher or lower depending on the trend of $H(l)$. The index of I which measures the local peak in the curves of $\dot{H}(l)$ is defined by the expression given in Equation (8):

$$I = \left(\dot{H}(l) - \dot{H}(l + \Delta l)\right) + \left(\dot{H}(l) - \dot{H}(l - \Delta l)\right) \tag{8}$$

With the increase of scale parameters, DTNP keeps increasing while MI and BSH keep decreasing. Therefore, when using MI and BSH to measure inter-segment heterogeneity, the lowest value of I represents the optimal segmentation parameter. When using DTNP to measure inter-segment heterogeneity, the largest value of I represents the optimal segmentation parameter.

In this paper, three inter-segment heterogeneity measures were evaluated by the optimal segmentation results obtained by the index of I in Section 3.1. To evaluate the effectiveness of FGS, we qualitatively analyzed its effectiveness through visual assessment, and qualitatively evaluate its effectiveness through the accuracy assessment measures in Section 3.2 by comparing with Johnson's method [33] and Wang's method [21].

3. Results

3.1. Effectiveness Analysis of DTNP

The Multi-resolution Segmentation algorithm in eCognition Developer 9.0 was used in four study areas to generate 31 segmentation results with 2 intervals, ranging from 10 to 70 for each study area. We used the method stated in Section 2.6 to compute the $\dot{H}$ and I of DTNP, MI, and BSH to verify the inter heterogeneity measure, as presented in Figure 4. Lower DTNP values indicate lower inter-segment heterogeneity, while lower MI and BSH indicate higher inter-segment heterogeneity. As segmentation scale parameters increase, the DTNP value was increasing, and the MI and BSH value was decreasing gradually in the four study areas. From the $\dot{H}$ value, we can find that the change trend of DTNP and MI was more stable than BSH. Because the three indices have different trends with scale, the maximum I value of DTNP indicates the optimal scale parameter, and the minimum I value of MI and BSH indicates the optimal scale parameter. For the T1 image, the optimal segmentation using the three inter-segment heterogeneity measures was obtained by using the scale at 30, 60, and 20, respectively. For the T2 image, the optimal segmentation using the three inter-segment heterogeneity measures was obtained by using the scale at 66, 66, and 50, respectively. For the T3 image, the optimal segmentation using the three inter-segment heterogeneity measures was obtained by using the scale at 52, 56, and 20, respectively. For the T4 image, the optimal segmentation using the three inter-segment heterogeneity measures was obtained by using the scale at 58, 62, and 32, respectively.

The local details of the optimal segmentation results of the four study areas obtained by DTNP, MI, and BSH are shown in Figure 5. In the first subset of T1, the playground was segmented well in the DTNP result, but it was under-segmented in the MI result and over-segmented in the BSH result. In the second subset of T1, buildings were effectively separated from shadows and some roads in the DTNP and BSH result, whereas there were some buildings mixed with shadows and some roads in the MI result. In the subset of T2, it was more under-segmented in the DTNP and MI result than in the BSH result. The buildings were not effectively separated from the surrounding buildings. In the first subset of T3, the results of DTNP and BSH were equivalent. Both can better separate buildings from buildings, and buildings from trees, but one object was under-segmented, which failed to effectively separate the two buildings. Each geographical object obviously exhibited serious over-segmentation phenomenon in the result of BSH. In the second subset of T3 and the first subset of T4, the farmland was segmented well in the DTNP and MI result, but it was over-segmented in the BSH result. In the second subset of T4, the farmland was segmented well in the DTNP result, but it was under-segmented in the MI result and over-segmented in the BSH result.

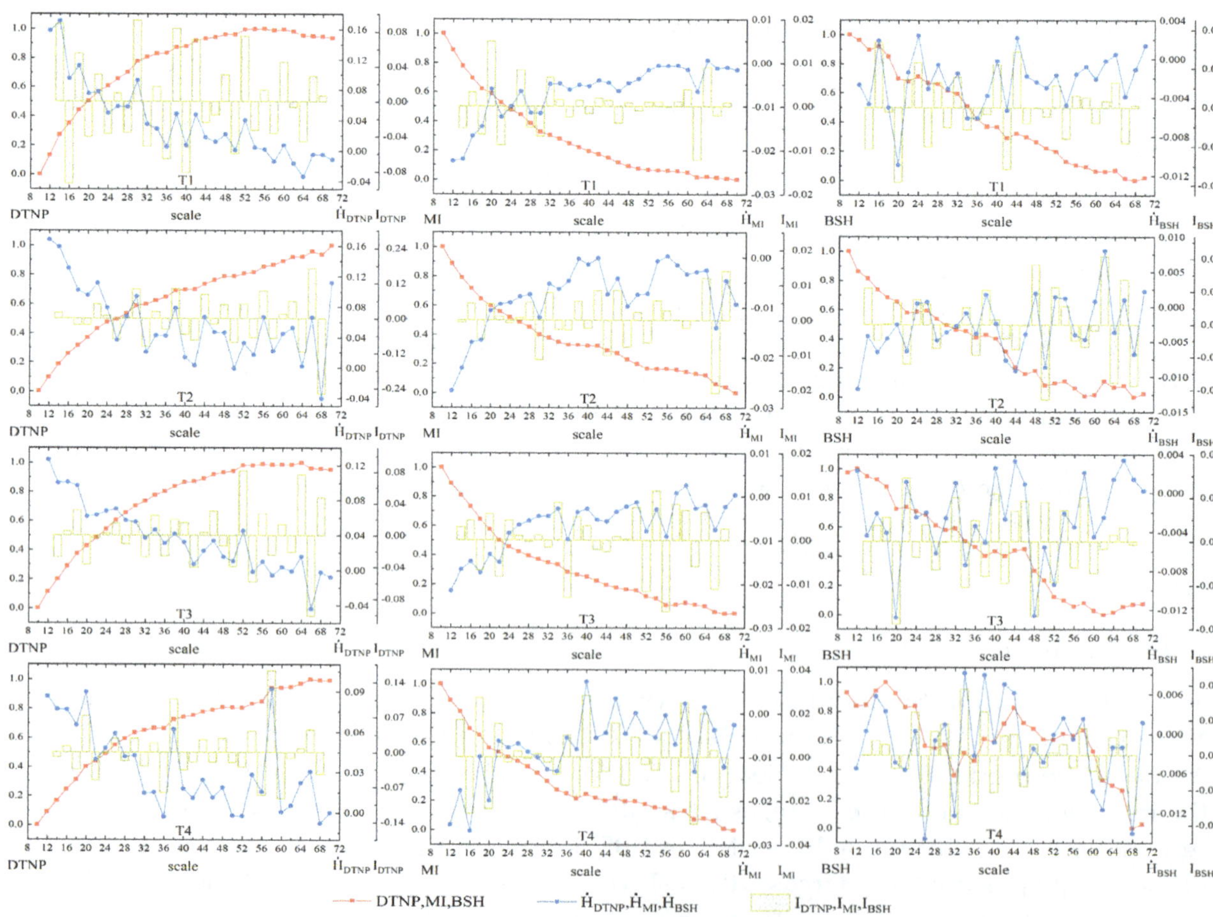

Figure 4. The $\dot{H}$, I, and normalized results of difference to neighbor pixels (DTNP), Moran's I (MI), and between-segment heterogeneity (BSH) of four test images.

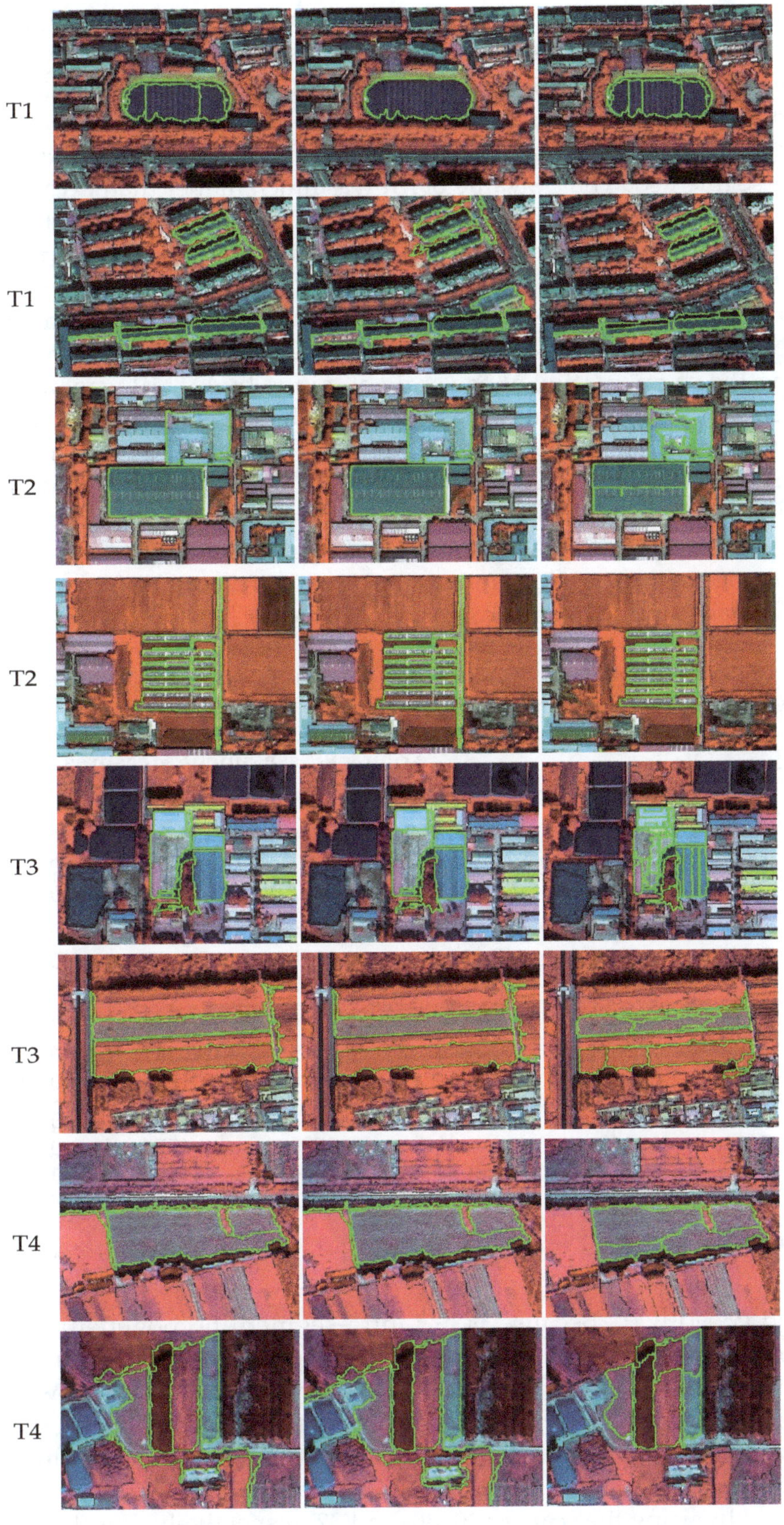

Figure 5. Subsets of the optimal segmentation for four test images obtained by the DTNP and other existing two inter-segment heterogeneity methods.

In order to quantitatively evaluate the effectiveness of DTNP, Table 1 shows the accuracy assessment results of the optimal segmentation for four study areas using the UE methods which only based on inter-segment heterogeneity. For the T1 image, the QR and D values of the DTNP methods were both lower than MI and BSH, thus, indicating that the DTNP had a better performance in the T1 result. Furthermore, the OS value of BSH was much higher than DTNP and MI, indicating that the result of BSH was more over-segmented. The US value of MI was much higher than DTNP and BSH, indicating that the result of MI was more under-segmented. For the T2 image, DTNP performed the same as MI, and the BSH had a better performance in the T2 result, because the QR and D values of the BSH methods were both lower than DTNP and MI. From the value of OS and US, we can find that the result of BSH was more over-segmented and the result of DTNP and MI was more under-segmented. For the T3 image, both QR and D values at the optimal scale obtained by DTNP were lower than MI and BSH, indicating the segmentation at the scale obtained by DTNP was better. The OS value of DTNP and MI was similar and much lower than that of BSH, indicating that the result of BSH was more over-segmented. From the value of US, we can find that the result of MI was most under-segmented and the result of BSH had less over-segmentation. For the T4 image, the QR and D values of the BSH methods were both lower than DTNP and MI, indicating that the BSH had a better performance in the T4 result. From the value of OS and US, we can find that the result of BSH was most over-segmented but had the best performance in the under-segmentation, and the result of MI was most under-segmented but had a best performance in the over-segmentation.

Table 1. The accuracy assessment measures results for the segmentations produced by the DTNP and other existing two inter-segment heterogeneity methods.

Test	Methods	Accuracy Assessment Measures			
		QR	OS	US	D
T1	**DTNP**	**0.2875**	**0.0849**	**0.2255**	**0.1914**
	MI	0.5636	0.0281	0.5512	0.3931
	BSH	0.3275	0.2282	0.1382	0.2188
T2	DTNP	0.4236	0.0338	0.3973	0.2961
	MI	0.4236	0.0338	0.3973	0.2961
	BSH	**0.3383**	**0.0708**	**0.2808**	**0.2338**
T3	**DTNP**	**0.3491**	**0.0589**	**0.3055**	**0.2363**
	MI	0.3806	0.0579	0.3517	0.2614
	BSH	0.4927	0.4528	0.1265	0.3470
T4	DTNP	0.4204	0.0533	0.3884	0.2886
	MI	0.4864	0.0508	0.4575	0.3368
	BSH	**0.3013**	**0.1906**	**0.1461**	**0.2025**

3.2. *Effectiveness Analysis of FGS*

To further prove the effectiveness of FGS, this paper compared it with Johnson's method (JM) [30] and Wang's method (WM) [24]. Because FGS's intra-segment homogeneity evaluation indicators are the same as them, the difference lies in the inter-segment heterogeneity evaluation indicators. Note that all three methods use the same normalization method, thus, the lower measure values in Johnson's method (JM) indicate higher segmentation quality and the higher measure values in the FGS and Wang's method (WM) indicate higher segmentation quality. The optimal results obtained by the three methods are shown in Figure 6. For the T1 image, the optimal segmentation using the three methods was obtained by using the scale at 30, 30, and 32, respectively. For the T2 image, the optimal segmentation using the three methods was obtained by using the scale at 24, 18, and 38, respectively. For the T3 image, the optimal segmentation using the three methods was obtained by using the scale at 34, 24, and 32, respectively. For the T4 image, the optimal segmentation using the three methods was obtained by using the scale at 30, 38, and 44, respectively.

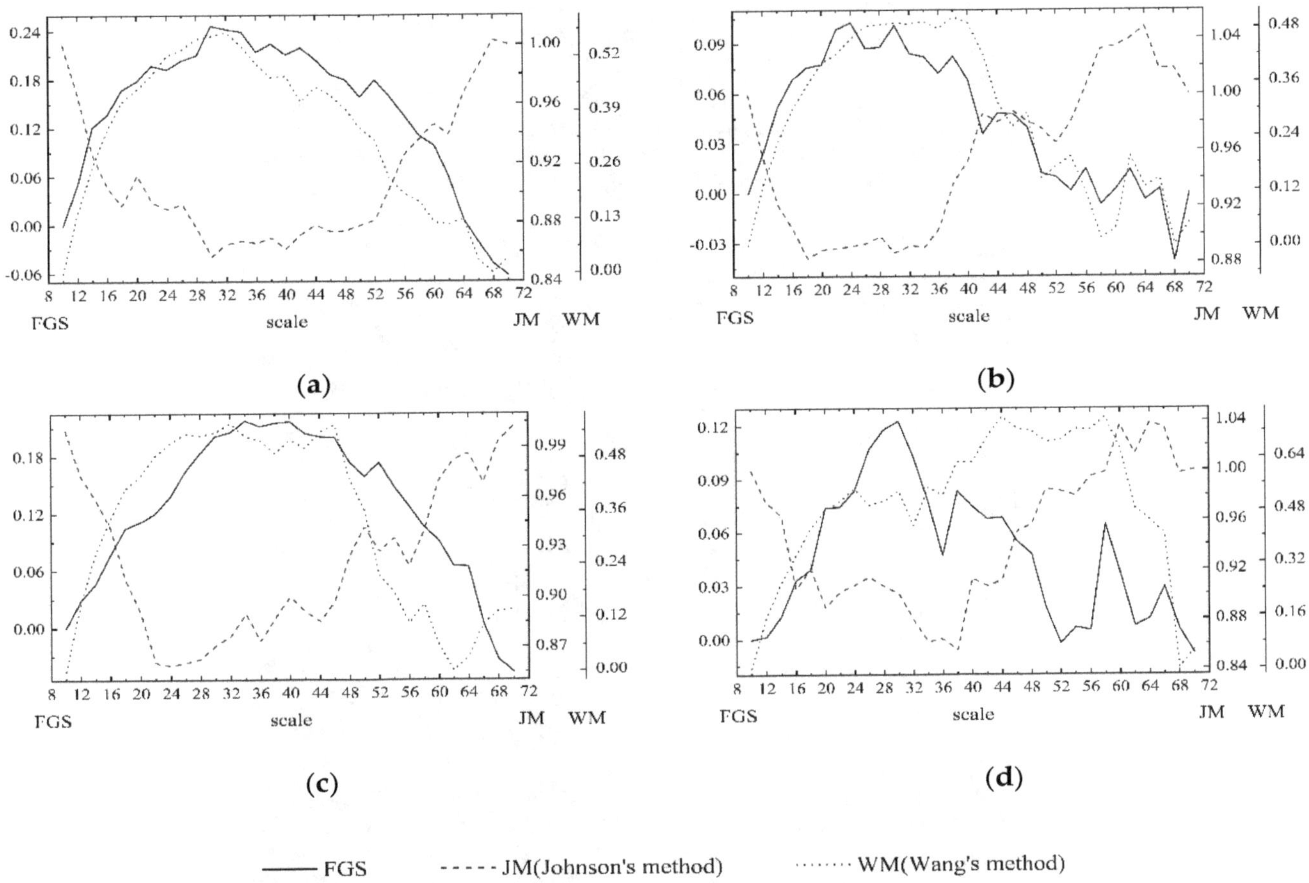

Figure 6. The unsupervised evaluation (UE) results of the segmentations produced by the different UE methods for the four study areas: (**a**) T1, (**b**) T2, (**c**) T3, and (**d**) T4.

The local details of the optimal segmentation results of the four study areas obtained by FGS, JM, and WM are shown in Figure 7. In the first subset of T1, the playground was segmented better in the FGS and JM result, and it was under-segmented in the WM result. In the second subset of T1, the results of the three methods were under-segmented, because the buildings were confused with some vegetation, and the result of WM was more under-segmented. In the first subset of T2, the FGS and JM had similar performance and the result of them was more over-segmented, but the result of WM showed the phenomenon of under segmentation locally. In the second subset of T2, the results were over-segmented in the FGS and JM results. In the WM result, buildings and vegetation were not effectively separated, and under-segmentation occurred. In the first subset of T3, the buildings were all over-segmented in the results of the three methods, but the FGS and WM performed better than JM. In the second subset of T3, the farmland was all over-segmented in the results of the three methods, but geographic objects in JM were more fragmented. In the two subsets of T4, the and WM performed better in segmenting farmland than FGS, although the results of the three methods were all over-segmented.

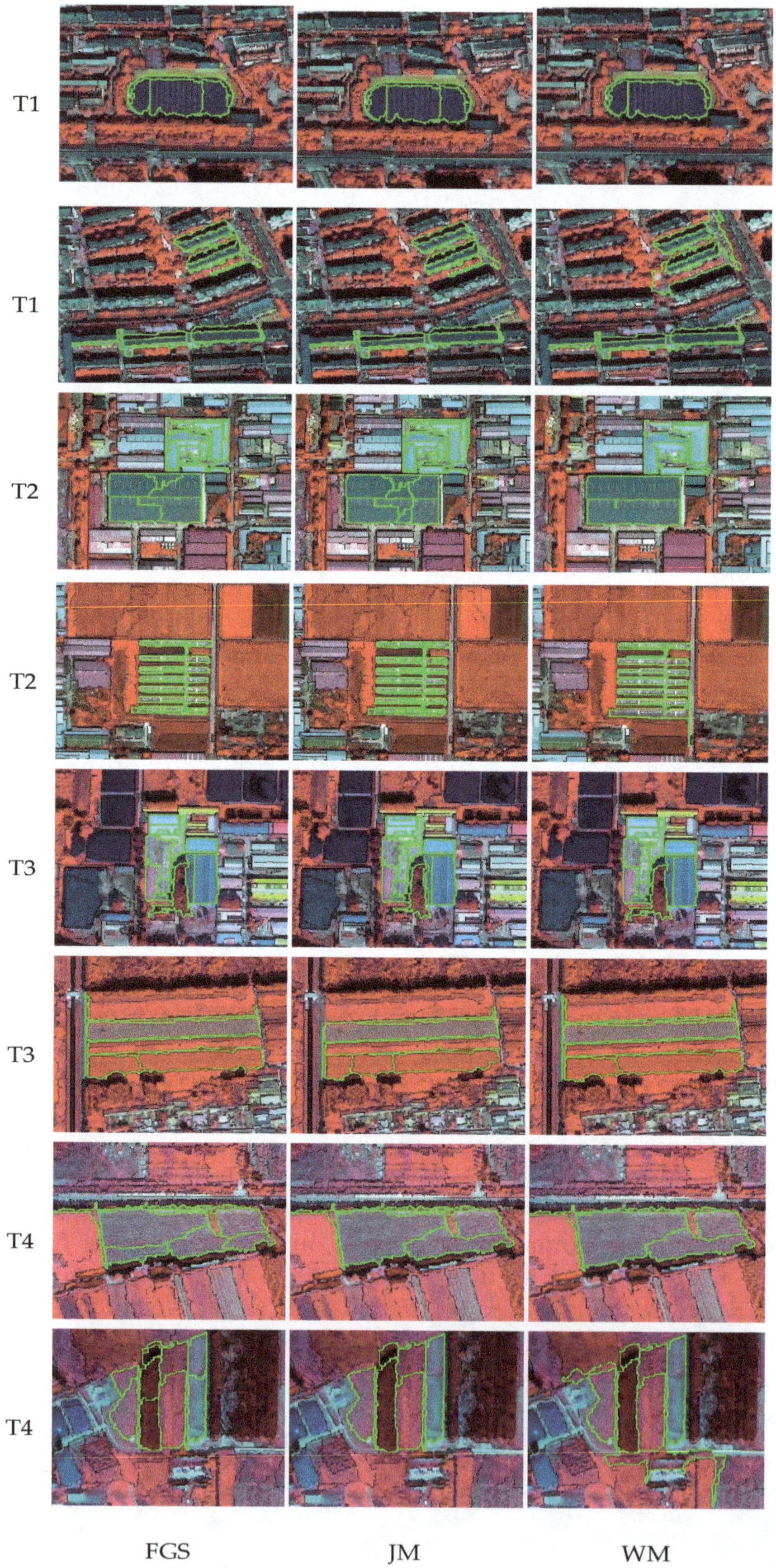

Figure 7. Subsets of the optimal segmentation for four test images obtained by the fast-global score (FGS) and other existing two UE methods.

In order to quantitatively evaluate the effectiveness of FGS, Table 2 shows the accuracy assessment results of the optimal segmentation for four study areas using the three methods. For the T1 image, the QR and D values of the WM methods were both higher than FGS and JM, thus, indicating that the FGS and JM had a better performance in the T1 result. From the value of OS and US, we can find that the result of WM was most under-segmented and the three methods were equivalent in over-segmentation. For the T2 image, the QR and D values of the WM methods were both lowest, indicating that the WM had the best performance. The QR and D values of the JM methods were both highest, indicating that the JM had the worst performance. The OS value of JM was much higher than FGS and WM, indicating that the result of JM was more over-segmented. The US value of WM was much higher than FGS and JM, indicating that the result of WM was more under-segmented. For the T3 image, FGS performed the same as WM, and the JM had the worst performance in the T2 result, because the QR and D values of the JM methods were both higher than FGS and WM. From the value of OS and US, we can find that the result of JM was more over-segmented and the result of FGS and WM was more under-segmented. For the T4 image, both QR and D values at the optimal scale obtained by WM were lower than FGS and JM, indicating the segmentation at the scale obtained by WM was better. From the value of OS, we can find that the result of FGS was most under-segmented and the result of WM had less over-segmentation. The US value of JM and WM was similar and much higher than that of FGS, indicating that the results of JM and WM were more under-segmented.

Table 2. The accuracy assessment measures results for the segmentations produced by the FGS and other existing two UE methods.

Test	Methods	Accuracy Assessment Measures			
		QR	OS	US	D
T1	**FGS**	**0.2875**	**0.0849**	**0.2255**	**0.1914**
	Johnson's method	**0.2875**	**0.0849**	**0.2255**	**0.1914**
	Wang's method	0.3016	0.0839	0.2414	0.2012
T2	FGS	0.2843	0.2293	0.0856	0.1944
	Johnson's method	0.4242	0.3913	0.0864	0.2974
	Wang's method	**0.2795**	**0.1319**	**0.1702**	**0.1909**
T3	**FGS**	**0.3281**	**0.1744**	**0.1894**	**0.2212**
	Johnson's method	0.3924	0.3271	0.1285	0.2697
	Wang's method	**0.3281**	**0.1744**	**0.1894**	**0.2212**
T4	FGS	0.3237	0.2487	0.1184	0.2194
	Johnson's method	0.2969	0.1171	0.2073	0.1978
	Wang's method	**0.2915**	**0.0830**	**0.2269**	**0.1929**

3.3. The Performance of FGS on Other Datasets

In order to evaluate the performance of FGS more effectively, Massachuetts Buildings Dataset was used [44], which can be obtained from the website (http://www.cs.toronto.edu/~vmnih/data/). The images in this dataset contain three multi-spectral bands (blue, green, and red) with a resolution of 1 m. In this paper, four images with the size of 1500 × 1500 were selected for the experiment (Figure 8). The optimal results obtained by the three methods are shown in Figure 9. For the T5 image, the optimal segmentation using the three methods was obtained by using the scale at 28, 24, and 34, respectively. For the T6 image, the optimal segmentation using the three methods was obtained by using the scale at 32, 32, and 28, respectively. For the T7 image, the optimal segmentation using the three methods was obtained by using the scale at 32, 44, and 28, respectively. For the T8 image, the optimal segmentation using the three methods was obtained by using the scale at 30, 28, and 26, respectively.

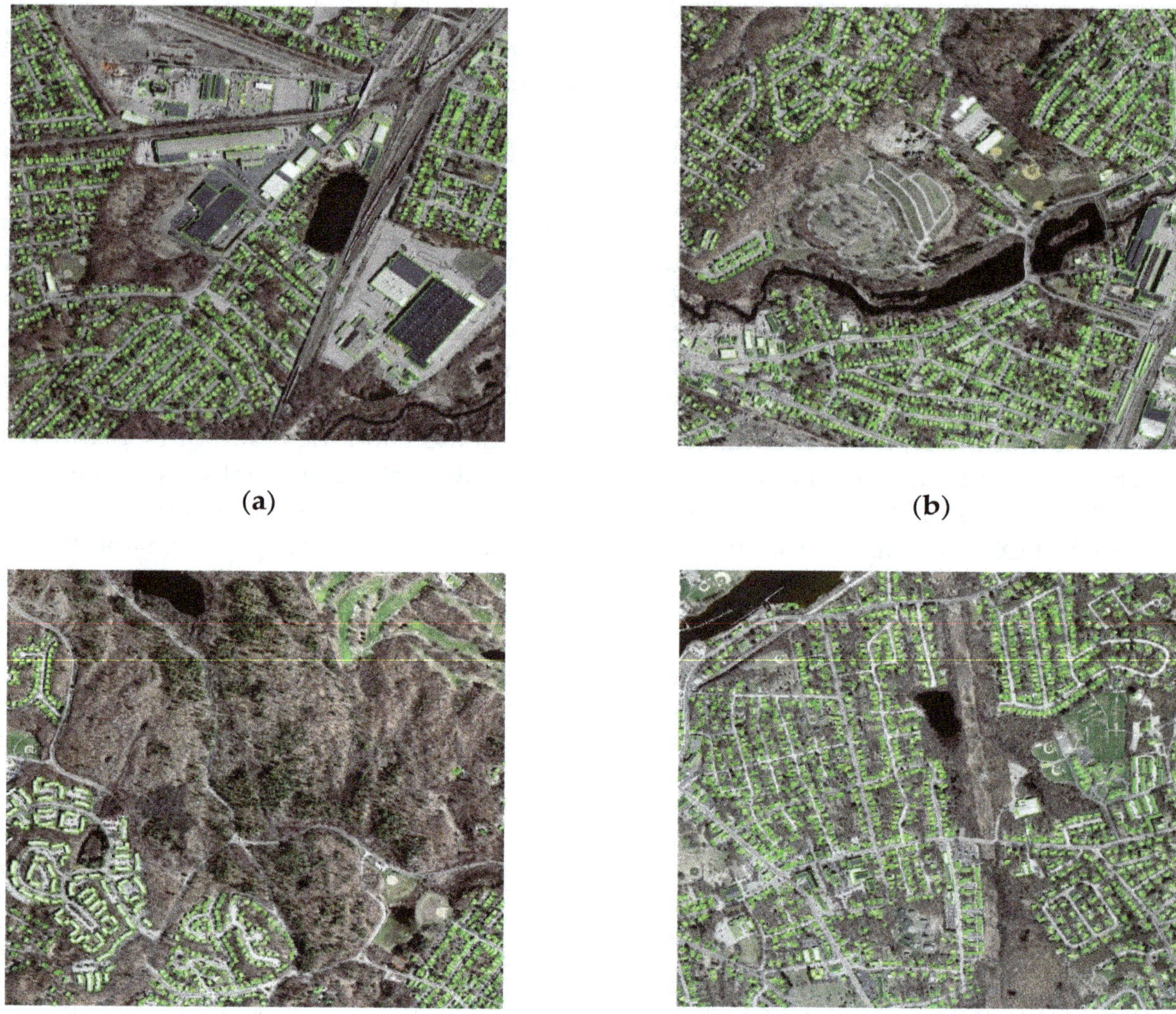

(a) (b) (c) (d)

Figure 8. Four images in Massachuetts Buildings Dataset: (**a**) T5, (**b**) T6, (**c**) T7, and (**d**) T8. Green polygons represent the reference of buildings.

The accuracy assessment results of the optimal segmentation for four study areas using the three methods are presented in Table 3. For the T5 image, the performance of FGS lied between WM and JM, indicated by the lower QR and D values. The OS value of JM was highest, indicating that the result of JM was more over-segmented. The US value of WM was highest, indicating that the result of WM was more under-segmented. For the T6 image, FGS performed the same as JM, and both performed worse than WM, indicated by the lower QR and D values. From the value of OS and US, we can find that WM performed worse in over-segmentation and better in under-segmentation. For the T7 image, the QR and D values of the FGS were lower than WM and higher than JM, thus, indicating that the FGS performed better than WM and worse than JM. From the value of OS and US, we can find that the result of JM was most under-segmented and the result of WM was most over-segmented. For the T8 image, both QR and D values at the optimal scale obtained by FGS were higher than JM and WM, indicating the segmentation at the scale obtained by FGS was worst. From the value of OS and US, we can find that the result of WM was most over-segmented and the result of FGS was most under-segmented. In addition, the QR and D values were not much different on Massachuetts Buildings Dataset, indicating that the performance of the three methods was very similar.

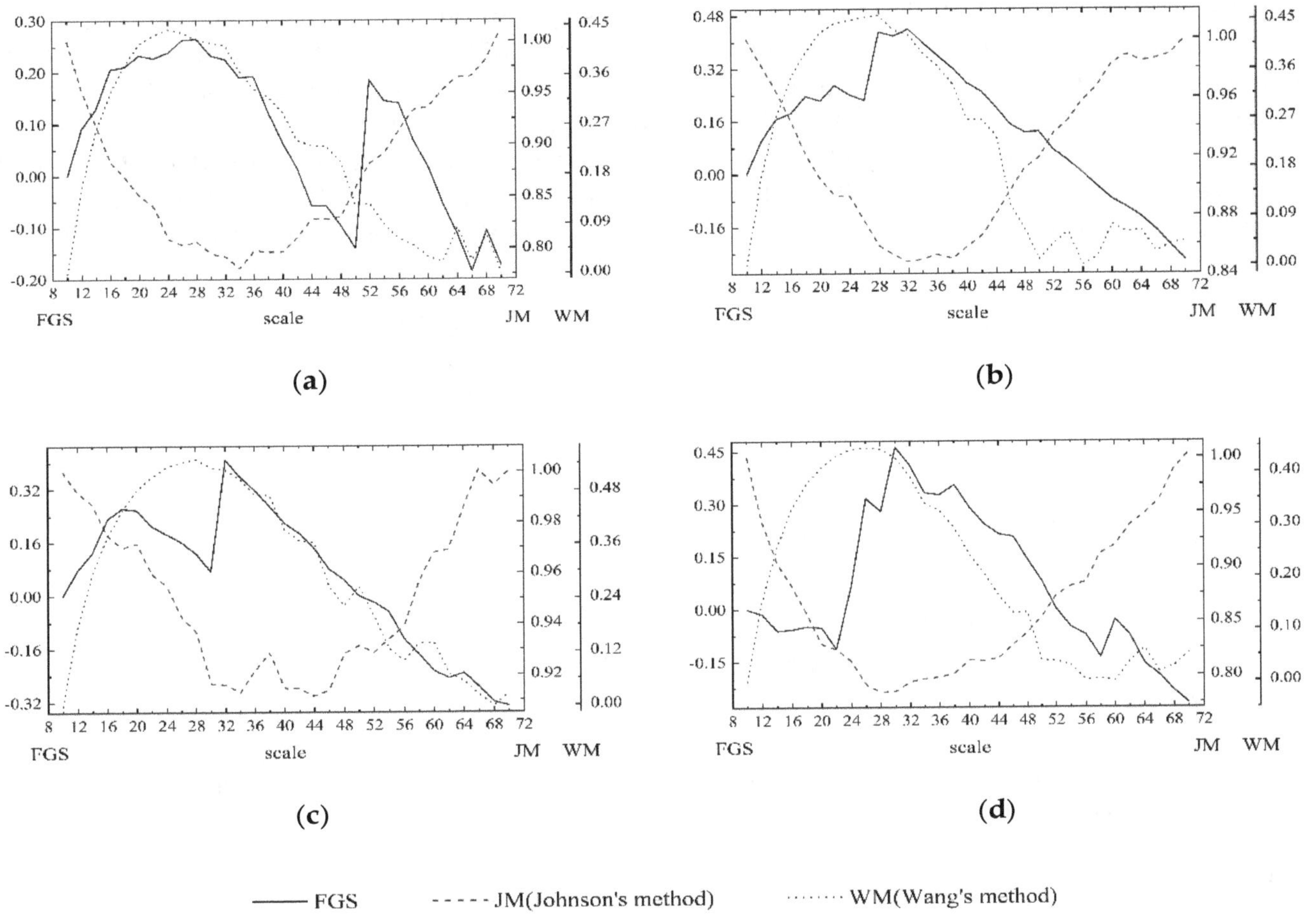

Figure 9. The UE results of the segmentations produced by the different UE methods for the Massachuetts Buildings Dataset: (**a**) T5, (**b**) T6, (**c**) T7, and (**d**) T8.

Table 3. The accuracy assessment measurement results for the segmentations produced by the FGS and other existing two UE methods, using the Massachuetts Buildings Dataset.

Test	Methods	Accuracy Assessment Measures			
		QR	OS	US	D
T5	FGS	0.6411	0.4502	0.3709	0.4589
	Johnson's method	0.6503	0.5059	0.3185	0.4650
	Wang's method	**0.6371**	**0.3877**	**0.4307**	**0.4564**
T6	FGS	0.6231	0.3828	0.4091	0.4437
	Johnson's method	0.6231	0.3828	0.4091	0.4437
	Wang's method	**0.6206**	**0.4276**	**0.3622**	**0.4419**
T7	FGS	0.6131	0.4192	0.3392	0.4351
	Johnson's method	**0.6116**	**0.3260**	**0.4345**	**0.4348**
	Wang's method	0.6276	0.4715	0.3036	0.4467
T8	FGS	0.6491	0.3881	0.4446	0.4658
	Johnson's method	**0.6439**	**0.4077**	**0.4181**	**0.4615**
	Wang's method	0.6465	0.4363	0.3924	0.4635

3.4. Computational Cost

This paper uses T1 as an example to analyze the computational cost of FGS, JM, and WM, and the relationship with the total number of objects. Because the intra-segment homogeneity measures of the three methods of FGS, JM, and WM are the same and the difference is the inter-segment heterogeneity measure, the computing efficiency of DTNP, MI, and BSH also reflects the efficiency of the three UE methods. Please note that all three measures of inter-segment heterogeneity were computed under the environment of eCognition Developer 9.0. DTNP, MI, BSH and the total number of objects at different scales of T1 are shown in Figure 10. As the scale increases, the number of objects obtained by segmentation decreases geometrically. The computational cost of DTNP, MI and BSH is positively correlated with the number of objects, and the correlation coefficients are 0.9906, 0.9999 and 0.9999 respectively, P values are all less than 0.05. Under the same scale, DTNP has the least computational cost, far less than MI and BSH. Therefore, FGS has the least computational cost compared with JM and WM.

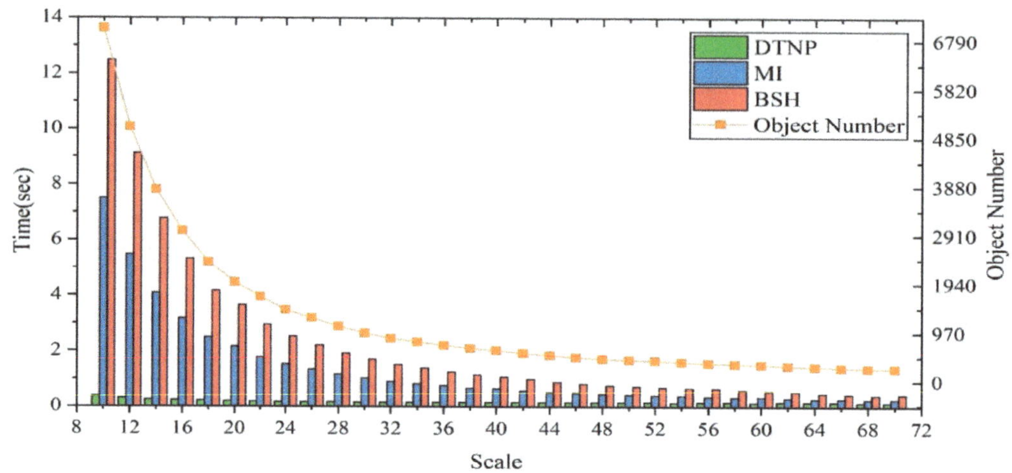

Figure 10. Execution time of three inter-segment heterogeneity methods and object number at different scales.

4. Discussion

With the development of aviation and aerospace remote sensing technology, high-resolution remote sensing images containing more detailed spatial information are increasingly used. In order to make better use of spatial details, image segmentation is an indispensable processing step. The quality of image segmentation directly affects the accuracy of subsequent analysis, so it is important to evaluate image segmentation. Although the SE method can provide more accurate results, it requires manual generation of reference objects, which makes it difficult to be widely used in practical applications. The UE method can objectively evaluate the segmentation results without prior knowledge. Compared with SE, UE can not only be used for comparison and selection of segmentation methods, and the setting of segmentation parameters, but also for segmentation parameters optimization. Therefore, the UE method has gained more attention.

This study proposed the FGS combining intra-heterogeneity and inter-homogeneity to evaluate image segmentation. Compared with the existing UE method, FGS uses the same inter-segment heterogeneity evaluation measure, but creatively uses DTNP as the inter-segment heterogeneity measure. As the scale increases, the inter-segment heterogeneity also increases. The DTNP curve shown in Figure 8 can clearly reveal the process of increasing heterogeneity. FGS is always low when over- or under-segmented. The FGS curve shown in Figure 6 can clearly reveals the change from under-segmentation to optimal segmentation results and under-segmentation.

DTNP was proposed to measure inter-segment heterogeneity quickly and effectively. Through qualitative analysis of three inter-segment heterogeneity measures (Figure 5), we found that the BSH had a higher separability and MI had a lower separability in segmentation results, but the result was also most over-segmented in BSH result. Through quantitative analysis of three inter-segment heterogeneity measures (Table 1), we found that the QR and D values of DTNP were the lowest in the T1 and T3, the same as MI and were both lower than BSH in the T3, and the highest in the T4, thus, indicating that DTNP has comparable performance to MI and BSH as the inter-segment heterogeneity measure. Furthermore, the OS value of BSH was highest and the US value of BSH was lowest in four study areas, indicating that BSH performs best in under-segmentation but worst in over-segmentation. The OS value of MI was lowest and the US value of MI was highest in four study areas, indicating that MI performs best in over-segmentation but worst in under-segmentation. From the value of OS and US, we can find that DTNP can perform better than BSH on over-segmentation and better than MI on under-segmentation. Therefore, compared to MI and BSH, which are more sensitive to over-segmentation or under-segmentation, DTNP can keep stability and remain sensitive to both over-segmentation and under-segmentation.

To further prove the effectiveness of the proposed FGS, we compared it with JM and WM. The qualitative results (Figure 7) show that the segmentation results determined by the three methods can distinguish different geographic objects, but cannot accurately distinguish all geographic objects, because a single segmentation scale is not applicable to all geographic objects. Therefore, it is necessary to further study the optimization of segmentation parameters. The QR and D in the qualitative results (Table 2) indicate that FGS does not always perform better than the JM and WM methods. However, in the quantitative results of T1 and T3 images, the QR and D values of FGS were the lowest, indicating that FGS performs best. The experiment on Massachuetts Buildings Dataset demonstrates that the performance of FGS is similar to that of the compared method, and it can effectively evaluate the segmentation quality.

From the value of QR and D in Tables 1 and 2, we can find that the quality of the segmentation results obtained by determining the optimal segmentation scale only through the evaluation of inter-segment heterogeneity was worse than that of the segmentation results obtained by determining the optimal segmentation scale through combining the evaluation of intra-segment homogeneity and inter-segment heterogeneity, which indicates the importance of intra-segment homogeneity and inter-homogeneity in segmentation quality evaluation.

From the result of computational cost, we can find that the computational cost of DTNP was much less than that of MI and BSH, and the computational cost of BSH was more than that of MI. DTNP need not construct RAG and computes with neighborhood objects, which lead the result of less computational cost. The computation process of BSH needs to use RAG twice, but it cannot be stored in the eCognition environment. Therefore, the computational cost of BSH is significantly increased by the two construction processes of RAG. If MI and BSH are computed by programming, the difference in computational cost will be greatly reduced. Note that the correlation coefficient between the computational cost of BSH and the number of objects is the lowest, because the number of objects is too few and the computational cost is too low, so it is difficult to make full use of computing resources. Since the computational cost of the heterogeneity in UE is more than the computational cost of other steps, we can infer from the computational cost of DTNP, MI, and BSH that the computational cost of FGS is much less than JM and WM.

5. Conclusions

The FGS based on WV and DTNP was proposed to evaluate the remote sensing image segmentation. The innovation of this study is to use the pixels of the object neighborhood as the inter-segment heterogeneity computational object. Compared with the existing methods, the performance is similar and the computational cost is greatly reduced. The experimental results show DTNP can clearly reveal the process of increasing heterogeneity and FGS can clearly reveal the change from over-segmentation to optimal segmentation results and under-segmentation. Furthermore, compared to MI and BSH, DTNP can keep stability and remain sensitive to both over-segmentation and under-segmentation. FGS has the similar performance with JM and WM but has the least computational cost. This advantage is more prominent when the UE method is applied to segmentation parameter optimization. The result in Sections 3.1 and 3.2 indicates both homogeneity and heterogeneity play an important role in segmentation evaluation. This method can be effectively used in GEOBIA. In future research, more effective and time-consuming unsupervised evaluation methods should be studied and effectively applied to segmentation parameter optimization.

Author Contributions: Conceptualization, M.Z. and Q.M.; methodology, M.Z.; validation, Y.Z. and L.Z.; formal analysis, L.Z. and D.H.; writing—original draft preparation, M.Z.; writing—review and editing, Q.M. and M.A.; visualization, M.Z.; project administration, Q.M.; and funding acquisition, Q.M. All authors have read and agreed to the published version of the manuscript.

References

1. Chen, F.; Chen, X.; Van de Voorde, T.; Roberts, D.; Jiang, H.; Xu, W. Open water detection in urban environments using high spatial resolution remote sensing imagery. *Remote Sens. Environ.* **2020**, *242*, 111706. [CrossRef]
2. Wang, S.; Yang, H.; Wu, Q.; Zheng, Z.; Wu, Y.; Li, J. An Improved Method for Road Extraction from High-Resolution Remote-Sensing Images that Enhances Boundary Information. *Sensors* **2020**, *20*, 2064. [CrossRef] [PubMed]
3. Wu, T.; Luo, J.; Zhou, Y.n.; Wang, C.; Xi, J.; Fang, J. Geo-Object-Based Land Cover Map Update for High-Spatial-Resolution Remote Sensing Images via Change Detection and Label Transfer. *Remote Sens.* **2020**, *12*, 174. [CrossRef]
4. Zhang, L.; Wu, J.; Fan, Y.; Gao, H.; Shao, Y. An Efficient Building Extraction Method from High Spatial Resolution Remote Sensing Images Based on Improved Mask R-CNN. *Sensors* **2020**, *20*, 1465. [CrossRef] [PubMed]
5. Thomas, N.; Hendrix, C.; Congalton, R.G. A comparison of urban mapping methods using high-resolution digital imagery. *Photogramm. Eng. Remote Sens.* **2003**, *69*, 963–972. [CrossRef]
6. Myint, S.W.; Gober, P.; Brazel, A.; Grossman-Clarke, S.; Weng, Q. Per-pixel vs. object-based classification of urban land cover extraction using high spatial resolution imagery. *Remote Sens. Environ.* **2011**, *115*, 1145–1161. [CrossRef]
7. Huang, L.; Ni, L. Object-oriented classification of high resolution satellite image for better accuracy. In Proceedings of the 8th International Symposium Spatial Accuracy Assessement Natural Resources Environironmental Science Vol Ii: Accuracy Geomaterials, Shanghai, China, 25–27 June 2008; pp. 211–218.
8. Hossain, M.D.; Chen, D. Segmentation for Object-Based Image Analysis (OBIA): A review of algorithms and challenges from remote sensing perspective. *ISPRS J. Photogramm. Remote Sens.* **2019**, *150*, 115–134. [CrossRef]
9. Blaschke, T. Object based image analysis for remote sensing. *ISPRS J. Photogramm. Remote Sens.* **2010**, *65*, 2–16. [CrossRef]
10. Karlson, M.; Reese, H.; Ostwald, M. Tree Crown Mapping in Managed Woodlands (Parklands) of Semi-Arid West Africa Using WorldView-2 Imagery and Geographic Object Based Image Analysis. *Sensors* **2014**, *14*, 22643–22669. [CrossRef]
11. Zhang, X.; Feng, X.; Xiao, P.; He, G.; Zhu, L. Segmentation quality evaluation using region-based precision and recall measures for remote sensing images. *ISPRS J. Photogramm. Remote Sens.* **2015**, *102*, 73–84. [CrossRef]
12. Hernandez, I.E.R.; Shi, W. A Random Forests classification method for urban land-use mapping integrating spatial metrics and texture analysis. *Int. J. Remote Sens.* **2018**, *39*, 1175–1198. [CrossRef]
13. Goncalves, J.; Pocas, I.; Marcos, B.; Mucher, C.A.; Honrado, J.P. SegOptim—A new R package for optimizing object-based image analyses of high-spatial resolution remotely-sensed data. *Int. J. Appl. Earth Obs. Geoinf.* **2019**, *76*, 218–230. [CrossRef]
14. Chen, L.C.; Papandreou, G.; Kokkinos, I.; Murphy, K.; Yuille, A.L. DeepLab: Semantic Image Segmentation with Deep Convolutional Nets, Atrous Convolution, and Fully Connected CRFs. *IEEE Trans. Pattern Anal. Mach. Intell.* **2018**, *40*, 834–848. [CrossRef] [PubMed]
15. Kemker, R.; Salvaggio, C.; Kanan, C. Algorithms for semantic segmentation of multispectral remote sensing imagery using deep learning. *ISPRS J. Photogramm. Remote Sens.* **2018**, *145*, 60–77. [CrossRef]
16. Li, D.R.; Zhang, G.F.; Wu, Z.C.; Yi, L.N. An Edge Embedded Marker-Based Watershed Algorithm for High Spatial Resolution Remote Sensing Image Segmentation. *IEEE Trans. Image Process.* **2010**, *19*, 2781–2787. [CrossRef] [PubMed]
17. Vincent, L.; Soille, P. Watersheds in Digital Spaces: An Efficient Algorithm Based on Immersion Simulations. *IEEE Trans. Pattern Anal. Mach. Intell.* **1991**, *13*, 583–598. [CrossRef]
18. Shafarenko, L.; Petrou, M.; Kittler, J. Automatic watershed segmentation of randomly textured color images. *IEEE Trans. Image Process.* **1997**, *6*, 1530–1544. [CrossRef]
19. Eddy, P.R.; Smith, A.M.; Hill, B.D.; Peddle, D.R.; Coburn, C.A.; Blackshaw, R.E. Hybrid

segmentation—Artificial Neural Network classification of high resolution hyperspectral imagery for Site-Specific Herbicide Management in agriculture. *Photogramm. Eng. Remote Sens.* **2008**, *74*, 1249–1257. [CrossRef]

20. Yang, J.; He, Y.H.; Caspersen, J. Region merging using local spectral angle thresholds: A more accurate method for hybrid segmentation of remote sensing images. *Remote Sens. Environ.* **2017**, *190*, 137–148. [CrossRef]
21. Liu, J.H.; Pu, H.; Song, S.R.; Du, M.Y. An adaptive scale estimating method of multiscale image segmentation based on vector edge and spectral statistics information. *Int. J. Remote Sens.* **2018**, *39*, 6826–6845. [CrossRef]
22. Wang, Y.J.; Qi, Q.W.; Liu, Y. Unsupervised Segmentation Evaluation Using Area-Weighted Variance and Jeffries-Matusita Distance for Remote Sensing Images. *Remote Sens.* **2018**, *10*, 1193. [CrossRef]
23. Shen, Y.; Chen, J.Y.; Xiao, L.; Pan, D.L. Optimizing multiscale segmentation with local spectral heterogeneity measure for high resolution remote sensing images. *ISPRS J. Photogramm. Remote Sens.* **2019**, *157*, 13–25. [CrossRef]
24. Wang, Y.J.; Qi, Q.W.; Liu, Y.; Jiang, L.L.; Wang, J. Unsupervised segmentation parameter selection using the local spatial statistics for remote sensing image segmentation. *Int. J. Appl. Earth Obs. Geoinf.* **2019**, *81*, 98–109. [CrossRef]
25. Zhang, H.; Fritts, J.E.; Goldman, S.A. Image segmentation evaluation: A survey of unsupervised methods. *Comput. Vis. Image Underst.* **2008**, *110*, 260–280. [CrossRef]
26. Clinton, N.; Holt, A.; Scarborough, J.; Yan, L.; Gong, P. Accuracy Assessment Measures for Object-based Image Segmentation Goodness. *Photogramm. Eng. Remote Sens.* **2010**, *76*, 289–299. [CrossRef]
27. Liu, Y.; Bian, L.; Meng, Y.H.; Wang, H.P.; Zhang, S.F.; Yang, Y.N.; Shao, X.M.; Wang, B. Discrepancy measures for selecting optimal combination of parameter values in object-based image analysis. *ISPRS J. Photogramm. Remote Sens.* **2012**, *68*, 144–156. [CrossRef]
28. Yang, J.; He, Y.H.; Weng, Q.H. An Automated Method to Parameterize Segmentation Scale by Enhancing Intrasegment Homogeneity and Intersegment Heterogeneity. *IEEE Geosci. Remote Sens. Lett.* **2015**, *12*, 1282–1286. [CrossRef]
29. Espindola, G.M.; Camara, G.; Reis, I.A.; Bins, L.S.; Monteiro, A.M. Parameter selection for region-growing image segmentation algorithms using spatial autocorrelation. *Int. J. Remote Sens.* **2006**, *27*, 3035–3040. [CrossRef]
30. Johnson, B.; Xie, Z.X. Unsupervised image segmentation evaluation and refinement using a multi-scale approach. *ISPRS J. Photogramm. Remote Sens.* **2011**, *66*, 473–483. [CrossRef]
31. Yang, J.; Li, P.J.; He, Y.H. A multi-band approach to unsupervised scale parameter selection for multi-scale image segmentation. *ISPRS J. Photogramm. Remote Sens.* **2014**, *94*, 13–24. [CrossRef]
32. Gao, H.; Tang, Y.; Jing, L.; Li, H.; Ding, H. A Novel Unsupervised Segmentation Quality Evaluation Method for Remote Sensing Images. *Sensors* **2017**, *17*, 2427. [CrossRef] [PubMed]
33. Haralick, R.M.; Shapiro, L.G. Image Segmentation Techniques. *Comput. Vis. Graph. Image Process.* **1985**, *29*, 100–132. [CrossRef]
34. Xiong, Z.Q.; Zhang, X.Y.; Wang, X.N.; Yuan, J. Self-adaptive segmentation of satellite images based on a weighted aggregation approach. *GIScience Remote Sens.* **2019**, *56*, 233–255. [CrossRef]
35. Levine, M.D.; Nazif, A.M. Dynamic Measurement of Computer Generated Image Segmentations. *IEEE Trans. Pattern Anal. Mach. Intell.* **1985**, *7*, 155–164. [CrossRef]
36. Woodcock, C.E.; Strahler, A.H. The Factor of Scale in Remote-Sensing. *Remote Sens. Environ.* **1987**, *21*, 311–332. [CrossRef]
37. Vamsee, A.M.; Kamala, P.; Martha, T.R.; Kumar, K.V.; Sankar, G.J.; Amminedu, E. A Tool Assessing Optimal Multi-Scale Image Segmentation. *J. Indian Soc. Remote Sens.* **2018**, *46*, 31–41. [CrossRef]
38. Yin, R.J.; Shi, R.H.; Gao, W. Automatic Selection of Optimal Segmentation Scales for High-resolution Remote Sensing Images. In Proceedings of the Conference on Remote Sensing and Modeling of Ecosystems for Sustainability X, San Diego, CA, USA, 24 September 2013. [CrossRef]
39. Johnson, B.A.; Bragais, M.; Endo, I.; Magcale-Macandog, D.B.; Macandog, P.B.M. Image Segmentation Parameter Optimization Considering Within—And Between-Segment Heterogeneity at Multiple Scale Levels: Test Case for Mapping Residential Areas Using Landsat Imagery. *ISPRS Int. J. Geo Inf.* **2015**, *4*, 2292–2305. [CrossRef]

40. eCognition Software. Available online: https://geospatial.trimble.com/products-and-solutions/ecognition (accessed on 11 September 2020).
41. Sun, W.; Chen, B.; Messinger, D.W. Nearest-neighbor diffusion-based pan-sharpening algorithm for spectral images. *Opt. Eng.* **2013**, *53*, 013107. [CrossRef]
42. ENVI Software. Available online: https://www.harris.com/solution/envi (accessed on 11 September 2020).
43. Benz, U.C.; Hofmann, P.; Willhauck, G.; Lingenfelder, I.; Heynen, M. Multi-resolution, object-oriented fuzzy analysis of remote sensing data for GIS-ready information. *ISPRS J. Photogramm. Remote Sens.* **2004**, *58*, 239–258. [CrossRef]
44. Mnih, V. *Machine Learning for Aerial Image Labeling*; University of Toronto: Toronto, ON, Canada, 2013.

3

Flood Inundation Mapping from Optical Satellite Images using Spatiotemporal Context Learning and Modest AdaBoost

Xiaoyi Liu [1,2], Hichem Sahli [3,4], Yu Meng [1,*], Qingqing Huang [1] and Lei Lin [1,2]

1 Institute of Remote Sensing and Digital Earth, Chinese Academy of Sciences, Beijing 100101, China; liuxy01@radi.ac.cn (X.L.); huangqq@radi.ac.cn (Q.H.); linlei@radi.ac.cn (L.L.)
2 University of Chinese Academy of Sciences, Beijing 100049, China
3 Department of Electronics and Informatics, Vrije Universiteit Brussel, 1050 Brussels, Belgium; hsahli@etrovub.be
4 Interuniversity Microelectronics Centre (IMEC), 3001 Heverlee, Belgium
* Correspondence: mengyu@radi.ac.cn.

Academic Editors: Qi Wang, Nicolas H. Younan, Carlos López-Martínez and Prasad S. Thenkabail

Abstract: Due to its capacity for temporal and spatial coverage, remote sensing has emerged as a powerful tool for mapping inundation. Many methods have been applied effectively in remote sensing flood analysis. Generally, supervised methods can achieve better precision than unsupervised. However, human intervention makes its results subjective and difficult to obtain automatically, which is important for disaster response. In this work, we propose a novel procedure combining spatiotemporal context learning method and Modest AdaBoost classifier, which aims to extract inundation in an automatic and accurate way. First, the context model was built with images to calculate the confidence value of each pixel, which represents the probability of the pixel remaining unchanged. Then, the pixels with the highest probabilities, which we define as 'permanent pixels', were used as samples to train the Modest AdaBoost classifier. By applying the strong classifier to the target scene, an inundation map can be obtained. The proposed procedure is validated using two flood cases with different sensors, HJ-1A CCD and GF-4 PMS. Qualitative and quantitative evaluation results showed that the proposed procedure can achieve accurate and robust mapping results.

Keywords: inundation mapping; flood; optical sensors; spatiotemporal context learning; Modest AdaBoost; HJ-1A/B CCD; GF-4 PMS

1. Introduction

Natural disasters are common phenomena in all parts of the world. There are many types of natural disasters [1], of which a flood is considered to be one of the most destructive, widespread and frequent disasters [2,3]. Every year, tremendous loss of life and property is caused by flooding [3]. Due to the changes in global climate and land use, floods are becoming more severe and more frequent all around the world [3,4]. Although it is difficult to prevent floods, it is possible to minimise their impact through proper rescue, relief and resource allocation for recovery and reconstruction. Therefore, accurate inundation mapping, especially near real time, is very important for establishing a fast response plan and mitigating the disaster [5–7].

Traditional methods for inundation mapping are based on ground survey and aerial observation. However, when the flood spreads to a large scale, these approaches are time- and resource-consuming, which cannot satisfy the need for a fast response to a disaster. Moreover, aerial observation can be unrealistic in some extreme weather conditions, and the density of gauging stations is not satisfactory in many countries [8]. An alternative choice is provided by satellite remote sensing (RS) techniques [6]. Due to their time availability and cost effectiveness, satellite data has played an important role in understanding inundation [9–13]. The availability of multi-date images makes it possible to monitor the progress of floods.

Satellites used for mapping floods can be divided into those that are optical and those that are microwave. Due to its capacity to penetrate the frequent clouds in a flood event, microwave remote sensing is all-weather and invaluable for flood monitoring. With multispectral images, the flood can be analysed in a more straightforward way with simpler pre-processing [14]. In this study, we mainly focus on methods using multispectral satellite images.

Numerous methods have been proposed for mapping inundation using multispectral remote sensing images. Among them, the one most frequently used is thresholding. Usually, indices are first calculated through different band combinations, such as the normalised difference water index (NDWI) created by McFeeters [15], which has been proven to produce good results for inundated areas [16]. Then a threshold is selected to determine the water range in the image. A manual threshold is accurate, but has difficulty satisfying the need for fast disaster response. Moreover, it is subjective, as different operators may produce different results. To overcome the problems, unsupervised thresholding methods have been proposed. For instance, Xie et al. [17] introduced Otsu's algorithm to implement automatic selection of the water threshold. But due to the common illumination differences and mixed pixels in satellite images, its effectiveness is reduced, especially for some complicated scenes.

The segmentation (semi-supervised) technique [18] has been proposed to minimise the involvement of the user. The user first selects some seed points, with which the connectivity map is generated using fuzzy logic. For example, in [19], a fast flood map and a detailed flood map were obtained using growing strategies with seed points. However, the detailed map result still depended on the correctness of the seed points.

Unsupervised strategies without any human involvement have attracted a lot of attention in recent years. There are several kinds of unsupervised inundation mapping methods. Besides the unsupervised threshold, unsupervised feature extraction methods have been utilised. Chignell et al. [20] combines the pre- and post-flood images and apply the independent component analysis (ICA) to them. Segmentation and threshold are used to extract the flood from the change components. The cloud and crop components help to refine the maximum flood extent. In the work by Rokni et al. [21], the multi-temporal NDWI images are composited into one file. Principal component analysis (PCA) is applied to the composited file. The principle components are classified by the thresholding technique, and the result of the change detection for the lake is obtained. But these methods are only based on spectral information. When they are applied to cases using different sensors, the ability of the method can vary with the changes of spectral characteristics.

Recently, context information, especially spatiotemporal context information, has attracted more attention and proven to bring much improvement in monitoring the water surface. It is combined with other techniques to generate chains of processing for better representation of an event. Chen et al. [22] proposed a water surface monitoring method using contextual information. First, permanent water/non-water pixels were detected by judging the statistical consistency between an image point and its neighbourhood. Then, a distance-based classifier was used to map the other pixels with the obtained permanent pixels. Experiments on Moderate Resolution Imaging Spectroradiometer (MODIS) proved its validity and superiority over other unsupervised methods. However, the proposed definition of the statistical equality depended on simple one-dimensional features, which were the means and mediums of temporally adjacent pixels. A pixel was considered to be permanent if it had more than five spatially adjacent and statistically equal pixels. This simple count strategy can reduce the robustness of the method. Moreover, the low spatial resolution of MODIS data also limited its performance in spatial dimension.

To resolve these issues, in this paper, we introduce a spatiotemporal context learning (STCL) method and propose a novel work flow for flood mapping. The main objective is to delineate the water surface in an accurate and automatic way. First, a statistical model is built for the contextual information of multi-temporal NDWI. Then, permanent pixels are extracted according to their contextual consistency confidence values calculated from the model. Finally, a Modest AdaBoost (MADB) classifier, trained with the permanent pixels and a variety of spectral characteristics, is adopted to map the image into water and non-water categories. Through making full use of the spatiotemporal and spectral information, the proposed approach improves the ability to map inundated surfaces. The uncertainty caused by the sensor and scene differences is also reduced. Two different multispectral datasets with medium resolution, HJ-1A CCD (30 m) and GF-4 PMS (50 m), are employed for the validation.

2. Experimental Set Up

2.1. Datasets

Several kinds of multispectral satellite data have been used for flood mapping, such as Advanced Very High Resolution Radiometer (AVHRR), MODIS, and Landsat TM/ETM+ data. However, most of these data do not have high spatial and temporal resolution at the same time [23]. This has limited their ability to map inundation, which changes complicatedly and rapidly over time. For example, AVHRR and MODIS have a frequent revisiting cycle, which can be even shorter than 1 day. Their high temporal resolution makes them useful for monitoring environmental changes, while the spatial resolution of AVHRR and MODIS is 1 km and 250 m, respectively, which is coarse. Only general extent, not accurate results, can be obtained using these data for flood mapping. On the contrary, Landsat TM/ETM+ data have a middle-to-high spatial resolution of 30 m, but the observation is repeated every 16 days, which cannot satisfy the needs for timely response.

On 6 September 2008, two optical satellites named HJ-1A/B (short for HuanJing-1A/B), also known as the Chinese Environment and Disaster Monitoring and Forecasting Small Satellite Constellation, were launched in China. The data can be downloaded from the website (http://www.cresda.com/) free of charge, and have been successfully applied in several applications such as land mapping, yield prediction, and environment assessment. The two satellites were equipped with CCD cameras, which take multispectral images on the earth surface with a spatial resolution of 30 m. For each satellite, the time interval is 4 days. The constellation of the two satellites theoretically has a higher revisiting frequency of 2 days. With both the advantages of spatial and temporal resolution, HJ-1A/B satellites are regarded as an effective tool for monitoring and post-flood assessment [24].

The recently emerged geostationary satellite GaoFen-4 (GF-4) also has a high application value in rapid assessment and emergency response of floods [25]. Due to its optical geostationary orbit, GF-4 shows a better performance in time resolution over other satellites. It is equipped with a camera for visible, near infrared and middle-wavelength infrared spectra. The spatial resolution is 50 m. To the best of our knowledge, research work using GF-4 imagery is limited, as it was only launched on 29 December 2015, and officially put in use on 13 June 2016. In this work, we also want to explore the potential of the multispectral GF-4 PMS data in flood mapping. The main parameters of the HJ-1A/B CCD data and GF-4 PMS data are listed in Tables 1 and 2. Slightly different from the HJ-1A/B CCD data, the GF-4 PMS data have an additional panchromatic band.

Table 1. Technical parameters of the HJ-1A/B CCD data.

Satellite Sensor	Band No.	Spectral Range (μm)	Spatial Resolution (m)	Revisiting Time
HJ-1A/B CCD	1	0.43–0.52	30	4 days
	2	0.52–0.60		
	3	0.63–0.69		
	4	0.76–0.90		

As can be seen, the HJ-1A/B CCD data and the GF-4 PMS data show balanced abilities in spatial and temporal resolutions. Due to the limitations stated above in other multispectral data, we decide to utilise these two datasets for verifying the methods, and for understanding more about the potential of these two datasets in flood mapping as well, as they are not so commonly used as MODIS or Landsat data. Certainly, another reason why the HJ-1A/B CCD and GF-4 PMS data are chosen is because of their free access.

Table 2. Technical parameters of the GF-4 PMS data.

Satellite Sensor	Band No.	Spectral Range (μm)	Spatial Resolution (m)	Revisiting Time
GF-4 PMS	1	0.45–0.90	50	20 s
	2	0.45–0.52		
	3	0.52–0.60		
	4	0.63–0.69		
	5	0.76–0.90		

2.2. *Study Area*

The Heilongjiang River is one of the largest rivers in Northeast Asia, flowing through four countries (Mongolia, China, Russia and North Korea). The main stream has a total length of 2821 km, and also forms the boundary between China and Russia. There are abundant water resources in the Heilongjiang River, with a yearly runoff of 346.5 billion cubic meters. The main climate type in that region is monsoon. The precipitation distribution varies with the season. From April to October, the precipitation accounts for 90–93% of the annual precipitation, and the period from June to August accounts for 60–70%. From December, the winter dry season starts and the precipitation mainly falls in the form of snow.

From 12 August 2013, several severe precipitation events continuously hit the northeastern part of Asia, leading to great flood in 39 rivers including part of the Heilongjiang River. Especially for the Tongjiang and Fuyuan Reaches of the Heilongjiang River, the flood had been the most serious one in the past 100 years. On 24 August 2013, more than 5 million people were affected in this disaster. The first case study analyses the event in this region. Two cloud-free scenes of HJ-1A CCD data are utilised. One image was obtained on 12 July 2013, which is around one month before the flood, and the other one was obtained on 27 August 2013 during the peak flow period. The dimension of the study region is 1082 × 1321 pixels (around 1286 square kilometres). Its location and extent are shown in Figure 1.

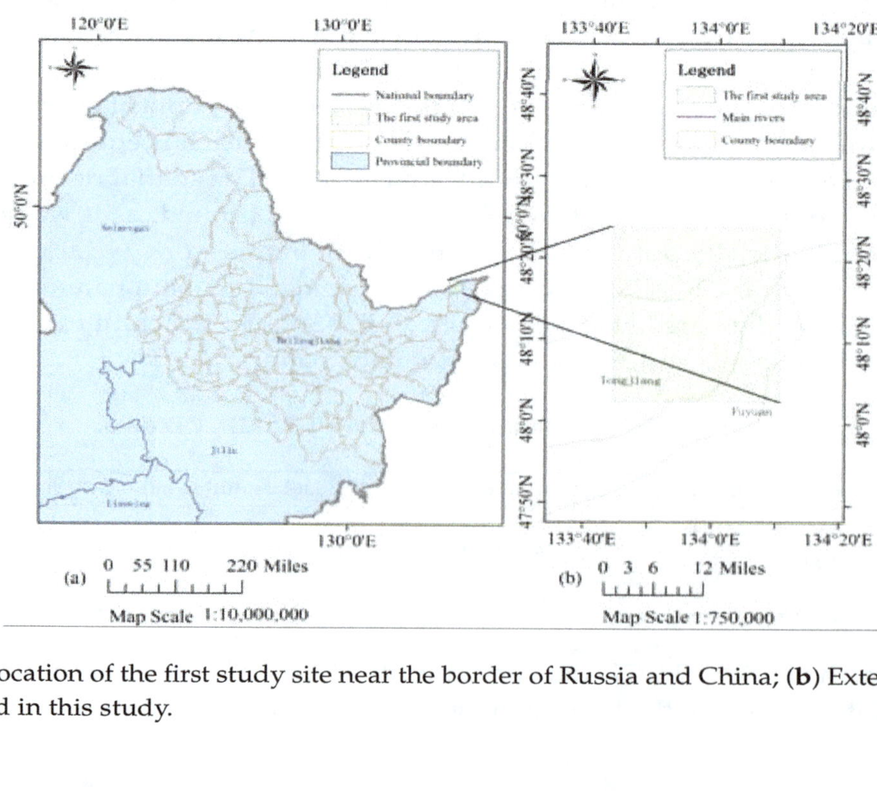

Figure 1. (**a**) Location of the first study site near the border of Russia and China; (**b**) Extent of the HJ-1A CCD data used in this study.

Dongting Lake is one of the most essential lakes in China, and one of the most important wetlands in the world as well. It is located on the southern bank of the Jinjiang section of the middle Yangtze River, and is one of the important dispatching lakes for the Yangtze River because of the strong ability of flood storage. The area of the lake is approximately 2690 square kilometres, across Hunan and Hubei provinces, and is roughly composed of East Dongting Lake, South Dongting Lake and West Dongting Lake. The water of Dongting Lake is clean and this area is one of the main freshwater fishery bases for commercial purposes. Due to its good environment and richness in water, soil and wildlife resources, it is one of the earliest birth places of Chinese rice raising agriculture. The basin area is of 262.8 thousand square kilometres, accounting for 14.6% of the Yangtze River basin area.

In June and July 2016, heavy rains hit the middle and lower reaches of the Yangtze River basin, causing a catastrophic and wide flood in southern China. Eleven provinces and more than 10 million people were affected. On 3 July 2016, the water at the Chenglingji station in Dongting Lake also surpassed the warning level 32.50 m. A regional flood occurred in Dongting Lake. The second case study focuses on this area during this flood. Two cloud-free GF-4 PMS images are selected as the experimental data, which were obtained on 17 June 2016 (before the flood occurred) and 23 July 2016 (during the flood period), with a dimension of 2534 × 2235 pixels (around 14,159 square kilometres). The location and extent of the second study site are shown in Figure 2.

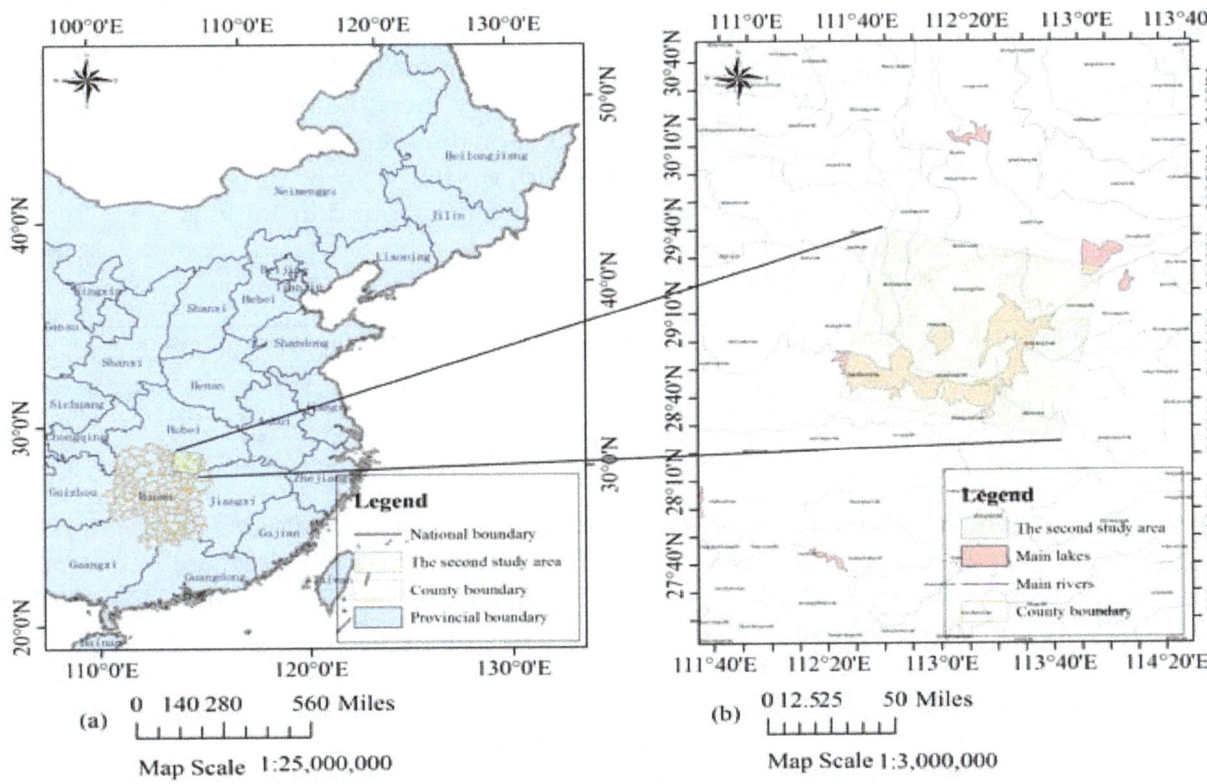

Figure 2. (**a**) Location of the second study site at the North of Hunan Province in China; (**b**) Extent of the GF-4 PMS data used in this study.

As can be seen from the figures, these two study sites are located in different geographic positions. The first case study mainly presents a river flood and the second one presents a lake flood. Studies on different kinds of floods can help validate the robustness of the method. Moreover, these two floods took place in 2013 and 2016. Each of them is one of the most severe flood events in that year, bringing about a large amount of damage and wide effects. The areas covered by the first and the second case studies are also areas with high flood risk every year, so it is significant to choose these two areas for study, which can help the government to make better decisions in disaster prevention in these areas.

Before the experiment, these two pairs of data are preprocessed. For HJ-1A/B CCD data, we download the absolute calibration coefficients from the data source website (http://www.cresda.com/), and apply the absolute radiometric corrections to the data. These coefficients are obtained through field experiment and authenticity testing by the China Centre for Resource Satellite Data and Application (CRESDA). For the GF-4 PMS data and the GF-1 WFV data used for validation, a relative radiometric correction is implemented before they are archived. We have not made further modifications to their radiation values. All the satellites images used in this study are geometrically registered using the software ERDAS IMAGEINE AutoSync. Specifically, in either of these two case studies, the experimental data before the flood is considered the reference data. Other images used in the same case are all registered to it. The co-registration technology adopts the cubic polynomial. The mean displacement error is 0.5 pixels. All the data are projected to the WGS 1984 UTM coordinate system.

2.3. Validation

The extent of the water surface during a flood process can have daily changes. It is almost impossible to obtain an accurate map of inundation regions on a particular day. In general, most of the flood products are a rough outline of the main inundated areas. In order to achieve the qualitative and quantitative evaluation, we produce two approximate reference maps for the first and second case studies. Either of them is based on a remote sensing image over the same site and taken on the same date as the corresponding experiment data. The spatial resolution of the data used for generating the reference map is necessarily higher than that of the experiment data. For the first case study using HJ-1A/B CCD data, there is a scene of GF-1 WFV data that can meet the requirements. The technical parameters of the GF-1 WFV data are listed in Table 3. For the second case using GF-4 PMS data, no corresponding GF-1 WFV data could be found. Instead, we find a scene of HJ-1B CCD data that is qualified. The technical parameters of the HJ-1B CCD data can be found in Table 1.

Table 3. Technical parameters of the GF-1 WFV data.

Satellite Sensor	Band No.	Spectral Range (μm)	Spatial Resolution (m)	Width
GF-1 WFV	1	0.45–0.52	16	800 km
	2	0.52–0.59		
	3	0.63–0.69		
	4	0.77–0.89		

For the process of how the reference map is made, we use a traditional water extraction method. We take the first case study as an example. First, the selected GF-1 WFV image is geometrically registered to the experimental HJ-1A CCD data. Then, we calculate the NDWI of the GF-1 WFV image. Compared with the ground information from Google Earth software, we manually select a threshold in NDWI to separate water and non-water pixels. Finally, the binary water mask is resampled to the spatial resolution of HJ-1A CCD data (30 m). Similar processes are applied to the second case study. Given that there is no detailed ground truth available, and that it is not feasible to get one by field investigation, we use the reference map in this study as an approximation of the real inundated extent, helping to evaluate and compare the detection results qualitatively and quantitatively.

3. Methods

3.1. Permanent Pixel Extraction Using Spatiotemporal Context Learning

The images before and after a flood are referred to as image #1 and image #2, respectively. The pixels with a constant land cover type, no matter what the type is, are defined as permanent pixels. The proposed method is divided into two steps. First, the permanent pixels in image #1 and image #2 are extracted based on the STCL strategy. This is a method that models the relative relationship between an object and its context. We introduce it to formulate the relationship between a satellite image pixel and its context. Through comparing the models at different time points, a confidence value for whether a pixel changes or not is calculated to extract the permanent pixels. Second, using these permanent pixels as a training set, a widely adopted machine learning classifier, Modest AdaBoost, is trained and implemented for mapping inundation in image #2. Modest AdaBoost is one of the derivations of the boosting algorithm, like the original AdaBoost algorithm. It combines the performance of a set of weak classifiers, and also proves better than other boosting algorithms for convergence ability. More details about these methods will be given below. In this section, we will first discuss the procedure in the first step.

Due to its capacity for targeting specific land cover type and reducing influence from inconstant band representation, spectral indices are commonly used in diverse remote sensing applications, such as disaster monitoring, land cover mapping and disease prevention [26–28]. For mapping different cover types in different applications, various indices have been proposed, including the normalised difference vegetation index (NDVI), the enhanced vegetation index (EVI), NDWI, and the normalised difference built-up index (NDBI) and so on. Among these indices, the NDWI has been successfully applied to mapping land surface water, and proved more effective than other general feature classification methods [29]. In this study, we calculate the NDWI in the experimental datasets (the HJ-1A CCD data for the first case study and the GF-4 PMS data for the second case study) first. Then, the steps for extracting permanent pixels will be executed on the NDWI data. The NDWI is calculated as:

$$\mathrm{NDWI} = \frac{\mathrm{Green} - \mathrm{NIR}}{\mathrm{Green} + \mathrm{NIR}} \tag{1}$$

where Green and NIR are the reflected green and near infrared radiance, respectively, which are replaced by band 2 and band 4 in the HJ1-A CCD data case, and band 3 and band 5 in the GF-4 PMS data case [15]. NDWI can eliminate the influence from the band value difference, but not the influence caused by the different weather conditions. However, as it is the relative relationship between neighbouring pixels that we use, influences from changes of overall brightness are limited in the proposed procedure.

In the visual tracking field, as the video frames usually change continuously, a strong spatiotemporal correlation is thought to exist between a target and its surroundings. In order to make better use of this relative relationship, Zhang et al. [30] proposed the STCL method. In this method, a rectangular contextual region was first built with the target in the centre. With the low-level features (including the image density and location) of the contextual region, the relative relationship between the target and its surroundings in contextual region was modelled. When a new frame came, it was put into the model to calculate a confidence map, indicating the location that best matched the contextual relationship of previous frames. This location was the inferred location of the target in the new frame. As it depended on a kind of relative relationship, the illumination difference during the frames cannot influence the result. Extensive experiments showed its effectiveness and good degree of precision. This method has also been further employed and extended in other visual trackers [31,32].

A remote sensing image time series shares many similar characteristics with video data, although video sequence images have a higher sampling rate. It can be inferred that there is also a relationship between a target pixel and its spatiotemporal neighbourhoods in a local scene of RS images, if the images are of good quality, without too many clouds and shadows. Due to the constant changes in weather and light conditions, radiation values of the same cover type can vary greatly in different scenes. Furthermore, it is rather difficult to calibrate the radiation of two RS images to absolute consistency. As a result, more false positives can be introduced in mapping the changes. While the relative relationship between unchanged pixels and their nearby pixels is relatively constant, the STCL method, which aims at modelling this kind of relationship, is supposed to be robust also to illumination variation in RS images. What is more, the STCL method provides a fast solution to online problems. In this work, we borrow the concept of STCL to build a procedure for extracting permanent pixels for flood mapping. The proposed flowchart for extracting permanent pixels is shown in Figure 3.

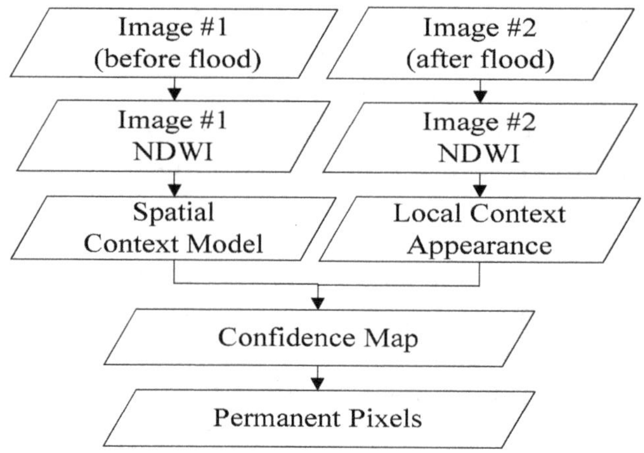

Figure 3. Flowchart of permanent pixel extraction procedure.

One core of the STCL method is its utilisation of the attention focus property in biological visual systems. In the mechanism of biological vision, assume that we are observing a point in a picture. Besides the point itself, which draws most of our attention, other points around the target point are the part we pay the second most attention to in the picture. The further one point is from the target point, the less concern it will get from the visual system. On the contrary, if someone tries to find a known point in an image, the visual mechanism first roughly figures out the background of the target, and then, on the basis of a correlation between the background and the point, the point can be easily targeted. But if only the feature of the target itself is considered, the search will be time and labour intensive. In the visual tracking field, it means the tracker may get lost.

According to this conception, the STCL proposed by Zhang et al. [30] in the visual tracking field uses the distribution of the attention focus, which is formulated as a curved surface function. In the function, the peak is located at the target point and its surroundings gradually decrease. With the weights from this function, the correlation of the target with its local background in image density and location are modelled. If the target location gradually changes, this context model will gradually change as well, and will be updated in each frame. When a new frame comes, although there are illumination variation and occlusion problems, the new location of the target can still be found by comparing the model with that of each pixel in the image. We borrow the concept and the formulation of the spatiotemporal context in STCL, and propose a method based on this context information for extracting permanent pixels. Details of the method are described as follows. The core of this problem is calculating the confidence map $c(x)$ between image #1 and image #2, which is also the probability of that the pixel is permanent. It can be formulated as

$$c(x) = P(x), \tag{2}$$

where $x \in R^2$ is a pixel location, and $P(x)$ is the probability. The higher $P(x)$ is, the more likely x will be permanent. After transformation, $P(x)$ can be given by

$$P(x) = \sum_{f(z) \in X_C} P(x|\,f(z))P(f(z)) \tag{3}$$

where $X_C = \{f(z) = (I(z), z)|z \in \Omega(x)\}$. $I(z)$ denotes the pixel value, i.e., the NDWI value, at location z, and $\Omega(x)$ is the neighbourhood of location x. $P(f(z))$ is the context prior probability that models the appearance of the local context, and $P(x|f(z))$ represents the relative relationship between x and its neighbourhood, which is defined as the spatial context model

$$g_{SC}(x - z) \;=\; P(x|\,f(z)) \tag{4}$$

In image #1, $x*$ and z are, respectively, the locations of the target pixel and its local context. For context prior probability, when $x*$ is permanent, if the values of $x*$ and z, as well as $I(x*)$ and $I(z)$, are closer, there is a higher probability that the pixel at location z will also be permanent. Different from the original solution for the tracking problem, we model the context prior probability $P(f(z))$ as

$$P(f(z)) = e^{-|I(x*)-I(z)|}\omega_\sigma(z - x*), \tag{5}$$

where $\omega_\sigma(z - x*)$ is a spatial weight function. With regard to the attention focus principle, if the local context pixel z is located closer to the object $x*$, z should make a greater contribution to the contextual characteristics of $x*$ in (5), and a higher weight should be given to it, and vice versa. Given that the weight should decrease smoothly with the increase of the distance to the object, ω_σ is defined as an exponential type as

$$\omega_\sigma(z) = ae^{-\frac{|z|^2}{2\sigma^2}}, \tag{6}$$

where a is a normalising constant that restricts $P(f(z))$ to a range from 0 to 1. $\sigma = 0.5$ is the scale parameter. As there are no changes occurring to $x*$ in image #1, we set its confidence value $c(x*) = 1$. According to the correlation between adjacent pixels, if the context pixel is located closer to the object, it should be more likely to be permanent. Therefore, the confidence function in image #1 can be modelled as

$$c(x) = P(x) = e^{-|\frac{x-x*}{\alpha}|^\beta}, \tag{7}$$

where α is a scale parameter and β is a shape parameter. The confidence value changes monotonically with the values of α and β. Therefore, these two parameters can be neither too large nor too small. For instance, if β is too large, the model can easily get over-fitted. While if β is too small, the smoothing may cause some errors. We empirically set $\alpha = 4.5$ and $\beta = 1$ for all the experiments here. Based on (2)–(5), it can be inferred that

$$\begin{aligned} c(x) = P(x) &= \sum_{f(z)\in X_C} P(x|\, f(z))P(f(z)) = \sum_{z\in\Omega(x')} g_{SC}(x-z)e^{-|I(x*)-I(z)|}\omega_\sigma(z - x*) \\ &= g_{SC}(x) \otimes (e^{-|I(x*)-I(x)|}\omega_\sigma(x - x*)) \end{aligned}, \tag{8}$$

where $\otimes$ denotes the convolution operation. According to (7), (8) can be transformed to the frequency domain as:

$$F(c(x)) = F(e^{-|\frac{x-x*}{\alpha}|^\beta}) = F(g_{SC}(x)) \odot F(e^{-|I(x*)-I(x)|}\omega_\sigma(x - x*)), \tag{9}$$

where F donates the Fourier transform function. $\odot$ is the element-wise product. So, for image #1, the spatial context model is

$$g_{SC}(x) = F^{-1}(\frac{F(e^{-|\frac{x-x*}{\alpha}|^\beta})}{F(e^{-|I(x*)-I(x)|}\omega_\sigma(x - x*))}) \tag{10}$$

With the spatial context model gained from image #1, according to (9), the confidence map of image #2 can be calculated by

$$c'(x) = F^{-1}(F(g_{SC}(x)) \odot F(P'(f'(x)))) = F^{-1}(F(g_{SC}(x)) \odot F(e^{-|I'(x*')-I'(x)|}\omega_\sigma(x - x*'))), \tag{11}$$

where $x*'$ is the location of the target pixel in image #2, and $f'(x)$ and $I'(x)$, respectively, represent the context prior probability and image intensity in image #2 [30]. After the permanence confidence map is calculated for image #2, obtained after a flood, we select the pixels with top $n\%$ confidence values as the final permanent pixels. In this work, we choose $n = 2$. More discussion on how n influences the result will be given later.

3.2. *Inundation Mapping Based on Modest AdaBoost*

3.2.1. Permanent Pixels Labelling

According to the previous section, we get the set of permanent pixels. In order to utilise the permanent pixels for training the classifier later, we need to label the permanent pixels into water and non-water categories. In keeping the whole process automatic, manual labelling should not be used. In this study, we adopt the openly accessible MODIS 250 m land-water mask, which is called MOD44W for short, to achieve this purpose. MOD44W is a constant product, which is derived from Terra MODIS data MOD44C 250 m 16-day composites. If a pixel is identified as water in more than 50% of the period May to September of years 2000–2002, this pixel is labelled as water in the MOD44W product. This method effectively smooths the short-term water surface changes caused by flood and drought. Therefore, although the MOD44W was produced years before the case study, it is widely accepted as the description of average water distribution [33]. Here, we adopt MOD44W to label the permanent pixels. It is acknowledged that there are most likely some mistakes, caused by small changes in water surface over the years. But, as the general condition changes little, and the permanent pixels have high probability of being unchanged, the labels from MOD44W are generally reliable.

For both the first and second study area, there is only one scene of MOD44W data. We resample the MOD44W data to the same spatial resolution as the experimental image, and then label the permanent pixels into permanent water pixels and permanent non-water pixels according to the MOD44W values. As the labels of the permanent pixels are used for classifier training, the proportion of the permanent pixels of each class will influence the classifier training result. However, as most changes happen inside or around the river regions, it can be inferred that the permanent confidence is generally higher in non-water regions than in water regions. Among the pixels of the highest confidence values, we selected the water and non-water permanent pixels with the same proportion as that in the same scene in MOD44W. The sum of water and non-water permanent pixels remained n% of the total. With the labelled permanent pixels, the Modest AdaBoost classifier is trained. Then it is applied to the testing set consisting of multiple features of image #2. The final inundation mapping result can be calculated.

3.2.2. Inundation Mapping

Boosting is a technique that combines several weak classifiers to generate a powerful one. The first proposed boosting algorithm, AdaBoost, was created by Freund and Schapire in 1996 [34], which is regarded as the basis for all other kinds of boosting method. Due to its good generalisation ability, low computational complexity and high execution efficiency, boosting has become one of the most popular and effective classification tools in computer vision [35] and pattern recognition [36]. A number of algorithms are derived from the boosting method, such as the Discrete AdaBoost (DADB), Real AdaBoost (RADB) and Gentle AdaBoost (GADB). DADB is a boosting method that mainly employs binary weak classifiers, and RADB is a generalisation version of the basic AdaBoost algorithm [37]. On the basis of RADB, GADB is designed with better performance and higher resistance to outliers [38]. Here, we adopt a different boosting method called Modest AdaBoost, which proves to outperform GADB in generalisation error and overfitting. Its natural stopping criterion is also an advantage, which other boosting techniques lack [36]. The flowchart of mapping inundation using Modest AdaBoost is shown in Figure 4.

Modest AdaBoost is a variant of boosting proposed by A. Vezhnevets et al. [36]. The basic idea of this method is that in every iteration for computing the new distribution, more importance is given to the samples that are misclassified in the previous step (with low margins). In every step, the method is committed to improve the lowest margins of samples. While those training samples that already have high margins may be misclassified with the new distribution and the margins are decreased, this forces the weak classifier to work only in its domain and be 'modest', which is the origin of the name MADB. Through this strategy, some regions of the input space have fewer chances to become overconfident, and the generalisation ability of the method benefits from this. The open source GML AdaBoost Matlab Toolbox [39] is used in the experiments to implement the Modest AdaBoost algorithm. It is a collection of classes and functions of several boosting algorithms. More details about the mapping procedure are presented below.

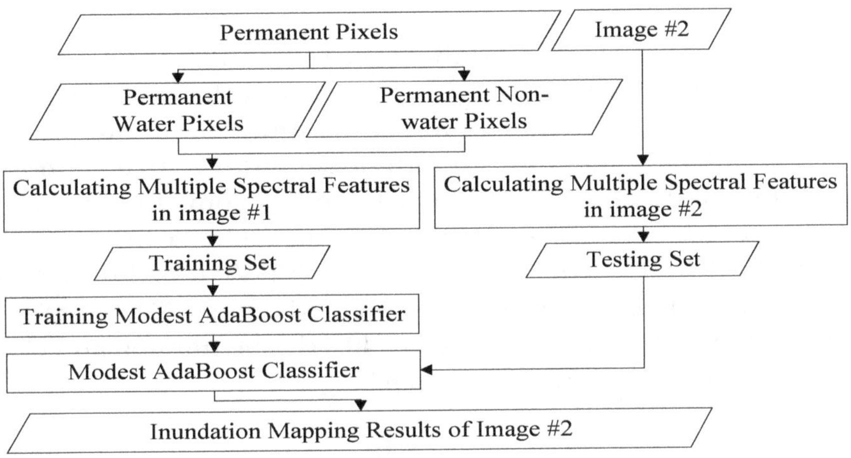

Figure 4. Flowchart of inundation mapping procedure.

First, each permanent pixel is set as a training sample point. Thus, the training dataset $(x_1, y_1),, (x_N, y_N)$ can be obtained. $x_i \in X$ is the input vector, which consists of several feature values of the permanent pixel, and $y_i \in \{-1, +1\}$, which is the corresponding class label of the permanent pixel. Here we define $y = +1$ when the pixel is water, and $y = -1$ when the pixel is non-water. N is the number of permanent pixels. At the beginning, we initialise the weight distribution on the input data as $D_0(i) = 1/N, i = 1, 2, \ldots, N$.

For each iteration $t = 1, \ldots, T$, with the weight distribution $D_t(i)$, the weak classifier $s_t(x) \in S$ can be trained by weighted least squares:

$$s_t = \underset{s}{\operatorname{argmin}} \left(\sum_{i=1}^{N} D_t(i) \cdot (y_i - s(x_i))^2 \right), \tag{12}$$

In addition, the 'inverted' distribution of the data weights is calculated by

$$D_t(i) = (1 - D_t(i))\overline{\alpha}_t, \tag{13}$$

where $\overline{\alpha}_t$ is the normalisation coefficient. Then we compute the probabilities:

$$P_t^{+1}(x) = P_{D_t}(y = +1 \cap s_t(x)), \tag{14}$$

$$\overline{P}_t^{+1}(x) = P_{\overline{D}_t}(y = +1 \cap s_t(x)), \tag{15}$$

$$P_t^{-1}(x) = P_{D_t}(y = -1 \cap s_t(x)), \tag{16}$$

$$\overline{P}_t^{-1}(x) = P_{\overline{D}_t}(y = -1 \cap s_t(x)). \tag{17}$$

Set

$$f_t(x) = (P_t^{+1}(1 - \overline{P}_t^{+1}) - P_t^{-1}(1 - \overline{P}_t^{-1}))(x), \tag{18}$$

and update the distribution by:

$$D_{t+1}(i) = D_t(i) \exp(-y_i f_t(x_i))\alpha_t, \tag{19}$$

where α_t is the normalisation coefficient. After T iterations or $f_t = 0$, the final classifier can be constructed by [36]

$$F(x) = sign[\sum_{t=1}^{T} f_t(s_t(x))] \tag{20}$$

The procedure of training the Modest AdaBoost classifier uses the permanent pixels extracted in previous steps, which contain the typical characteristics of water and non-water. After the strong classifier is obtained, it is applied to image #2 to get the inundation mapping results. Due to the difference in the bands of different satellites, the individual index of fixed band combination cannot always be effective in different flood scenarios. In order to overcome this shortcoming and make the method more robust, we set the components of the training vector using several bands and indices: (1) original bands; (2) NIR − Red; (3) NIR/Red; (4) EVI; (5) NDVI; (6) NDWI. All of these indices can be applied to optical satellite images. The computing method of NDWI is described in (1). For the EVI and the NDVI, the computing methods are as follows [40,41]. The training and classification processes are performed individually on each pixel of the image.

$$\mathrm{EVI} = \frac{2.5 \times (\mathrm{NIR} - \mathrm{Red})}{6 \times \mathrm{Red} + \mathrm{NIR} - 7.5 \times \mathrm{Blue} + 1}, \tag{21}$$

$$\mathrm{NDVI} = \frac{\mathrm{NIR} - \mathrm{Red}}{\mathrm{NIR} + \mathrm{Red}}, \tag{22}$$

For comparison, the commonly used unsupervised classification method K-MEANs, and another two different permanent pixel extraction methods combined with Modest AdaBoost, are also applied to the same experimental datasets. K-MEANs is implemented using the ENVI 5.0 software. The change threshold is set as 5.0%. One permanent pixel extraction method is from [22], which determines the permanent pixels through the means and mediums of the spatial neighbouring pixels. In another permanent pixel extraction method, a similar judgment rule using mediums and means, but extended to spatiotemporal field, is utilised. Specifically, each pixel has 8 spatial neighbouring pixels, and in the spatial neighbourhood-based permanent pixel extraction method (SP) in [22], if more than 5 among the 8 spatial neighbouring pixels are statistically equal (having the same medium or mean) to the target, the target pixel is considered permanent. For the spatiotemporal neighbourhood-based permanent pixel extraction method (STP), not only more than five spatially neighbouring pixels, but also more than five among the nine temporally neighbouring pixels, need to be statistically equal to the target for the target to be considered permanent. These permanent pixels are utilised with MADB in the same way as the proposed method. The inundation mapping results from each comparison method is then obtained. Here we call these two comparison methods utilising different permanent pixel strategies SP-MADB and STP-MADB, for short. All of these comparison methods are applied on the same multiple feature set of the classification step to the proposed method.

4. Results

4.1. Inundation Mapping Using HJ-1A CCD Data

As described above, in the first case study, two images, acquired on 12 July 2013 (before the flood) and 27 August 2013 (after the flood), are selected for the analysis. After pre-processing, the false colour composite images of the study area and the corresponding MOD44W product used for labelling the training data are shown in Figure 5.

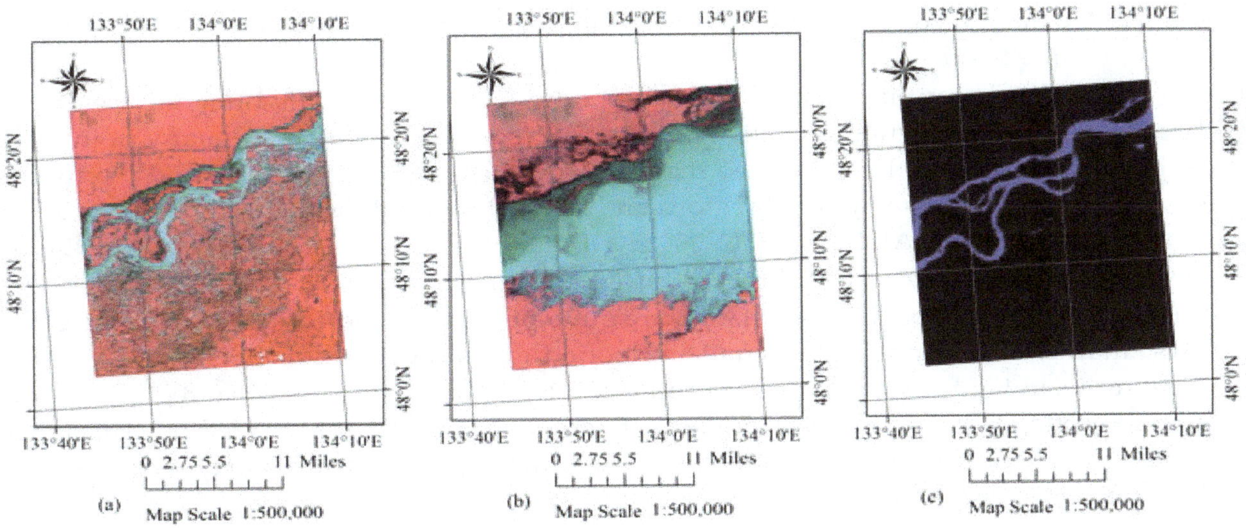

Figure 5. False colour composites (R4G3B2) of HJ-1A CCD images acquired on (**a**) 12 July 2013 (before the flood) and (**b**) 27 August 2013 (after the flood) for the first case study. (**c**) Corresponding MOD44W water mask product with water in blue and land in black.

The proposed spatiotemporal-context-learning-based permanent pixels-MADB (STCLP-MADB) method and three other comparison methods (K-MEANs, SP-MADB and STP-MADB) are each respectively applied to the experimental data. The final inundation mapping result for the individual method is shown in Figure 6. For a better visualization of the obtained results, we select four sub-regions and make a detailed zoom in. The location and size of the four sub-regions are shown in Figure 6e. The enlarged view of the small regions and their corresponding false colour composite, flood extraction result and the reference map are shown in Figure 7. Table 4 lists the number of inundated pixels derived by different methods in the full scene and sub-regions.

Table 4. The first case study in 2013—Number of inundated pixels.

Method	Full Region	Sub-Region A	Sub-Region B	Sub-Region C	Sub-Region D
K-MEANs	528,238	1833	16,236	7778	1442
SP-MADB	710,587	96	8470	5207	0
STP-MADB	843,570	1750	16,060	7009	1345
STCLP-MADB	812,610	1308	13,884	7011	1047
Reference Map	764,470	1176	11,861	6267	873

From the above figures and table, some comments can be made:

(1) The proposed inundation mapping method, based on STCL permanent pixel extraction and MADB, successfully extracts most of the flood regions in the first case study. In each column of the Table 4, the STCLP-MADB method achieves the closest number of inundated pixels to the reference, except in sub-region C. It is the second best among the methods, and has almost the same number of inundated pixels as the best. All of this evidence proves the effectiveness of the HJ-1A CCD data and the proposed procedure for mapping wide inundated areas in a river flood event.

(2) On the whole, it can be seen that the main regions of the flood are mostly well-delineated by each inundation mapping algorithm, except for small tributaries—for example the tributaries near the sub-region A and D—which are omitted by the SP-MADB method, and are shown in yellow. The STCLP-MADB performs better than the three other methods from the visual effect. In these regions, the K-MEANs and STP-MADB results present more false alarms, and the SP-MADB method makes more omissions. The result derived from STCLP-MADB is most consistent with

the reference map. Its effectiveness for precision mapping is significant for inferring the future evolution of the flood.

(3) From the detailed mapping results, it can be found that the inundated regions are delineated differently by different methods. In the map derived using K-MEANs, many points of false positive can be found in the unflooded regions. However, the SP-MADB method produces more false negatives in some small flood regions and half-submerged regions. More advanced results are obtained by STP-MADB and STCLP-MADB methods. Further comparisons of the details show that the results from STCLP-MADB provide finer outlines and are slightly better.

(4) Although the results from STCLP-MADB are quite promising, there are still some false positive errors, mainly occurring in the small unflooded areas surrounded by large flooded areas. For example, in Figure 6d, we can find some pixels in blue inside the main region of the flood, which are unflooded areas but determined as flood by the STCLP-MADB method. This is because these areas mostly comprise mixed pixels. Different proportions and locations of water in one mixed pixel influence what class the pixel is distributed to.

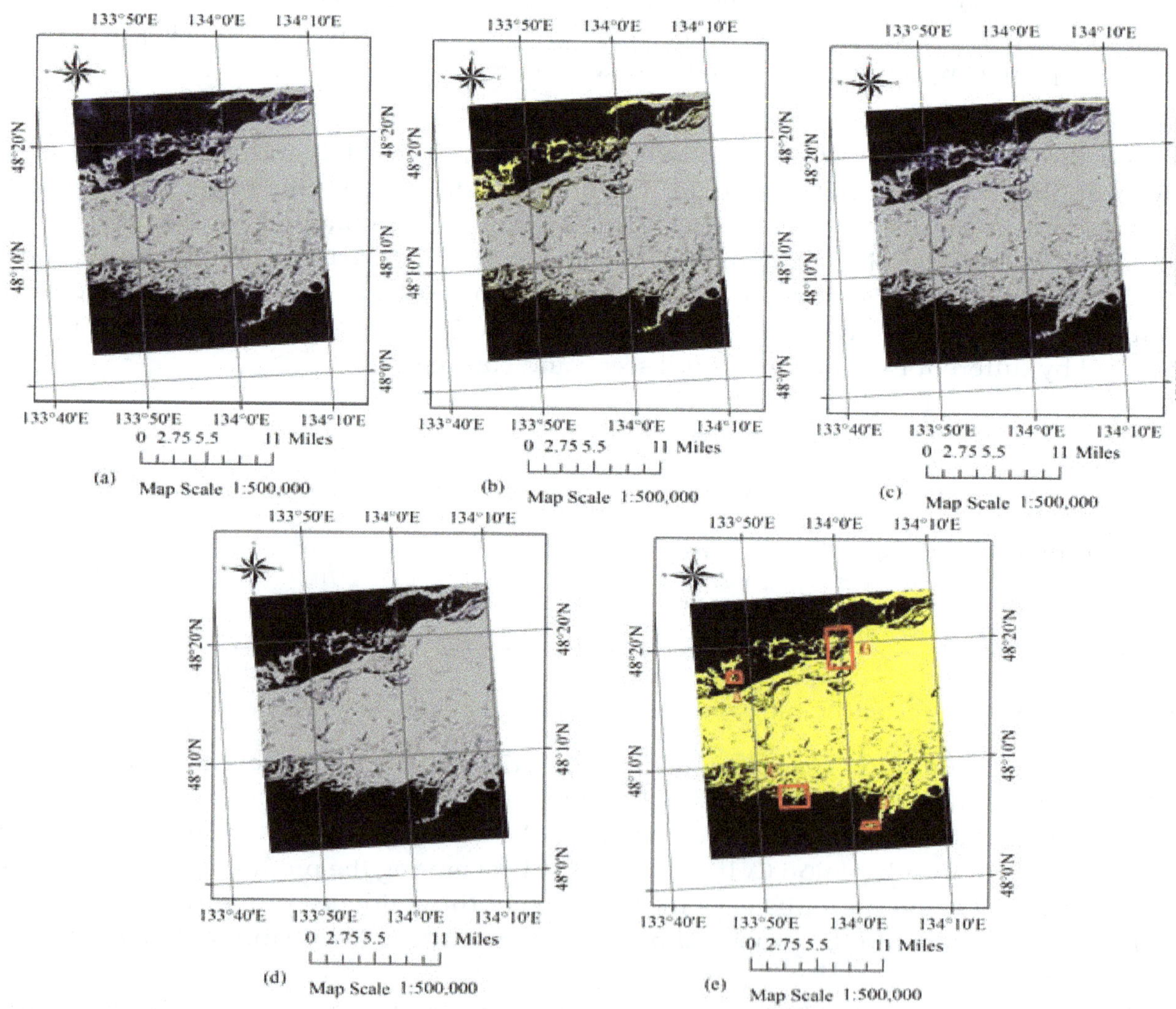

Figure 6. **(a–d)** Flood inundation mapping results for the first case study using K-MEANs, SP-MADB, STP-MADB and STCLP-MADB methods. (Gray: flood pixels in both the detection and reference maps; Blue: flood pixels only in the detection map; Yellow: flood pixels only in the reference map; Black: the background). **(e)** The locations of the four sub-regions in red rectangle, shown on the reference map of inundation derived from the GF-1 WFV data, with the water in yellow and the background in black.

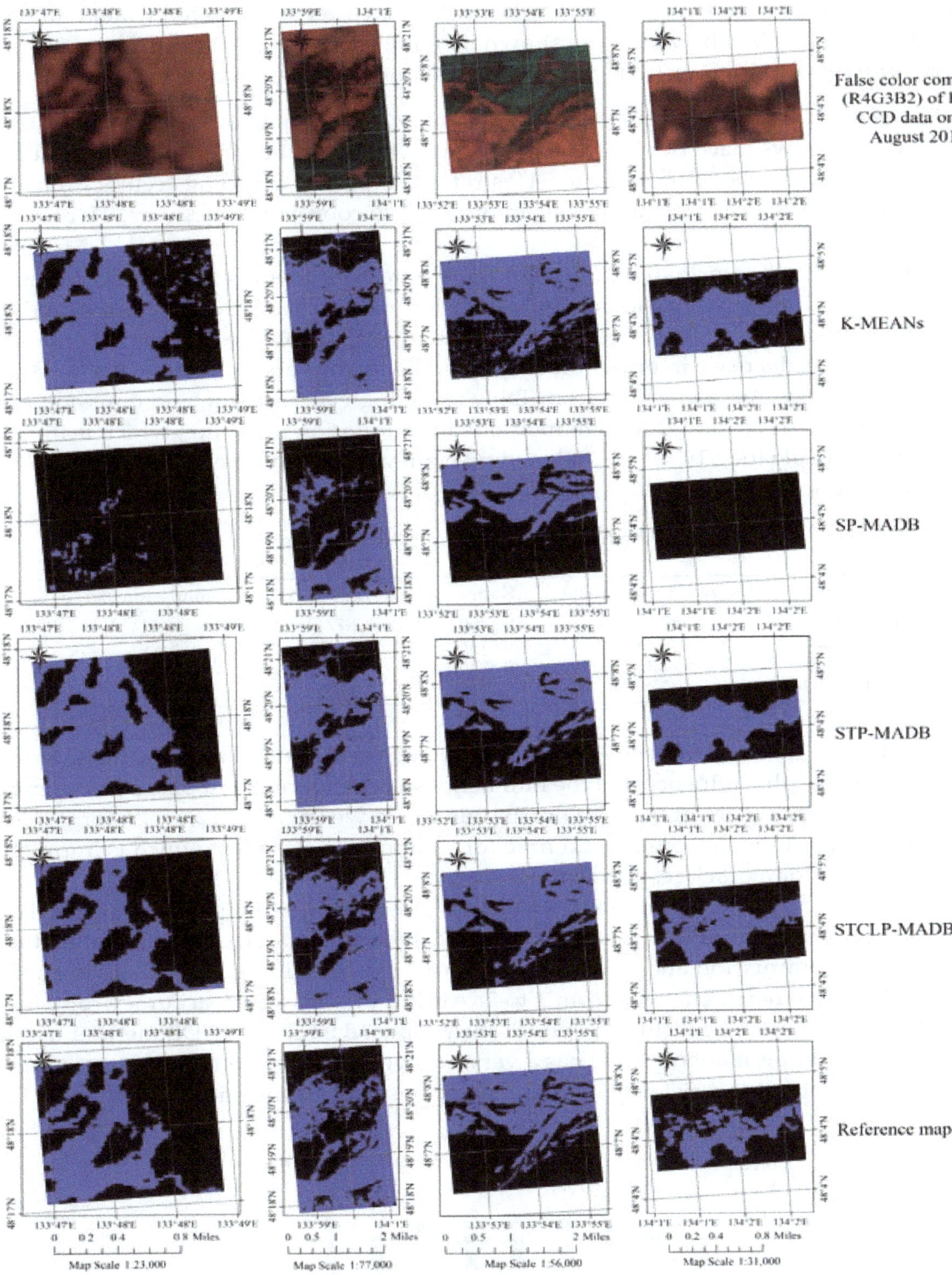

Figure 7. From the left to the right column: the regions of interest A, B, C and D for the first case study. From the second to the fifth row: corresponding detection and reference maps with the flood in blue and the background in black.

Besides the qualitative evaluation, a quantitative evaluation is also made for the test. For classification, the confusion matrix is one of the most commonly used methods for calculating accuracy. In this study, the reference and the detection results are all binary maps with two categories, water and non-water. Then, the confusion matrix can be produced. The accuracies of each method can be calculated, among which the overall accuracy is the rate of correctly classified water and non-water pixels among the total pixels. The values are shown in Table 5.

From the numbers reported in Table 5, a few further conclusions can be summarised:

(1) STCLP-MADB achieves the highest overall accuracy and kappa coefficient among these four methods, which shows that STCLP-MADB performs better than the others in terms of quantitative evaluation. Extending the SP strategy to STP strategy improves mapping accuracy. Furthermore, utilising STCL confidence calculation instead of a simple counting strategy in STP also enhances the mapping results.

(2) With incomplete flood information, different flood detectors produce different commission and omission errors. The best omission and commission rates are achieved by the K-MEANs and SP-MADB methods, respectively. However, there is always a balance between the omission and the commission. A decrease in omission errors usually brings about an increase in commission errors and vice versa. As can be seen from Table 5, the high commission and omission rates limit the ability of K-MEANs and SP-MADB methods in inundation mapping, which is illustrated in Figures 6 and 7, while the STCLP-MADB method achieves a balance between these two rates and provides a more acceptable result.

Table 5. The first case study in 2013—Accuracy.

Method	Overall Accuracy (%)	Kappa	Omission (%)	Commission (%)
K-MEANs	87.48	0.7450	2.77	17.51
SP-MADB	90.73	0.8146	12.18	5.53
STP-MADB	91.22	0.8220	3.03	12.13
STCLP-MADB	92.25	0.8435	4.10	9.78

We also discuss the relation between accuracy of the proposed method and the n in the permanent pixels extraction step, i.e., the influence that the number of selected permanent pixels has over mapping precision. The result is shown in Figure 8 below. In this work, we try $n = 1, 2, \ldots, 9$, for if n is too big, it will cost a lot of computation resources and time for training the classifier, which is impractical and cannot satisfy the need for a quick response to a disaster.

From the figure, it can be seen that more permanent pixels leads to an increase in commission and decrease in omission, but this only happens when $n \leq 2$. When the value of n gets higher, there is no significant change in commission and omission. Similarly, for overall accuracy, there is only a slight decrease (around 0.1%) when n changes from 1 to 2. After that, the overall accuracy remains almost unchanged. Therefore, it can be concluded that for the proposed STCLP-MADB method using HJ-1A CCD data, the number of permanent pixels has a very limited influence on mapping precision. Given the importance of computation efficiency in disaster response, it is quite enough to set n as 1 or 2.

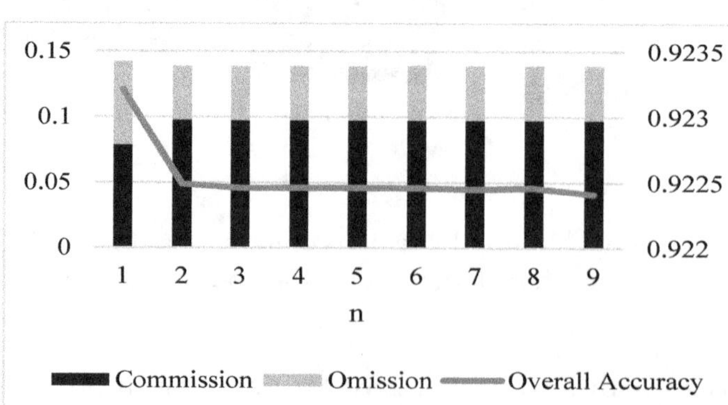

Figure 8. The first case study: commission, omission and overall accuracy in function of n, the percentage of permanent pixels.

4.2. Inundation Mapping Using GF-4 PMS Data

The second case study aims to analyse the GF-4 PMS data for the 2016 flood event at Dongting Lake. Figure 9 shows the two images selected for this case, which were acquired on 17 June 2016 (before the flood) and 23 July 2016 (after the flood). The corresponding MOD44W product used as the ancillary data is also shown in Figure 9.

In the second case study, inundation mapping results using different strategies are shown in Figure 10. Similar to the first case study, four sub-regions located at different positions are selected and shown in Figure 11, which aims to visually compare the results in a more detailed way. With regards to quantitative evaluation, Tables 6 and 7 report the number of inundated pixels and the final accuracy values, respectively. Figure 12 illustrates the relation between the permanent pixel proportion and detection accuracy.

As can be seen from the figures and the table, many similarities exist between the results of the second and the first case studies, and several slight differences as well. They are described as follows:

(1) In terms of the performance in categorisation, results in the second test are similar to that in the first test. The proposed STCLP-MADB method still achieves the best overall accuracy and kappa coefficient. K-MEANs and STP-MADB methods achieve the best omission and commission, respectively, while STCLP-MADB shows an average performance of these two rates. As the two test datasets are from different sensors, locations and inundation cases, this experiment further proves the good robustness of the proposed method.

(2) K-MEANs makes use of the statistical properties of the whole image, which causes high commission because the inundated pixels can have a different appearance in different contextual situations. SP-MADB and STP-MADB draw more attention to the local characteristics, but they make the determination of permanent pixels by counting, which lacks a theoretical foundation and can be easily disturbed. This can be found by comparing Figures 6c and 10c. In the first case, using HJ-1A CCD data, the STP-MADB method produces more commission, while in the second case study, using GF-4 PMS data, more omission than commission is introduced in the STP-MADB result. In the proposed method, a spatiotemporal context confidence calculating model is adopted to overcome the limitation of counting. With the formulised combination of local spatiotemporal and spectral information, we achieve a more accurate and robust inundation map than other methods.

(3) The changing curves of accuracy with n are more unstable than those in the first test. The influence of n on result precision does not change monotonically. It is difficult to find any rules in the curves at all. This could be because the outline of the inundation is more complicated in the second case study than in the first. Moreover, the spatial resolution of the GF-4 PMS data is sparser than that of the HJ-1A CCD data, which brings out more mixed pixels. With the increase in these uncertainties, the variation in accuracy becomes more unpredictable. Nevertheless, the fluctuation is still within a limited range. The effectiveness of the proposed method is rather stable.

(4) As the GF-4 satellite was officially put into service not long ago (in June 2016), research on GF-4 PMS data is rare. Our work explores the applied value of this new dataset and proves its effectiveness for inundation mapping. More promising research about GF-4 PMS data could be carried out in the future.

Table 6. The second case study in 2016—Number of inundated pixels.

Method	Full Region	Sub-Region A	Sub-Region B	Sub-Region C	Sub-Region D
K-MEANs	2,417,535	74,610	21,785	10,363	13,001
SP-MADB	1,103,408	23,946	1488	5884	6163
STP-MADB	873,738	16,908	1468	5233	3848
STCLP-MADB	1,215,464	30,502	2116	6668	6961
Reference Map	1,357,670	37,529	3654	7963	7510

Table 7. The second case study in 2016—Accuracy.

Method	Overall Accuracy (%)	Kappa	Omission (%)	Commission (%)
K-MEANs	79.73	0.5613	3.24	45.66
SP-MADB	92.88	0.7877	26.43	4.29
STP-MADB	91.01	0.7190	36.58	1.46
STCLP-MADB	93.66	0.8195	18.47	8.93

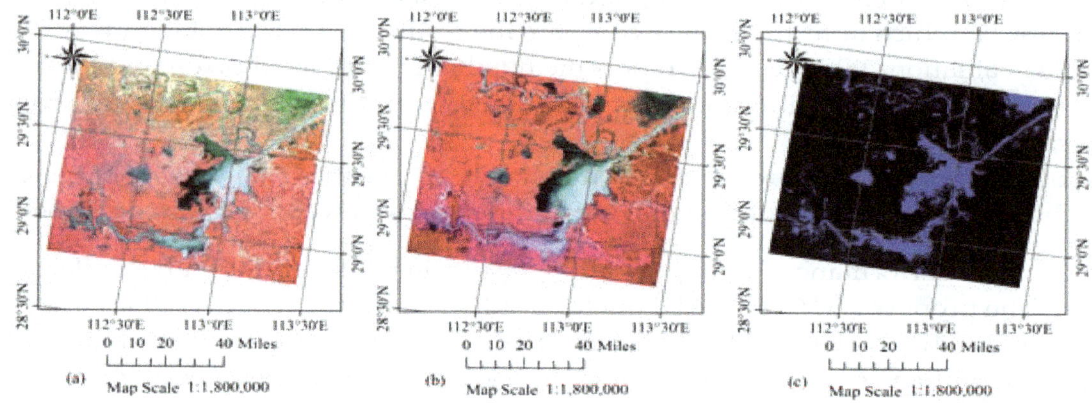

Figure 9. False colour composites (R: 5, G: 4, B: 3) of GF-4 PMS images acquired on (**a**) 17 June 2016 (before the flood) and (**b**) 23 July 2016 (after the flood) for the second test case. (**c**) Corresponding MOD44W water mask product with water in blue and land in black.

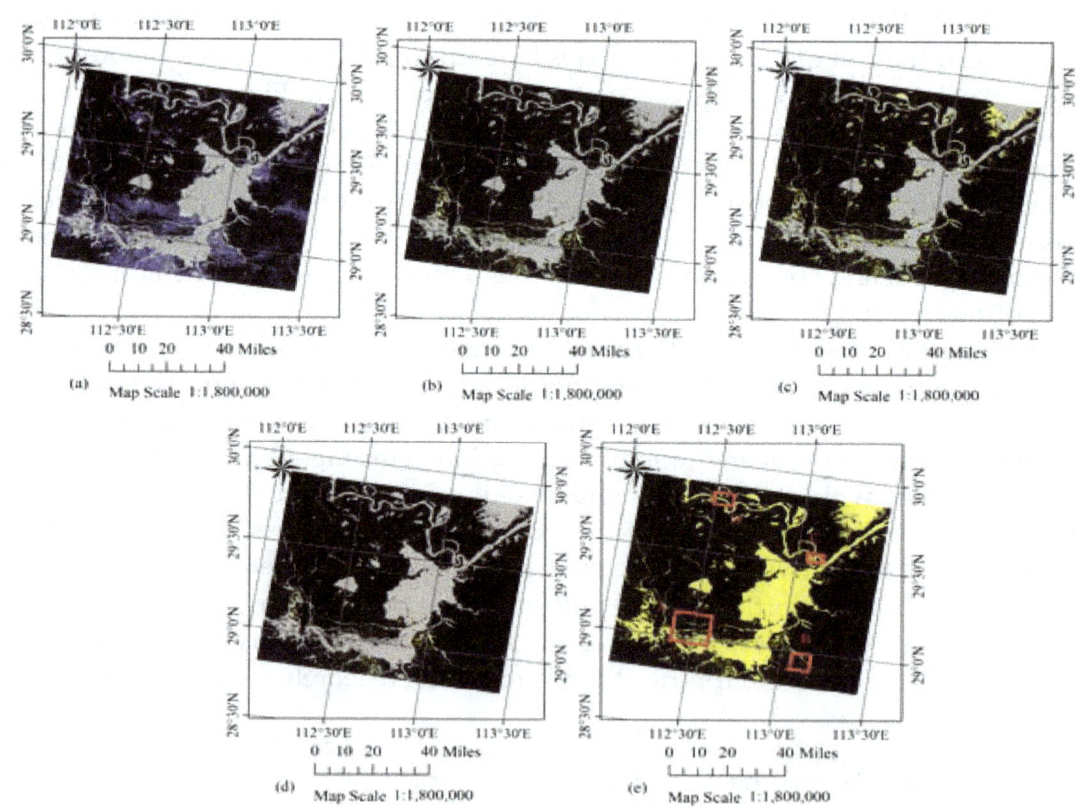

Figure 10. (**a–d**) Flood inundation mapping results for the second test area using K-MEANs, SP-MADB, STP-MADB and STCLP-MADB methods. (Gray: flood pixels in both the detection and reference maps; Blue: flood pixels only in the detection map; Yellow: flood pixels only in the reference map; Black: the background). (**e**) The locations of the four sub-regions in red rectangle, shown on the reference map of inundation derived from the HJ-1B CCD data, with the water in yellow and the background in black.

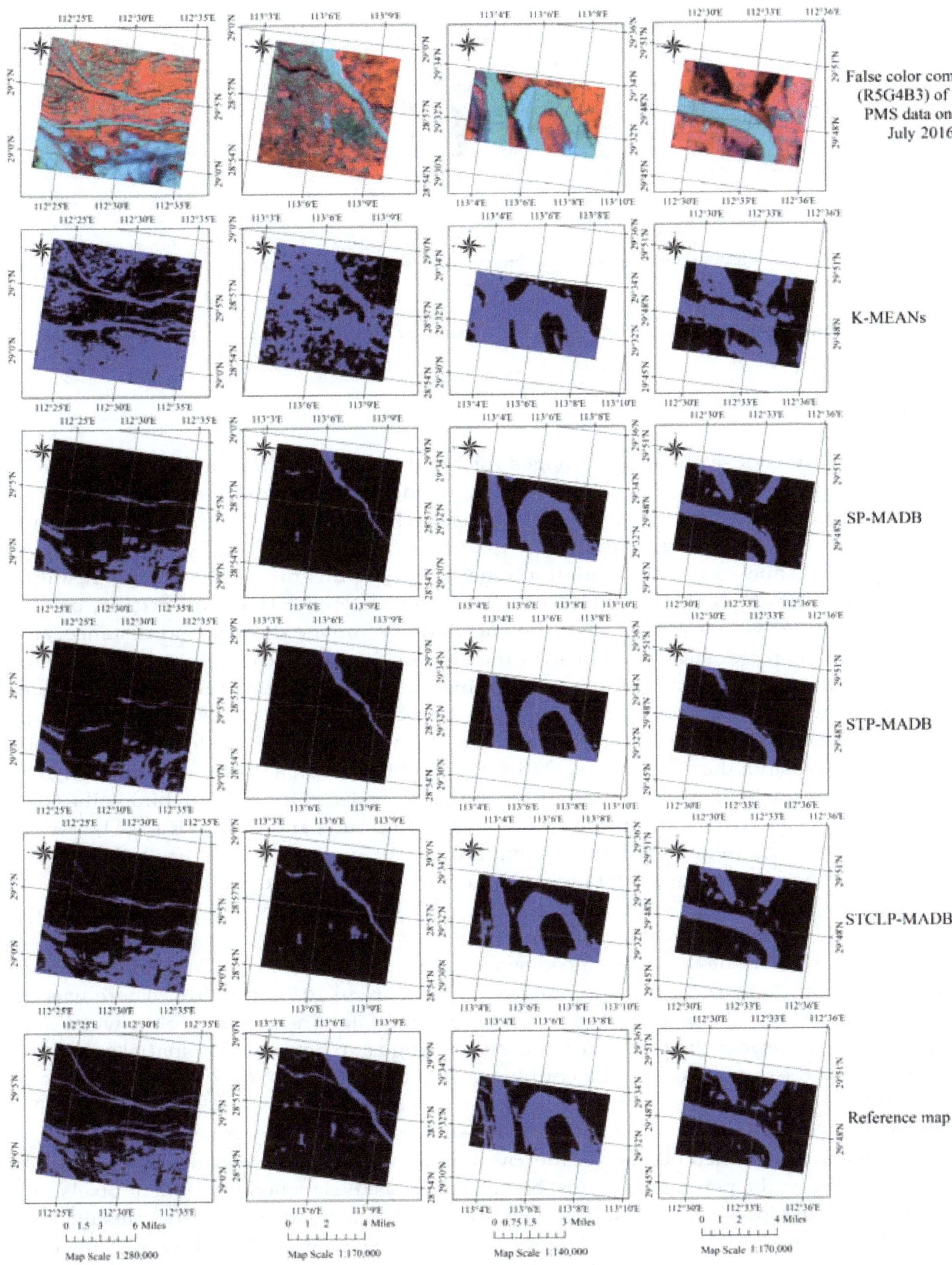

Figure 11. From the left to the right column: the regions of interest A, B, C and D for the second case study. From the second to the fifth row: corresponding detection and reference maps with the flood in blue and the background in black.

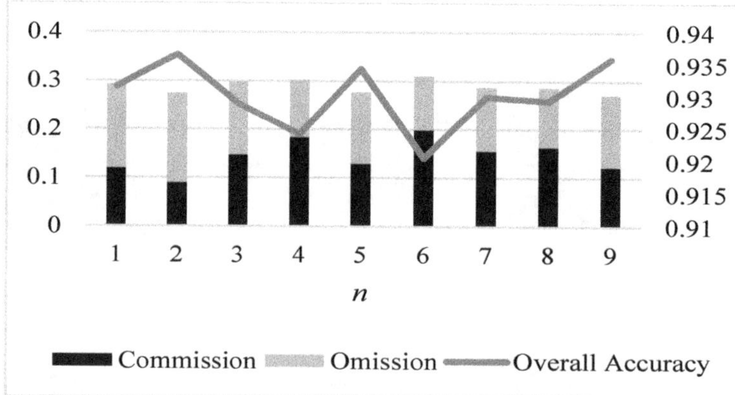

Figure 12. Second test—commission, omission and overall accuracy in function of n, the percentage of permanent pixels.

5. Discussion

In this study, we choose cloud-free images for the experiment. In practice, clouds and their shadows have been a critical issue for flood mapping using multispectral images, especially the as flood is usually accompanied by rainy and cloudy weather. This is because the visible and near infrared spectra cannot penetrate the cloud, so the image quality is frequently affected during flood periods. We put forward some analysis and speculation regarding how this may influence the result of the proposed method. First, the confidence value calculated in the step of STCL will certainly be affected by the clouds. As the STCL method models the correlation of image density and distance, and the cloud has different characteristics with those of the land or the water, the relative relationship will change a lot with the interference of clouds. According to the description in Section 3.1, if there are some clouds present nearby, the confidence value will decrease. However, since we only extract pixels with high confidence values for training the classifier, its impact on the final flood mapping may be limited. On the other hand, the GF-4 is a geostationary satellite. When a disaster happens, it can take images of the same region with a very high time resolution if needed. Through combining the common region of multiple images over a short time, data hidden by clouds and shadows may be recovered. Anyhow, it is a deficiency in our work that no experiment using cloudy data has been carried out. More explorations concerning cloudy data will be made in a future work.

In the proposed process for flood mapping, the MOD44W product plays a role in separating permanent pixels into permanent water pixels and permanent non-water pixels. With the introduction of this water mask, some issues are introduced as well. One is that this product is obtained based on the MODIS data from 2000 to 2002, while our case studies are in 2013 and 2016. There was over a decade between the MOD44W product and the experimental data. The outlines of the water are very likely to have altered. Besides the difference in time, the huge gap between the spatial resolutions of the water mask and the experimental data could also lead to problems. The spatial resolution of the MOD44W product is 250 m, which is much lower than that of the HJ-1A/B CCD and GF-4 PMS data. Many jagged edges can be found in the resampled result of the MOD44W product. Moreover, some small water surfaces are omitted because of the low spatial resolution. Both of these issues will bring about errors in labelling the permanent pixels. Nevertheless, as the labelled permanent pixels serve as the training set for the Modest AdaBoost classifier, not the final detailed classification result, we think a certain number of errors can be tolerated. Figures 5c and 9c also show that, in the experimental areas, from the visual effect, the MOD44W product is able to provide a general outline of the water before the flood comes. From another perspective, for the areas near the edges of the rivers and lakes, where most of the differences between the MOD44W product and the study data exist, the confidence value is generally low because of the changes induced by the flood. Therefore, the pixels at these areas are less likely to be selected as permanent pixels, and their corresponding MOD44W labels would have little influence on the final result.

According to the demand for automation and details in disaster assessment, this study aims to explore a novel solution for flood mapping that can achieve precise results with minimal human intervention. After two experiments on different regions and data, the proposed method shows better performance than other automatic methods. Several important reasons we infer are as follows. The first is the introduction of the machine learning classification method. Extensive literature shows that the precision of supervised classification methods is generally better than that of unsupervised classification methods. Unsupervised flood mapping methods, like K-MEANs, can bring about more errors in scenes of large area or complicated distribution. Because in these situations, the radiation value of water may vary a lot at different locations. Without a learning strategy, some non-water pixels with similar features to the water at other locations could be identified as water, as can be seen in Figure 10a of the Dongting Lake case. With the aid of the samples, the supervised classifier can learn and adapt itself better to different land cover characteristics in different scenes, resulting in higher accuracy. But the samples usually need to be selected manually, which limits their applicability in disaster response. Another essential advantage of the proposed method is that it proposes an automatic sample selection method, and combines it with a learning method. With the advantages of these two methods, both good precision and automation can be achieved.

The utilisation of local information is also a factor bringing improvement to the result. On one hand, it is more robust to utilise both contextual and global information than to utilise global information only. On the other hand, the experiment results show that the proposed method outperforms (qualitatively and quantitatively) the SP-MADB, STP-MADB methods. The only distinction among these three methods is the permanent pixel extraction strategy. All three methods utilise the local relationship between a pixel and its surroundings. The SP-MADB and STP-MADB methods count the number of 8-neighbourhood or 17-neighbourhood pixels with equal mean or medium to the object pixel. Noise and radiation variation, which exist all the time, can easily change the count result. Moreover, if a pixel and its neighbouring pixels simultaneously change from one cover type to another, the mean and medium will still remain the same, leading to errors in the permanent pixel set. From the experiment results we can see that the STP-MADB method obviously makes more commissions than omissions in the first case study, but makes more omissions in the second case, which proves its lack of robustness. Whereas, the proposed method builds a model between the pixel and its neighbouring regions, instead of counting the few adjacent pixels. Even if there are some noise pixels, the general structure of the model will not change. With better selection of the permanent pixels and the training set, the STCLP-MADB method produces a more precise outline of the inundated areas.

This study is proposed for floods, which is a practical problem. Hence it makes sense that this proposed method can be applied operationally, and that it can help when a real flood comes. Here we propose some suggestions for implementation, which may help the STCL-MADB method to be effectively applied in a real application. The whole workflow can be divided into three steps: extracting permanent pixels using STCL, training the classifier and mapping the inundation. The first step, especially the STCL algorithm, accounts for most of the time consumption in the whole process. Not only because the STCL algorithm has higher computation complexity than other steps, but the operations need to be performed pixel by pixel. For example, in the second case study at Dongting Lake, the size of the data is 2534 $\times$ 2235 pixels. The first step takes around 3 days, while the second and third steps take around 15 min and a few dozen seconds, respectively. All these experiments are implemented by MATLAB 2013 on a laptop with an i7-4710HQ CPU and 8 GB RAM. In practice, a library of permanent pixels can be built in advance for regions with high flood risk. With the accumulation of time series data, the library can be updated continuously. Then, the classifier can also be trained and updated with the new library. As soon as the latest scene of remote sensing data arrives, the ready classifier can be directly applied to it. In addition, implementing the process in other programming languages and utilising high-performance processors could also help promote efficiency.

Above all, remote sensing data with middle-high spatial and temporal resolution are recommended in the proposed method, as the spatiotemporal contextual information in the image is important for the method. If the spatial or the temporal resolution is rather low in the data, the correlation between neighbouring pixels could be unremarkable.

6. Conclusions

Due to its vast coverage in spatial and temporal scales, flood is considered to be one of the most complex disasters in the world. A novel inundation mapping approach based on spatiotemporal context learning and Modest AdaBoost is proposed and verified in this paper. The proposed method is implemented and evaluated in two different flooding cases using images from different sensors, HJ-1A CCD and GF-4 PMS. The experimental results show that the proposed approach is effective, and is able to produce more accurate mapping results than other state-of-the-art methods and, more importantly, without any artificial samples and thresholds.

On one hand, compared with the traditional global-based unsupervised flood mapping methods (such as K-MEANs), the SP-MADB, STP-MADB and the proposed method combine an automatic sample selection strategy with a machine learning classifier, leading to higher accuracies in an automatic way. With the samples extracted using local information, each of these three methods achieves an overall accuracy of more than 90% in both of the first and second case studies. By comparing the results of the SP-MADB and STP-MADB methods, it can be seen that only extending the neighbouring region to the temporal domain cannot significantly improve the performance of the SP-MADB method. With a formulised model of the spatiotemporal context information instead of simple counting, the proposed approach achieves a more accurate and robust result than other methods. As a result of mixed pixels, there are still some inaccuracies in the result. Moreover, the effect of the proposed method on cloudy data needs to be explored. Future work will focus on these aspects and validate the proposed method with more kinds of data.

Acknowledgments: This work is supported by the National Natural Science Foundation of China (Grant No. 41401474) and the National Key Research and Development Plan (No. 2016YFB0502502). The HJ-1A/B CCD data, GF-1 WFV data and GF-4 PMS data were provided by the China Centre for Resources Satellite Data and Application (CRESDA, http://www.cresda.com). The MODIS water mask (MOD44W) was obtained through the online Data Pool at the NASA Land Processes Distributed Active Archive Centre (LP DAAC), USGS/Earth Resources Observation and Science (EROS) Centre (https://lpdaac.usgs.gov). We thank the above organisations for providing the data. We thank the University of Chinese Academy of Sciences for scholarship support of Xiaoyi Liu.

Author Contributions: Xiaoyi Liu conceived and designed the experiments, performed the experiments and wrote the paper under the guidance of Hichem Sahli and Yu Meng. Qingqing Huang and Lei Lin analysed the data and contributed analysis tools.

References

1. O'Keefe, P.; Westgate, K.; Wisner, B. Taking the naturalness out of natural disasters. *Nature* **1976**, *260*, 566–567. [CrossRef]
2. Sanyal, J.; Lu, X.X. Application of remote sensing in flood management with special reference to monsoon Asia: A review. *Nat. Hazards* **2004**, *33*, 283–301. [CrossRef]
3. Berz, G.; Kron, W.; Loster, T.; Rauch, E.; Schimetschek, J.; Schmieder, J.; Siebert, A.; Smolka, A.; Wirtz, A. World map of natural hazards—A global view of the distribution and intensity of significant exposures. *Nat. Hazards* **2001**, *23*, 443–465. [CrossRef]
4. Akıncı, H.; Erdoğan, S. Designing a flood forecasting and inundation-mapping system integrated with spatial data infrastructures for Turkey. *Nat. Hazards* **2014**, *71*, 895–911. [CrossRef]
5. Smith, L.C. Satellite remote sensing of river inundation area, stage, and discharge: A review. *Hydrol. Process.* **1997**, *11*, 1427–1439. [CrossRef]
6. Brivio, P.A.; Colombo, R.; Maggi, M.; Tomasoni, R. Integration of remote sensing data and GIS for accurate mapping of flooded areas. *Int. J. Remote Sens.* **2002**, *23*, 429–441. [CrossRef]

7. Wang, Y.; Colby, J.D.; Mulcahy, K.A. An efficient method for mapping flood extent in a coastal floodplain using Landsat TM and DEM data. *Int. J. Remote Sens.* **2002**, *23*, 3681–3696. [CrossRef]
8. Rahman, M.S.; Di, L. The state of the art of spaceborne remote sensing in flood management. *Nat. Hazards* **2017**, *85*, 1223–1248. [CrossRef]
9. Li, L.; Chen, Y.; Yu, X.; Liu, R.; Huang, C. Sub-pixel flood inundation mapping from multispectral remotely sensed images based on discrete particle swarm optimization. *ISPRS J. Photogramm. Remote Sens.* **2015**, *101*, 10–21. [CrossRef]
10. Brakenridge, R.; Anderson, E. MODIS-based flood detection, mapping and measurement: The potential for operational hydrological applications. In *Transboundary Floods: Reducing Risks through Flood Management*; Marsalek, J., Stancalie, G., Balint, G., Eds.; Springer: Dordrecht, The Netherlands, 2006; Volume 72, pp. 1–12.
11. Ticehurst, C.J.; Chen, Y.; Karim, F.; Dutta, D.; Gouweleeuw, B. Using MODIS for mapping flood events for use in hydrological and hydrodynamic models: Experiences so far. In Proceedings of the 20th International Congress on Modelling and Simulation, Adelaide, Australia, 1–6 December 2013; pp. 1–6.
12. Kwak, Y.; Park, J.; Yorozuya, A.; Fukami, K. Estimation of flood volume in Chao Phraya River basin, Thailand, from MODIS images couppled with flood inundation level. In Proceedings of the 2012 IEEE International Geoscience and Remote Sensing Symposium (IGARSS), Munich, Germany, 22–27 July 2012.
13. Khan, S.I.; Hong, Y.; Wang, J.; Yilmaz, K.K.; Gourley, J.J.; Adler, R.F.; Brakenridge, G.R.; Policelli, F.; Habib, S.; Irwin, D. Satellite remote sensing and hydrologic modeling for flood inundation mapping in Lake Victoria basin: Implications for hydrologic prediction in ungauged basins. *IEEE Trans. Geosci. Remote Sens.* **2011**, *49*, 85–95. [CrossRef]
14. Schumann, G. Preface: Remote Sensing in Flood Monitoring and Management. *Remote Sens.* **2015**, *7*, 17013–17015. [CrossRef]
15. McFeeters, S.K. The use of the Normalized Difference Water Index (NDWI) in the delineation of open water features. *Int. J. Remote Sens.* **1996**, *17*, 1425–1432. [CrossRef]
16. Jain, S.K.; Singh, R.D.; Jain, M.K.; Lohani, A.K. Delineation of flood-prone areas using remote sensing techniques. *Water Resour. Manag.* **2005**, *19*, 333–347. [CrossRef]
17. Xie, H.; Luo, X.; Xu, X.; Pan, H.; Tong, X. Evaluation of Landsat 8 OLI imagery for unsupervised inland water extraction. *Int. J. Remote Sens.* **2016**, *37*, 1826–1844. [CrossRef]
18. Giordano, F.; Goccia, M.; Dellepiane, S. Segmentation of coherence maps for flood damage assessment. In Proceedings of the IEEE International Conference on Image Processing, Genova, Italy, 14 September 2005; Volume 2, p. II-233.
19. Dellepiane, S.; Angiati, E.; Vernazza, G. Processing and segmentation of COSMO-SkyMed images for flood monitoring. In Proceedings of the 2010 IEEE International Geoscience and Remote Sensing Symposium, Honolulu, HI, USA, 25–30 July 2010.
20. Chignell, S.M.; Anderson, R.S.; Evangelista, P.H.; Laituri, M.J.; Merritt, D.M. Multi-temporal independent component analysis and Landsat 8 for delineating maximum extent of the 2013 Colorado front range flood. *Remote Sens.* **2015**, *7*, 9822–9843. [CrossRef]
21. Rokni, K.; Ahmad, A.; Selamat, A.; Hazini, S. Water feature extraction and change detection using multitemporal Landsat imagery. *Remote Sens.* **2014**, *6*, 4173–4189. [CrossRef]
22. Chen, X.C.; Khandelwal, A.; Shi, S.; Faghmous, J.H.; Boriah, S.; Kumar, V. Unsupervised method for water surface extent monitoring using remote sensing data. In *Machine Learning and Data Mining Approaches to Climate Science*; Springer International Publishing: Cham, Switzerland, 2015; pp. 51–58.
23. Huang, C.; Chen, Y.; Wu, J. Dem-based modification of pixel-swapping algorithm for enhancing floodplain inundation mapping. *Int. J. Remote Sens.* **2014**, *35*, 365–381. [CrossRef]
24. Lu, S.; Wu, B.; Yan, N.; Wang, H. Water body mapping method with hj-1a/b satellite imagery. *Int. J. Appl. Earth Obs. Geoinf.* **2011**, *13*, 428–434. [CrossRef]
25. Gu, X.; Tong, X. Overview of china earth observation satellite programs. *IEEE Geosci. Remote Sens. Mag.* **2015**, *3*, 113–129.
26. Gu, Y.; Hunt, E.; Wardlow, B.; Basara, J.B.; Brown, J.F.; Verdin, J.P. Evaluation of MODIS NDVI and NDWI for vegetation drought monitoring using oklahoma mesonet soil moisture data. *Geophys. Res. Lett.* **2008**, *35*, 1092–1104. [CrossRef]
27. George, C.; Rowland, C.; Gerard, F.; Balzter, H. Retrospective mapping of burnt areas in central siberia using a modification of the normalised difference water index. *Remote Sens. Environ.* **2006**, *104*, 346–359. [CrossRef]

28. Mcfeeters, S.K. Using the normalized difference water index (NDWI) within a geographic information system to detect swimming pools for mosquito abatement: A practical approach. *Remote Sens.* **2013**, *5*, 3544–3561. [CrossRef]
29. Li, W.; Du, Z.; Ling, F.; Zhou, D.; Wang, H.; Gui, Y.; Sun, B.; Zhang, X. A comparison of land surface water mapping using the normalized difference water index from TM, ETM+ and ALI. *Remote Sens.* **2013**, *5*, 5530–5549. [CrossRef]
30. Zhang, K.; Zhang, L.; Yang, M.H.; Zhang, D. Fast tracking via spatio-temporal context learning. *arXiv*, **2013**, preprint arXiv:1311.1939.
31. Zuo, Z.Y.; Tian, S.; Pei, W.Y.; Yin, X.C. Multi-strategy tracking based text detection in scene videos. In Proceedings of the IEEE 13th International Conference on Document Analysis and Recognition, Tunis, Tunisia, 23–26 August 2015.
32. Xu, J.; Lu, Y.; Liu, J. Robust tracking via weighted spatio-temporal context learning. In Proceedings of the 2014 IEEE International Conference on Image Processing (ICIP), Paris, France, 27–30 October 2014.
33. Muster, S.; Heim, B.; Abnizova, A.; Boike, J. Water body distributions across scales: A remote sensing based comparison of three arctic tundra wetlands. *Remote Sens.* **2013**, *5*, 1498–1523. [CrossRef]
34. Freund, Y.; Schapire, R.E. A desicion-theoretic generalization of on-line learning and an application to boosting. In *European Conference on Computational Learning Theory*; Springer: Berlin/Heidelberg, Germany, 1995; pp. 23–37.
35. Viola, P.; Jones, M. Robust real-time object detection. *Int. J. Comput. Vis.* **2001**, *57*, 34–37.
36. Vezhnevets, A.; Vezhnevets, V. Modest AdaBoost-teaching AdaBoost to generalize better. *Graphicon* **2005**, *12*, 987–997.
37. Sam, K.T.; Tian, X.L. Rapid license plate detection using Modest AdaBoost and template matching. In Proceedings of the 2nd International Conference on Digital Image Processing, Singapore, 26 February 2010.
38. Qahwaji, R.; Al-Omari, M.; Colak, T.; Ipson, S. Using the real, gentle and modest AdaBoost learning algorithms to investigate the computerised associations between coronal mass ejections and filaments. In Proceedings of the 2008 IEEE Communications, Computers and Applications, Amman, Jordan, 8–10 August 2008.
39. MSU Graphics & Media Lab, Computer Vision Group. Available online: http://graphics.cs.msu.ru (accessed on 3 June 2016).
40. Gao, B.C. NDWI—A normalized difference water index for remote sensing of vegetation liquid water from space. *Remote Sens. Environ.* **1996**, *58*, 257–266. [CrossRef]
41. Huete, A.; Didan, K.; Miura, T.; Rodriguez, E.P.; Gao, X.; Ferreira, L.G. Overview of the radiometric and biophysical performance of the MODIS vegetation indices. *Remote Sens. Environ.* **2002**, *83*, 195–213. [CrossRef]

4

Supervised and Semi-Supervised Multi-View Canonical Correlation Analysis Ensemble for Heterogeneous Domain Adaptation in Remote Sensing Image Classification

Alim Samat [1,2], Claudio Persello [3], Paolo Gamba [4], Sicong Liu [5], Jilili Abuduwaili [1,2,*] and Erzhu Li [6]

1. State Key Laboratory of Desert and Oasis Ecology, Xinjiang Institute of Ecology and Geography, Chinese Academy of Sciences, Urumqi 830011, China; alim.smt@gmail.com
2. CAS Research Center for Ecology and Environment of Central Asia, Chinese Academy of Sciences, Urumqi 830011, China
3. Department of Earth Observation Science, Faculty of Geo-Information Science and Earth Observation (ITC), University of Twente, 7500 AE Enschede, The Netherlands; c.persello@utwente.nl
4. Department of Electrical, Computer and Biomedical Engineering, University of Pavia, 27100 Pavia, Italy; paolo.gamba@unipv.it
5. College of Surveying and Geoinformatics, Tongji University, Shanghai 200092, China; sicongliu.rs@gmail.com
6. Department of Geographical Information Science, Nanjing University, Nanjing 210000, China; lierzhu2008@126.com

* Correspondence: jilil@ms.xjb.ac.cn

Academic Editors: Qi Wang, Nicolas H. Younan, Carlos López-Martínez, Xiaofeng Li and Prasad S. Thenkabail

Abstract: In this paper, we present the supervised multi-view canonical correlation analysis ensemble (SMVCCAE) and its semi-supervised version (SSMVCCAE), which are novel techniques designed to address heterogeneous domain adaptation problems, i.e., situations in which the data to be processed and recognized are collected from different heterogeneous domains. Specifically, the multi-view canonical correlation analysis scheme is utilized to extract multiple correlation subspaces that are useful for joint representations for data association across domains. This scheme makes homogeneous domain adaption algorithms suitable for heterogeneous domain adaptation problems. Additionally, inspired by fusion methods such as Ensemble Learning (EL), this work proposes a weighted voting scheme based on canonical correlation coefficients to combine classification results in multiple correlation subspaces. Finally, the semi-supervised MVCCAE extends the original procedure by incorporating multiple speed-up spectral regression kernel discriminant analysis (SRKDA). To validate the performances of the proposed supervised procedure, a single-view canonical analysis (SVCCA) with the same base classifier (Random Forests) is used. Similarly, to evaluate the performance of the semi-supervised approach, a comparison is made with other techniques such as Logistic label propagation (LLP) and the Laplacian support vector machine (LapSVM). All of the approaches are tested on two real hyperspectral images, which are considered the target domain, with a classifier trained from synthetic low-dimensional multispectral images, which are considered the original source domain. The experimental results confirm that multi-view canonical correlation can overcome the limitations of SVCCA. Both of the proposed procedures outperform the ones used in the comparison with respect to not only the classification accuracy but also the computational efficiency. Moreover, this research shows that canonical correlation weighted voting (CCWV) is a valid option with respect to other ensemble schemes and that because of their ability to balance diversity and accuracy, canonical views extracted using partially joint random view generation are more effective than those obtained by exploiting disjoint random view generation.

Keywords: heterogeneous domain adaptation; transfer learning; multi-view canonical correlation analysis ensemble; semi-supervised learning; canonical correlation weighted voting; ensemble learning; image classification

1. Introduction

Supervised learning algorithms predominate over all other land cover mapping/monitoring techniques that use remote sensing (RS) data. However, the performance of supervised learning algorithms varies as a function of labeled training data properties, such as the sample size and the statistically unbiased and discriminative capabilities of the features extracted from the data [1]. As monitoring requires multi-temporal images, radiometric differences, atmospheric and illumination conditions, seasonal variations, and variable acquisition geometries can affect supervised techniques, potentially causing a distribution shift in the training data [2,3]. Regardless of the cause, any distribution change or domain shift that occurs after learning a classifier can degrade performance.

In the pattern recognition (PR) and RS image classification communities, this challenge is commonly referred to as covariate shift [4] or sample selection bias [5]. Many solutions have been proposed to resolve this problem, including image-to-image normalization [6], absolute and relative image normalization [7,8], histogram matching [9], and a multivariate extension of the univariate matching [10]. Recently, domain adaptation (DA) techniques, which attempt to mitigate performance the degradation caused by a distribution shift, has attracted increasing attention and is widely considered to provide an efficient solution [11–16].

According to the technical literature in PR and machine learning (ML), DA is a special case of transductive transfer learning (TTL). Its goal is to learn a function that predicts the label of a novel test sample in the target domain [12,15]. Depending on the availability of the source and the target domain data, the DA problem can result into supervised domain adaptation (SDA), semi-supervised domain adaptation (SSDA), unsupervised domain adaptation (UDA), multisource domain adaptation (MSDA) and heterogeneous domain adaption (HDA) [14–19].

Moreover, according to the "knowledge" transferred across domains or tasks, classical approaches to DA can be grouped into parameter adapting, instance transferring, feature representation, and relational knowledge transfer techniques.

Parameter adapting approaches aim to transfer and adapt a classification model and/or its parameters to the target domain; the model and/or parameters are learned from the source domain (SD) [20]. The seminal work presented by Khosla et al. [5] and Woodcock et al. [7], which features parameter adjustment for a maximum-likelihood classifier in a multiple cascade classifier system by retraining, can be categorized into this group.

In instance transferring, the samples from the SD are reweighted [21] or resampled [22] for their use in the TD. In the RS community, active learning (AL) has also been applied to address DA problems. For example, AL for DA in the supervised classification RS images is proposed by Persello and Bruzzone [23] via iteratively labeling and adding to the training set the minimum number of the most informative samples from the target domain, while removing the source-domain samples that do not fit with the distributions of the classes in the TD.

For the third group, feature representation-based adaptation searches for a set of shared and invariant features using feature extraction (FE), feature selection (FS) or manifold alignment to reduce the marginal, conditional and joint distributions between the domains [16,24–26]. Matasci et al. [14] investigated the semi-supervised transfer component analysis (SSTCA) [27] for both hyperspectral and multispectral high resolution image classification, whereas Samat et al. [16] analyzed a geodesic Gaussian flow kernel based support vector machine (GFKSVM) in the context of hyperspectral image classification, which adopts several unsupervised linear and nonlinear subspace feature transfer techniques.

Finally, relational knowledge transfer techniques address the problem of how to leverage the knowledge acquired in SD to improve accuracy and learning speed in a related TD [28]. Among these four groups, it is easy to recognize the importance of RS image classification of adaptation strategies based on feature representation. However, most previous studies have assumed that data from different domains are represented by the same types of features with the same dimensions. Thus, these techniques cannot handle the problem of data from source and target domains represented by heterogeneous features with different dimensions [18,29]. One example of this scenario is land cover updating using current RS data; each time, there are different features with finer spatial resolution and more spectral bands (e.g., Landsat 8 OLI with nine spectral bands at 15–30 m spatial resolution, and Airborne Visible/Infrared Imaging Spectrometer (AVIRIS) with 224 spectral bands at 20 m spatial resolution), when the training data are only available at coarser spatial and spectral resolutions (e.g., MSS with four spectral bands and 60 m spatial resolution).

One of the simplest feature-based DA approaches is the feature augmentation proposed in [17], whose extended versions, called heterogeneous feature augmentation (HFA) and semi-supervised HFA (SHFA), were recently proposed in [18]. Versions that consider the intermediate domains as being manifold-based were proposed in [30,31]. However, none of these approaches have been considered in RS image classification.

Finding a joint feature representation between the source and target domains requires FS [12,19] or FE [16] to select the most effective feature set. To accomplish this aim, canonical correlation analysis (CCA), which aims to maximize the correlation between two variable sets (in this case, the different domains) could be a very effective technique. Indeed, CCA and kernel CCA (KCCA) have already been applied with promising results in object recognition and text categorization [29], action recognition and image-to-text classification [32]. However, existing joint optimization frameworks such as [32] are limited to scenarios in which the labeled data from both domains are available. This is not the case in many practical situations. To solve this problem, CTSVM was proposed in [29], incorporating the DA ability into the classifier design for a cross-domain recognition scenario of labeled data that is available only in the SD. However, the CTSVM might fail to balance the possible mismatches between the heterogeneous domains.

One solution might be to multi-view learning (MVL), a procedure that implies the splitting of high-dimensional data into multiple "views" [33,34]. If multiple views are available, then multiple classification results must be reconciled, and this step is efficiently performed using Ensemble Learning (EL) [35,36]. Accordingly, this work introduces an EL technique based on supervised multi-view CCA, which is called supervised multi-view canonical correlation analysis ensemble (SMVCCAE), and we prove its effectiveness for DA (and specifically heterogeneous DA) problems.

Additionally, in real applications, it is typical to experience situations in which there are very limited or even no labeled samples available. In this case, a semi-supervised learning (SSL) technique (e.g., [37]), which uses of unlabeled data to improve performance using a small amount of labeled data from the same domain, might be an appropriate solution. As a matter of fact, many SSDAs have been proposed. However, most existing studies, such as asymmetric kernel transforms (AKT) [38], domain-dependent regularization (DDR) [32], TCA, SSTCA [14,27], and co-regularization based SSDA [39], were designed for homogeneous DA. Very recently, Li et al. [18] proposed a semi-supervised heterogeneous DA by convex optimization of standard multiple kernel learning (MKL) with augmented features. Unfortunately, this optimization is quite challenging in real-world applications. This work instead proposes a semi-supervised version of the above-mentioned multi-view canonical correlation analysis ensemble (called SSMVCCAE), incorporating multiple speed-up spectral regression kernel discriminant analysis (SRKDA) [40] into the original supervised algorithm.

2. Related Work

2.1. Notation for HDA

According to the technical literature, feature-based approaches to HDA can be grouped into the following three clusters, depending on the features used to connect the target and the SD:

(1) If data from the source and target domains share the same features [41–43], then latent semantic analysis (LSA) [44], probabilistic latent semantic analysis (pLSA) [45], and risk minimization techniques [46] may be used.

(2) If additional features are needed, "feature augmentation" approaches have been proposed, including the method in [37], HFA and SHFA [18], manifold alignment [31], sampling geodesic flow (SGF) [47], and geodesic flow kernel (GFK) [16,30]. All of these approaches introduce a common subspace for the source and target data so that heterogeneous features from both domains.

(3) If features are adapted across domains through learning transformations, feature transformation-based approaches are considered. This group of approaches includes the HSMap [48], the sparse heterogeneous feature representation (SHFR) [49], and the correlation transfer SVM (CTSVM) [29]. The algorithms that we propose fit into this group.

Although all of the approaches reviewed above have achieved promising results, they also have some limitations of all the approaches reviewed above. For example, the co-occurrence features assumption used in [41–43] may not hold in applications such as object recognition, which uses only visual features [32]. For the feature augmentation based approaches discussed in [18,30,31], the domain-specific copy process always requires large storage space, and the kernel version requires even more space and computational complexity because of the parameter tuning. Finally, for the feature transformation based approaches proposed in [29,32,48], they do not optimize the objective function of a discriminative classifier directly, and the computational complexity is highly dependent on the total number of samples or features used for adaptation [12,19].

In this work, we assume that there is only one SD (S_D) and one TD (T_D). We also define $\mathbf{X}_S = \left[x_1^S, \ldots, x_{n_S}^S\right]^\dagger \in \Re^{d_S \times n_S}$ and $\mathbf{X}_T = \left[x_1^T, \ldots, x_{n_T}^T\right]^\dagger \in \Re^{d_T \times n_T}$ as the feature spaces in the two domains, with the corresponding marginal distributions $p(\mathbf{X}_S)$ and $p(\mathbf{X}_T)$ for S_D and T_D, respectively. The parameters d_S and d_T represent the size of $x_i^S, i = 1, \ldots, n_S$ and $x_j^T, j = 1, \ldots, n_T$, n_S and n_T are the sample sizes for $\mathbf{X}_S$ and $\mathbf{X}_T$, and we have $S_D = \{\mathbf{X}_S, P(\mathbf{X}_S)\}$, $T_D = \{\mathbf{X}_T, P(\mathbf{X}_T)\}$. The labeled training samples from the SD are denoted by $\left\{ \left(x_j^S, y_j^S\right)\Big|_{j=1}^{n_S} \right\}, y_j^S \in \Omega = \{\omega_l\}_{l=1}^{c}$, and they refer to c classes. Furthermore, let us consider as "task" Y the task to assign to each element of a set a label selected in a label space by means of a predictive function f, so that $v = \{y, f\}$.

In general, if the feature sets belong to different domains, then either $\mathbf{X}_S \neq \mathbf{X}_T$ or $p(\mathbf{X}_S) \neq p(\mathbf{X}_T)$, or both. Similarly, the condition $v_S \neq v_T$ implies that either $Y_S \neq Y_T$ ($Y_S = [y_1^S, \ldots, y_{n_S}^S]$, $Y_T = [y_1^T, \ldots, y_{n_T}^T]$) or $p(Y_S|\mathbf{X}_S) \neq p(Y_T|\mathbf{X}_T)$, or both. In this scenario, a "domain adaptation algorithm" is an algorithm that aims to improve the learning of the predictive function f_T in the TD T_D using the knowledge available in the SD S_D and in the learning task v_S, when either $S_D \neq T_D$ or $v_S \neq v_T$. Moreover, in heterogeneous problems, the additional condition $d_S \neq d_T$ holds.

2.2. Canonical Correlation Analysis

Let us now assume that $n_S = n_T$ for the feature sets (called "views" here) in the source and target domains. The CCA is the procedure for obtaining the transformation matrices $\boldsymbol{\omega}_S$ and $\boldsymbol{\omega}_T$ which maximize the correlation coefficient between the two sets [50]:

$$\max_{\boldsymbol{\omega}_S, \boldsymbol{\omega}_T} \rho = \frac{\boldsymbol{\omega}_S^\dagger \Sigma_{ST} \boldsymbol{\omega}_T}{\sqrt{\boldsymbol{\omega}_S^\dagger \Sigma_{SS} \boldsymbol{\omega}_S} \sqrt{\boldsymbol{\omega}_T^\dagger \Sigma_{TT} \boldsymbol{\omega}_T}} \tag{1}$$

where $\Sigma_{ST} = \mathbf{X}_S\mathbf{X}_T^{\dagger}$, $\Sigma_{SS} = \mathbf{X}_S\mathbf{X}_S^{\dagger}$, $\Sigma_{TT} = \mathbf{X}_T\mathbf{X}_T^{\dagger}$, $\rho \in [0,1]$, and "†" means the matrix transpose. In practice, $\boldsymbol{\omega}_S$ can be obtained by a generalized eigenvalue decomposition problem:

$$\Sigma_{ST}(\Sigma_{TT})^{-1}\Sigma_{ST}^{\dagger}\boldsymbol{\omega}_S = \eta(\Sigma_{SS})\boldsymbol{\omega}_S \tag{2}$$

where η is a constraint factor. Once $\boldsymbol{\omega}_S$ is obtained, $\boldsymbol{\omega}_T$ can be obtained by $\Sigma_{TT}{}^{-1}\Sigma_{ST}\boldsymbol{\omega}_S/\eta$. By adding the regularization terms $\lambda_S\mathbf{I}$ and into Σ_{SS} and Σ_{TT} to avoid overfitting and singularity problems, Equation (2) becomes:

$$\Sigma_{ST}(\Sigma_{TT} + \lambda_T\mathbf{I})^{-1}\Sigma_{ST}^{\dagger}\boldsymbol{\omega}_S = \eta(\Sigma_{SS} + \lambda_S\mathbf{I})\boldsymbol{\omega}_S \tag{3}$$

As a result, the source and target view data can be transformed into correlation subspaces by:

$$\mathbf{X}_S^C = \mathbf{X}_S{\cdot}\boldsymbol{\omega}_S, \boldsymbol{\omega}_S \in \Re^{d_S \times d} \tag{4}$$

$$\mathbf{X}_T^C = \mathbf{X}_T{\cdot}\boldsymbol{\omega}_T, \boldsymbol{\omega}_T \in \Re^{d_T \times d} \tag{5}$$

Note that one can derive more than one pair of transformation matrices $\{\omega_i^S\}_{i=1}^d$ and $\{\omega_i^T\}_{i=1}^d$, where $d = \min\{d_S, d_T\}$ is the dimension of the resulting CCA subspace. Once the correlation subspaces $\mathbf{X}_S^C$ and $\mathbf{X}_T^C$ spanned by $\boldsymbol{\omega}_S$ and $\boldsymbol{\omega}_T$ are derived, test data in the target view can be directly labeled by any model M_S^C that is trained using the source features $\mathbf{X}_S^C$.

2.3. Fusion Methods

If multiple "views" are available, then for each view, a label can be associated with each pixel used, for instance, CCA. If multiple labels are present, then they must be fused to obtain a single value using a so-called decision-based fusion procedure. Decision-based fusion aims to provide the final classification label for a pixel by combining the labels obtained, in this case, by multiple view analysis. This usually is obtained using two classes of procedures: weighted voting methods and meta-learning methods [51].

For weighted voting, the labels are combined using the weights assigned to each result. Many variants have been proposed in past decades. For the sake of comparison and because we must consider these options to evaluate the performance of the canonical correlation weighted voting (CCWV) scheme proposed in this paper, here, we consider only the following state-of-the-art techniques:

- Accuracy weighted voting (AWV), in which the weight of each member is set proportionally to its accuracy performance on a validation set [51]:

$$w_i = \frac{a_i}{\sum_{j=1}^{T} a_j} \tag{6}$$

  where a_i is a performance evaluation of the i-th classifier on a validation set.
- Best–worst weighted voting (BWWV), in which the best and the worst classifiers are given a weight of 1 or 0, respectively [51], and for the ones the weights are compute according to:

$$\alpha_i = 1 - \frac{e_i - \min_i(e_i)}{\max_i(e_i) - \min_i(e_i)} \tag{7}$$

  where e_i is the error of the i-th classifier on a validation set.
- Quadratic best–worst weighted voting (QBWWV), that computes the intermediate weights between 0 and 1 via squaring the above-mentioned BWWV:

$$\alpha_i = \left(\frac{\max_i(e_i) - e_i}{\max_i(e_i) - \min_i(e_i)}\right)^2 \tag{8}$$

3. The (Semi) Supervised Canonical Correlation Analysis Ensemble

3.1. Supervised Procedure

The idea of this procedure is to adopt MVL to decompose the target domain data into multiple disjoint or partial joint feature subsets (views), where each view is assumed to bring complementary information [52]. Next, these multiple views are used for DA, providing multiple matches between the source and the target domains. Eventually, the labeling task in the SD is transferred into the target domain through CCA, and the results of this "multi-view" CCA are combined to achieve a more efficient heterogeneous DA.

Specifically, without loss of generality, let us assume a heterogeneous DA from a low-dimensional $\mathbf{X}_S$ to a high-dimensional $\mathbf{X}_T$, with $d_S < d_T$, which requires that $\mathbf{X}_T$ is decomposed into N views, i.e., $\mathbf{X}_T = \{\mathbf{X}_T^i\}_{i=1}^N, \mathbf{X}_T^i \in \Re^{d_i \times n_T}, d_T = \sum_{i=i}^N d_i$. In this case, the implementation of MVCCA corresponds to searching for the following:

$$\underset{(\boldsymbol{\omega}_S^i, \boldsymbol{\omega}_T^i),\ldots,(\boldsymbol{\omega}_S^N,\ldots,\boldsymbol{\omega}_T^N)}{\operatorname{argmax}} (\rho_1,\ldots,\rho_N) = \sum_{i=1}^{N} \frac{(\boldsymbol{\omega}_S^i)^{\dagger} \Sigma_{ST}^i \boldsymbol{\omega}_T^i}{\sqrt{(\boldsymbol{\omega}_S^i)^{\dagger} \Sigma_{SS}^i \boldsymbol{\omega}_S^i} \sqrt{(\boldsymbol{\omega}_T^i)^{\dagger} \Sigma_{TT}^i \boldsymbol{\omega}_T^i}} \tag{9}$$

where $\Sigma_{ST}^i = \mathbf{X}_S (\mathbf{X}_T^i)^{\dagger}, \Sigma_{SS}^i = \mathbf{X}_S \mathbf{X}_S^{\dagger}$ and $\Sigma_{TT}^i = \mathbf{X}_T^i (\mathbf{X}_T^i)^{\dagger}$. Generalizing the standard CCA, Equation (9) can be rewritten as:

$$\begin{gathered} \underset{(\boldsymbol{\omega}_S^i, \boldsymbol{\omega}_T^i),\ldots,(\boldsymbol{\omega}_S^N,\ldots,\boldsymbol{\omega}_T^N)}{\operatorname{argmax}} (\rho_1,\ldots,\rho_N) = \sum_{i=1}^{N} (\boldsymbol{\omega}_S^i)^{\dagger} \Sigma_{ST}^i \boldsymbol{\omega}_T^i \\ s.t. (\boldsymbol{\omega}_S^1)^{\dagger} \Sigma_{ST}^1 \boldsymbol{\omega}_T^1 = 1, \ldots, (\boldsymbol{\omega}_S^N)^{\dagger} \Sigma_{ST}^N \boldsymbol{\omega}_T^N = 1 \end{gathered} \tag{10}$$

As a result, by using the solutions $\omega_S^i|_{i=1}^N$ and $\omega_T^i|_{i=1}^N$, we will have multiple transformed correlation subspaces, each one considering the SD and one of the target "views":

$$\mathbf{X}_S^{Ci} = \mathbf{X}_S \cdot \boldsymbol{\omega}_S^i, \boldsymbol{\omega}_S^i \in \Re^{d_S \times \hat{d}_i} \tag{11}$$

$$\mathbf{X}_T^{Ci} = \mathbf{X}_T^i \cdot \boldsymbol{\omega}_T^i, \boldsymbol{\omega}_T^i \in \Re^{d_T \times \hat{d}_i} \tag{12}$$

For any new instance of the target domain, i.e., $\mathrm{x} = \{\mathrm{x}_i\}|_{i=1}^N, \mathrm{x}_i \in \mathbf{X}_T^{Ci}$, the decision function of this SMVCCAE, trained with labeled training samples $\left\{ \left(\mathrm{x}_j^{SC}, y_j^S \right) \Big|_{j=1}^{n_S} \right\}, \mathrm{x}_j^{SC} \in \mathbf{X}_S^{Ci}, i = \forall N$, can be implemented via majority voting (MV):

$$\begin{aligned} H(\mathrm{x}) &= sign\left(\sum_{i=1}^{N} h_i(\mathrm{x}_i) \right) \\ &= \begin{cases} \varpi_l, \text{ if} \sum_{i=1}^{N} h_i^l(\mathrm{x}_i) \succ \frac{1}{2} \sum_{k=1}^{c} \sum_{i=1}^{N} h_i^k(\mathrm{x}_i) \\ reject, \quad \text{otherwise} \end{cases} \end{aligned} \tag{13}$$

However, to further optimize the ensemble results, one can also recall that the canonical correlations $\boldsymbol{\rho} = \left\{ \{\rho_1,\ldots,\rho_j\}|_{j=1}^{\hat{d}_1}, \ldots, \{\rho_1,\ldots,\rho_j\}|_{j=1}^{\hat{d}_N} \right\}$ obtained together with the transformation matrices $\boldsymbol{\omega}_S^i$ and $\boldsymbol{\omega}_T^i$ provide information about correlation between the SD and each target view. Since larger values of $\forall \{\rho_j\}|_{j=1}^{\hat{d}_i} \in \{\rho_i\}|_{i=1}^N$ show a greater correlation, this can also be considered a hint to obtain a better domain transfer ability for the corresponding view. We expect that poor correlation values (i.e., low values of $\sum_{j=1}^{\hat{d}_i} \rho_j$) will result in poor domain transfer abilities. Therefore, $\sum_{j=1}^{\hat{d}_i} \rho_j$ may be used to quantitatively evaluate the domain transfer ability of the transformation matrices $\boldsymbol{\omega}_S^i$ and

ω_T^i. Accordingly, we propose to include the following canonical correlation coefficient in the voting strategy of Equation (13):

$$H(x) = sign\left(\sum_{i=1}^{N} \sum_{j=1}^{\hat{d}_i} \rho_j h_i(x_i)\right) \tag{14}$$

The algorithmic steps of the new algorithm (called Supervised MVCCA Ensemble, or SMVCCAE for short) are summarized in Algorithm 1.

Algorithm 1. Algorithmic details of SMVCCAE.

---

1. ***Inputs:*** SD $\mathbf{X}_S = [x_1^S, \ldots, x_{ns}^S] \in \Re^{d_S \times n_S}$; TD $\mathbf{X}_T = [x_1^T, \ldots, x_{nT}^T] \in \Re^{d_T \times n_T}$; *id* for labeled training samples
2. $\left\{ \left(x_j^S, y_j^S\right)\Big|_{j=1}^{n_S} \right\}, y_j^S \in \Omega = \{\omega_l\}_{l=1}^c$ from $\mathbf{X}_S$, where the superscript C represents the number of class types;
3. Supervised classifier ζ; $\boldsymbol{N}$ the number of views of the TD; and $\min(d_S, d_T) \le \left\lfloor \frac{\max(d_S, d_T)}{N} \right\rfloor$.
4. ***Train:*** *for i = 1 to N*
5. generate the target domain view $\mathbf{X}_T^i \in \Re^{d_i \times n_T}, d_T = \sum_{i=i}^{N} d_i$;
6. return the transformation matrices $\boldsymbol{\omega}_S^i$ and $\boldsymbol{\omega}_T^i$ according to Equation (10);
7. obtain the correlation subspaces $\mathbf{X}_S^{Ci}$ and $\mathbf{X}_T^{Ci}$ according to Equations (11) and (12);
8. compute the transformed training samples $\left\{ \left(x_j^{SC}, y_j^S\right)\Big|_{j=1}^{n_S} \right\}$ from $\mathbf{X}_S^{Ci}$ according to *id;*
9. train the classifier $h_i = \zeta(x^{SC}, y^S)$;
10. *end*
11. ***Output:*** return the classifier pool $\{h_1, \ldots, h_N\}$;
12. ***Classification:*** For a given new instance $x = \{x_i\}|_{i=1}^N, x_i \in \mathbf{X}_T^{Ci}$, predict the label according to Equation (14).

3.2. Semi-Supervised Version

To implement a semi-supervised version of the proposed algorithm, the multiple speed-up SRKDA approach has been incorporated into the supervised procedure. SRDKA essentially improves the original idea of the spectral regression proposed in [53] for linear discriminant analysis (LDA), by transforming the eigenvector decomposition based discriminant analysis into a regression framework via spectral graph embedding [40]. For the sake of clarity, we briefly recall here the SRKDA notation before formalizing its implementation in the new procedure.

Given the labeled samples $\left\{ \left(x_j^S, y_j^S\right)\Big|_{j=1}^{n_S} \right\}, y_j^S \in \Omega = \{\omega_l\}_{l=1}^c$, the LDA objective function is:

$$\begin{aligned} \boldsymbol{a}_{LDA} &= \operatorname{argmax} \frac{\boldsymbol{a}^\dagger \boldsymbol{\psi}_b \boldsymbol{a}}{\boldsymbol{a}^\dagger \boldsymbol{\psi}_w \boldsymbol{a}} \\ \boldsymbol{\psi}_b &= \sum_{k=1}^{c} n_k \left(\boldsymbol{u}^{(k)} - \boldsymbol{u}\right)\left(\boldsymbol{u}^{(k)} - \boldsymbol{u}\right)^\dagger \\ \boldsymbol{\psi}_w &= \sum_{k=1}^{c} \left(\sum_{q=1}^{n_k} (\mathbf{x}_q^{(k)} - \boldsymbol{u}^{(k)})(\mathbf{x}_q^{(k)} - \boldsymbol{u}^{(k)})^\dagger \right) \end{aligned} \tag{15}$$

where $\boldsymbol{u}$ is the global centroid, n_k is the number of samples in the k-th class, $\boldsymbol{u}^{(k)}$ is the centroid of the k-th class, $\mathbf{x}_q^{(k)}$ is the q-th sample in the k-th class, and $\boldsymbol{\psi}_w$ and $\boldsymbol{\psi}_b$ represent the within-class scatter matrix and the between-class scatter matrix respectively, so that the total scatter matrix is computed as $\boldsymbol{\psi}_t = \boldsymbol{\psi}_b + \boldsymbol{\psi}_w$. The best solutions for Equation (15) are the eigenvectors that correspond to the nonzero eigenvalues of:

$$\boldsymbol{\psi}_b \boldsymbol{a}_{LDA} = \lambda \boldsymbol{\psi}_t \boldsymbol{a}_{LDA} \tag{16}$$

To address the nonlinearities, the kernel extension of this procedure maps the input data to a kernel Hilbert space through nonlinear positive semi-definite kernel functions, such as the Gaussian kernel $K(\mathbf{x}, y) = \exp\left(-\|\mathbf{x} - y\|^2/2\sigma^2\right)$, the polynomial kernel $K(\mathbf{x}, y) = \left(1 + \mathbf{x}^{\dagger}y\right)^d$ and the sigmoid kernel $K(\mathbf{x}, y) = \tanh\left(\mathbf{x}^{\dagger}y + a\right)$. Generalizing Equation (15), the projective function of KDA is therefore:

$$\begin{aligned}
\boldsymbol{\upsilon}_{KDA} &= \operatorname{argmax} \frac{\boldsymbol{\upsilon}^{\dagger}\boldsymbol{\psi}_b^{\phi}\boldsymbol{\upsilon}}{\boldsymbol{\upsilon}^{\dagger}\boldsymbol{\psi}_t^{\phi}\boldsymbol{\upsilon}} \\
\boldsymbol{\psi}_b^{\phi} &= \sum_{k=1}^{c} n_k \left(\boldsymbol{u}_{\phi}^{(k)} - \boldsymbol{u}_{\phi}\right)\left(\boldsymbol{u}_{\phi}^{(k)} - \boldsymbol{u}_{\phi}\right)^{\dagger} \\
\boldsymbol{\psi}_w^{\phi} &= \sum_{k=1}^{c} \left(\sum_{q=1}^{n_k} (\phi(\mathbf{x}_q^{(k)}) - \boldsymbol{u}_{\phi}^{(k)})(\phi(\mathbf{x}_q^{(k)}) - \boldsymbol{u}_{\phi}^{(k)})^{\dagger} \right) \\
\boldsymbol{\psi}_t^{\phi} &= \boldsymbol{\psi}_b^{\phi} + \boldsymbol{\psi}_w^{\phi}
\end{aligned} \tag{17}$$

where $\boldsymbol{\psi}_b^{\phi}$, $\boldsymbol{\psi}_w^{\phi}$, and $\boldsymbol{\psi}_t^{\phi}$ denote the between-class, within-class and total scatter matrices in the kernel space, respectively.

Because the eigenvectors of $\boldsymbol{\psi}_b^{\phi}\boldsymbol{\upsilon}_{KDA} = \lambda\boldsymbol{\psi}_t^{\phi}\boldsymbol{\upsilon}_{KDA}$ are linear combinations of $\phi(\mathbf{x}_q)$ [54], there is always a coefficient $\boldsymbol{\varepsilon}_q$ such as $\boldsymbol{\upsilon}_{KDA} = \sum_{q=1}^{n_k} \varepsilon_q \phi(\mathbf{x}_q)$. This constrain makes Equation (17) equivalent to:

$$\boldsymbol{\varepsilon}_{KDA} = \operatorname{argmax} \frac{\boldsymbol{\varepsilon}^{\dagger}\mathbf{KWK}\boldsymbol{\varepsilon}}{\boldsymbol{\varepsilon}^{\dagger}\mathbf{KK}\boldsymbol{\varepsilon}} \tag{18}$$

where $\boldsymbol{\varepsilon}_{KDA} = \left[\varepsilon_1, \ldots, \varepsilon_{n_k}\right]^{\dagger}$. Then, the corresponding eigenproblem becomes:

$$\mathbf{KWK}\boldsymbol{\varepsilon}_{KDA} = \lambda\mathbf{KK}\boldsymbol{\varepsilon}_{KDA} \tag{19}$$

where $\mathbf{K}$ is the kernel matrix, and the affinity matrix $\mathbf{W}$ is defined using either HeatKernel [55] or the binary weight mode:

$$W_{i,j} = \begin{cases} 1/n_k, \text{if } \mathbf{x}_i \text{ and } \mathbf{x}_j \text{ both belong to the } k^{\text{th}} \text{ class;} \\ 0, \qquad \text{otherwise.} \end{cases} \tag{20}$$

To efficiently solve the KDA eigenproblem in Equation (19), let us consider $\boldsymbol{\vartheta}$ to be the solution of $\mathbf{W}\boldsymbol{\vartheta} = \lambda\boldsymbol{\vartheta}$. Replacing $\mathbf{K}\boldsymbol{\varepsilon}_{KDA}$ on the left side of Equation (19) by $\boldsymbol{\vartheta}$, we have:

$$\mathbf{KWK}\boldsymbol{\varepsilon}_{KDA} = \mathbf{KW}\boldsymbol{\vartheta} = \mathbf{K}\lambda\boldsymbol{\vartheta} = \lambda K\boldsymbol{\vartheta} = \lambda\mathbf{KK}\boldsymbol{\varepsilon}_{KDA} \tag{21}$$

To avoid singularities, a constant matrix δI is added to $\mathbf{K}$ to keep it positive definite:

$$\boldsymbol{\varepsilon}_{KDA} = (\mathbf{K} + \delta I)^{-1}\boldsymbol{\vartheta} \tag{22}$$

where $\boldsymbol{I}$ is the identity matrix, and $\delta \geq 0$ represents the regularization parameter. It can be easily verified that the optimal solution given by Equation (22) is the optimal solution of the following regularized regression problem [56]:

$$\min_{f \in F} \sum_{j=1}^{n_S} \left(f(\mathbf{x}_j) - y_j\right)^2 + \delta\|f\|_K^2 \tag{23}$$

where F is the kernel space associated with the kernel K, and $\|f\|_K$ is the corresponding norm.

According to Equations (19) and (21), the solution can be reached in two steps: (1) solve the eigenproblem $\mathbf{W}\boldsymbol{\vartheta} = \lambda\boldsymbol{\vartheta}$ to obtain $\boldsymbol{\vartheta}$; and (2) find a vector ε_{KDA} that satisfies $\mathbf{K}\varepsilon_{KDA} = \boldsymbol{\vartheta}$. For Step 1, it is easy to check that the involved affinity matrix $\mathbf{W}$ has a block-diagonal structure:

$$\mathbf{W} = \begin{bmatrix} \mathbf{W}^{(1)} & 0 & \cdots & 0 \\ 0 & \mathbf{W}^{(2)} & \cdots & 0 \\ \vdots & \vdots & \ddots & \vdots \\ 0 & 0 & \cdots & \mathbf{W}^{(c)} \end{bmatrix} \quad (24)$$

where $\left\{\mathbf{W}^{(k)}\right\}_{k=1}^{c}$ is an $n_k \times n_k$ matrix with all of the elements defined in Equation (19), and it is straightforward to show that $\mathbf{W}^{(k)}$ has the eigenvector $\mathbf{e}^{(k)}$ associated with $e^{(k)} = [1, 1, ..., 1]^{\dagger}$. In addition, there is only one nonzero eigenvalue of $\mathbf{W}^{(k)}$ because the rank of $\mathbf{W}^{(k)}$ is always 1. Thus, there are exactly c eigenvectors of $\mathbf{W}$ with the same eigenvalue 1:

$$\boldsymbol{\vartheta}_k = [\underbrace{0, ..., 0}_{\sum_{i=1}^{k-1} n_i}, \underbrace{1, ..., 1}_{n_k}, \underbrace{0, ..., 0}_{\sum_{i=k+1}^{c} n_i}]^{\dagger} \quad (25)$$

According to the theorem proven by Cai and He in [57], the kernel matrix is positive definite, and the $\boldsymbol{c}$-1 projective function of KDA gives exactly the same solutions as the $\boldsymbol{c}$-1 linear equations systems $\mathbf{K}\varepsilon^{k}_{KDA} = \overline{\boldsymbol{\vartheta}}^{k}$. Then let $\boldsymbol{\Theta} = [\varepsilon_1, ..., \varepsilon_{c-1}]$ be the KDA transformation matrix which embeds the data into the KDA subspace:

$$\boldsymbol{\Theta}^{\dagger}\left[K(:, x_1), ..., K(:, x_{n_k})\right] = \overline{Y}^{\dagger} \quad (26)$$

where the columns of $\overline{Y}^{\dagger}$ are the embedding results. Accordingly, the data with the same label correspond to the same point in the KDA subspace when the kernel matrix is positive definite.

To perform SRKDA in a semi-supervised way, one straightforward solution is to use the label information to guide the construction of the affinity matrix $\mathbf{W}$, as in [57–59]. Let $G = (V, E)$ be a graph with set of vertices V, which is connected by a set of edges E. The vertices of the graph are the labeled and unlabeled instances $\left(x_j^S, y_j^S\right)\Big|_{j=1}^{n_S} \cup \left\{\left(x_j^T\right)\Big|_{j=1}^{n_T}\right\}$. An edge between two vertices (or labeled and unlabeled samples) i, j represents the similarity of two instances with an associated weight $\{W_{i,j}\}$. Then, the affinity matrix W is built using both labeled and unlabeled samples. To achieve this goal, $\boldsymbol{p}$-nearest neighbors, ε-neighbors, or fully connected graph techniques can be adopted, where 0–1 weighting, Gaussian kernel weighting, Polynomial kernel weighting and Dot-product weighting can be considered to establish the graph weights [57,58]. Usually, graph-based SSL methods compute the normalized graph Laplacian:

$$L = I - D^{-1/2} W D^{-1/2} \quad (27)$$

where D denotes a diagonal matrix defined by $D_{ii} = \sum_j W_{i,j}$ (see [59,60] (Chapter 5) for more details on different families of graph based SSL methods).

According to this procedure, and inserting the notation for DA using multiple view CCA, the new semi-supervised procedure follows the steps reported in Algorithm 2.

Algorithm 2. Algorithmic details of SSMVCCAE.

Inputs: SD $\mathbf{X}_S = \left[x_1^S, \ldots, x_{ns}^S\right] \in \Re^{d_S \times n_S}$; TD $\mathbf{X}_T = \left[x_1^T, \ldots, x_{nT}^T\right] \in \Re^{d_T \times n_T}$; id_S^L for labeled training
samples $\left\{ \left(x_j^S, y_j^S\right)\Big|_{j=1}^{n_S} \right\}, y_j^S \in \Omega = \{\omega_l\}_{l=1}^{C}$ from $\mathbf{X}_S$, where superscript C represents the number of class
types; id_T^U for unlabeled candidates $\left\{ \left(x_j^T\right)\Big|_{j=1}^{n_T} \right\}$ from $\mathbf{X}_T$ Semi-supervised classifier ζ_{SRKDA}; $\boldsymbol{N}$ =
Number of views of the target domain; and $\min(d_S, d_T) \leq \left\lfloor \frac{\max(d_S, d_T)}{N} \right\rfloor$.
Train: *for i = 1 to N*
  generate the target domain view $\mathbf{X}_T^i \in \Re^{d_i \times n_T}, d_T = \sum_{i=i}^{N} d_i$;
  return the transformation matrices $\boldsymbol{\omega}_S^i$ and $\boldsymbol{\omega}_T^i$ according to Equation (10);
  obtain the correlation subspaces $\mathbf{X}_S^{Ci}$ and $\mathbf{X}_T^{Ci}$ according to Equations (11) and (12);
  compute the transformed training samples $\left\{ \left(x_j^{SC}, y_j^S\right)\Big|_{j=1}^{n_S} \right\}$ from $\mathbf{X}_S^{Ci}$ according to id_S^L and the
  transformed unlabeled samples $\left\{ \left(x_j^{TC}\right)\Big|_{j=1}^{n_T} \right\}$ from $\mathbf{X}_T$ according to id_T^U;
  build the graph Laplacian L_i according to Equation (27) using $\left(x_j^{SC}, y_j^{SC}\right)\Big|_{j=1}^{n_S} \cup \left\{ \left(x_j^{TC}\right)\Big|_{j=1}^{n_T} \right\}$;
  obtain the KDA transformation matrix $\boldsymbol{\Theta}_i$ according to the solutions of Equation (26) and Equation (22);
  return the embedded results $\overline{Y}_i^{\dagger}$;
  end
Output: return the KDA transformation matrices $\{\boldsymbol{\Theta}_i\}_{i=1}^{N}$ and the full KDA subspace embedded results $\left\{\overline{Y}_i^{\dagger}\right\}_{i=1}^{N}$;
Classification: For a given new instance $\mathrm{x} = \{\mathrm{x}_i\}\big|_{i=1}^{N}, \mathrm{x}_i \in \mathbf{X}_T^{Ci}$
for i = 1 to N
  first map $\mathbf{x}_i$ into RKHS with the specified kernel function $\phi(\mathbf{x}_i)$;
  obtain the embedded results $\overline{Y}_{iT}$ in KDA space according to Equation (26);
  return the decision function $h_i(x) = \operatorname{argmin} \sum_{j=1}^{C} \left(\|\overline{\mathbf{y}}_{iT} - \boldsymbol{u}_j\|^2\right), \overline{\mathbf{y}}_{iT} \in \overline{Y}_{iT}$, and $\boldsymbol{u}_j = \sum_{\mathbf{x} \in c_i} \mathbf{x} / \left|c_j\right|$, which represents the class center of c_i in the KDA embedded space.
end
obtain the final predicted label by a majority voting ensemble strategy using Equation (14).

Summing up algorithmic details of the SMVCCAE and SSMVCCAE as described in Sections 3.1 and 3.2, Figure 1 illustrate the general flowchart for the proposed heterogeneous DA algorithms for RS image classification.

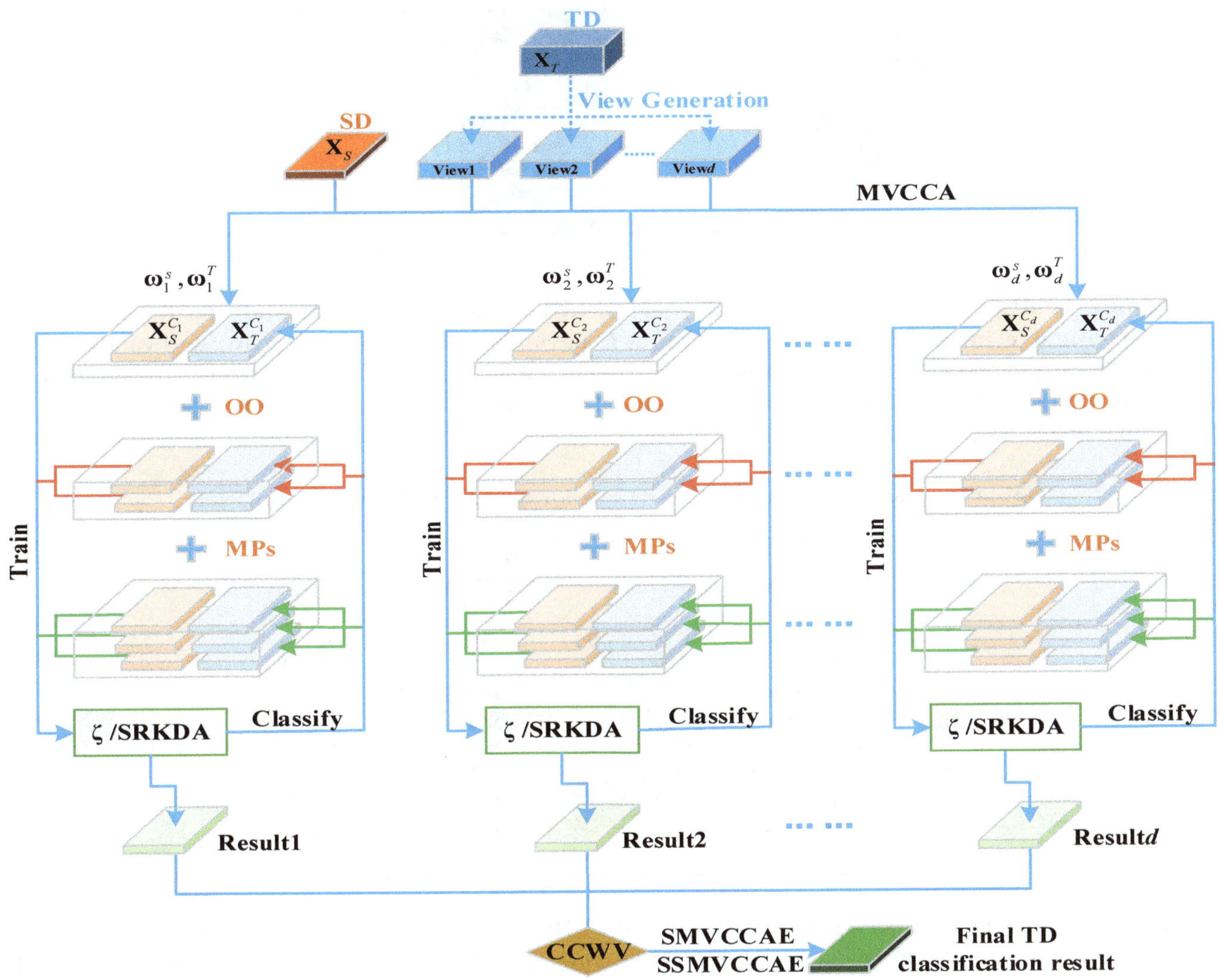

Figure 1. General flowchart for the proposed heterogeneous DA algorithms SMVCCAE and SSMVCCAE for RS image classification.

4. Data Sets and Setups

4.1. Datasets

For our analyses and evaluations, we consider two datasets, with different spatial and spectral resolutions. The first dataset is a 1.3 m spatial resolution image collected by the Reflective Optics Spectrographic Image System (ROSIS) sensor over the University of Pavia, with a size of 610 × 340 pixels (Figure 2). A total of 103 spectral reflectance bands that cover a region of the spectrum between 430 and 860 nm were retained for the analyses. The captured scene primarily represents a built-up setting with these thematic classes: asphalt, meadows, gravel, trees, metal sheets, bitumen, bare soil, bricks and shadows, as listed in Table 1. As described earlier, the main purpose of this article is to investigate the proposed methods in a heterogeneous DA problem. In this sense, the low-dimensional image is simulated by clustering the spectral space of the original ROSIS image. Specifically, the original bands of the original ROSIS image are clustered into seven groups using the K-Means algorithm, and the mean value of each cluster is considered as a new spectral band, providing a total of seven new bands. In the experiments, the new synthetic image is considered as the SD, whereas the original ROSIS image is considered as the TD.

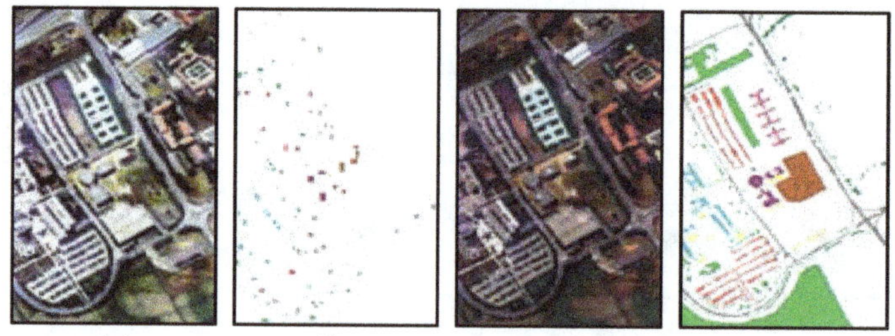

(a) (b) (c) (d)

Figure 2. (**a**–**d**) False color composite of the: synthetic low spectral resolution (**a**); and the original hyperspectral (**c**) images of the University campus in Pavia, together with: training (**b**); and validation (**d**) data sets (legend and sample details are reported in Table 1). False color composites are obtained and are displayed as R, G, and B bands 7, 5, and 4 for the synthetic, and bands 60, 30, and 2 for the original image, respectively.

Table 1. Class legend and sample details for the ROSIS University data set.

No.	Class	Code	Source	Target
			Train	Test
1	Asphalt		548	6631
2	Meadows		540	18649
3	Gravel		392	2099
4	Trees		524	3064
5	Metal sheets		265	1345
6	Bare soil		532	5029
7	Bitumen		375	1330
8	Bricks		514	3682
9	Shadows		231	947

The second dataset was gathered by the AVIRIS sensor over the Indian Pines test site in North-western Indiana in 1992, with 224 spectral reflectance bands in the wavelength range of 0.4 to 2.5 μm. It consists of 145 × 145 pixels with moderate spatial resolution of 20 m per pixel, and a 16-bit radiometric resolution. After an initial screening, the number of bands was reduced to 200 by removing bands 104–108, 150–163, and 220, due to noise and water absorption phenomena. This scene contains two-thirds agriculture, and one-third forest or other natural perennial vegetation. For the other Pavia data set, K-Means is used to simulate a low dimensional image with 10 bands. For illustrative purposes, Figure 3a,b shows false color composition of the simulated low dimensional and the original AVIRIS Indian Pines scene, whereas Figure 3b shows the ground truth map that is available for the scene, which is displayed in the form of a class assignment for each labeled pixel. In the experimenting stage, this ground truth map is subdivided into two parts for training and validation purposes, as detailed in Table 2.

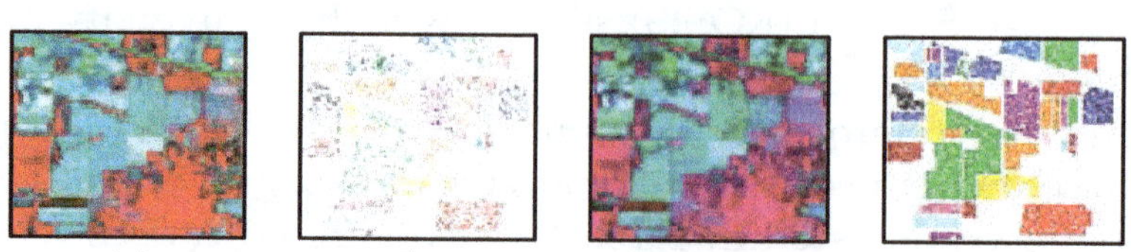

(a) (b) (c) (d)

Figure 3. (**a**–**d**) False color composites of the: simulated low spectral resolution (**a**); and original hyperspectral (**c**) images of Indian Pines data, together with: training (**b**); and validation (**d**) data sets (color legend and sample details are reported in Table 2). False color composites are obtained displaying as R, G, and B bands, 6, 4, and 5 for the synthetic, and bands 99, 51, and 21 for the original image, respectively.

Table 2. Class legend and sample details for the AVIRIS Indian Pines data set.

No.	Class	Code	Source	Target
			Train	Test
1	Alfalfa		23	23
2	Corn-notill		228	1200
3	Corn-mintill		130	700
4	Corn-notill		57	180
5	Grass-pasture		83	400
6	Grass-trees		130	600
7	Grass-pasture-mowed		14	14
8	Hay-windrowed		78	400
9	Oats		10	10
10	Soybean-notill		172	800
11	Soybean-mintill		255	2200
12	Soybean-clean		93	500
13	Wheat		55	150
14	Woods		265	1000
15	Buildings-grass-trees-drives		86	300
16	Stone-steel-towers		43	50

4.2. Experiment Setups

All of the experiments were performed using Matlab™ on a Windows 10 64-bit system with Intel® Core™ i7-4970 CPU, @3.60 GHz, 32GB RAM. For the sake of evaluation and comparison, a Random Forest classifier (RaF) is considered as benchmark classifier for both the SMVCCAE and SVCCA approaches, because of its proven velocity, and its generalized and easy-to-implement properties [61,62]. The number of decision trees in RaF is set by default to 100, whereas the number of features is set by default to the floor of the square root of the original feature dimensionality.

For both the ROSIS and Indian Pines data sets, all of the initial and derived features have been standardized to a zero mean and unit variance. For incorporated object oriented (OO), five statistics are utilized, including the pixels' mean and standard deviation, area, orientation and major axis length of the segmented objects via K-Means clustering algorithm, whereas the spatial feature morphology profiles (MPs) are applied to the three transferred features that have the highest canonical correlation coefficients. Specifically, MPs are constructed by applying closing by reconstruction (CBR) with a circular element with a radius of 3–11 pixels, and opening by reconstruction (OBR) with an element with a radius of 3–6 pixels, refer to works carried out in [63,64]. Therefore, the feature dimensionality set in the experiments is 7 (10) vs. 103 (200) when using spectral features only for ROSIS (Indian Pines), 7 + 5 (10 + 5) vs. 103 + 5 (200 + 5) when using spectral features stacked with OO ones, 7 + 39 (10 + 39) vs. 103 + 39 (200 + 39) when using spectral features stacked with MPs features, and finally 7 + 5 + 39 (10 + 5 + 39) vs. 103 + 5 + 39 (200 + 5 + 39) when using all spectral, OO, and MPs features.

To assess the classification performances of the proposed semi-supervised approach, two state-of-the-art semi-supervised classifiers, Logistic label propagation (LLP) [65] and Laplacian support vector machine (LapSVM) [66] were considered. For the critical parameters of the semi-supervised technique (SRKDA), such as the regularization parameter δ and the number of neighbors NN used to construct the graph Laplacian L with HeatKernel [40], their values are obtained by a heuristic search in the (0.01–1) and (1–15) ranges, respectively. The parameter settings for LLP and LapSVM are instead reported in Table 3. Because LapSVM was originally proposed for binary classification problems, a one-against-all (OAA) scheme was adopted to handle the multiclass classification in our experiments.

Table 3. Parameter details for LLP and LapSVM.

Classifier	Parameters	Meanings	Values
LLP	g	graph complete type	KNN
	τ	neighborhood type	Supervised
	N	neighbor size for constructing graph	5
	ω	weights for edge in graph	Heat Kernel
	σ	parameter for Heat Kernel	1
	C	regularization scale	0.001
	M	maxim iteration number	1000
	η	weight function for labeled samples	mean
LapSVM	γa	regularization parameter (ambient norm)	10^{-5}
	γi	regularization parameter (intrinsic norm)	1
	α	the initial weights	0
	κ	kernel type	RBF
	σ	RBF kernel parameter	0.01
	M	maximum iteration number	200
	c	LapSVM training type	primal
	η	Laplacian normalization	TRUE
	N	neighbor size for constructing graph	6

5. Experimental Results and Discussion

5.1. Domain Transfer Ability of MVCCA

As discussed in Section 3.1, each dimension in the derived CCA subspace is associated with a different canonical correlation coefficient which is a measure of its transfer ability. Moreover, in the MVCCA scenario, the transfer ability of each view and dimension is controlled not only by the number of views but also by the view generation technique. In this sense, Figure 4 presents the results of the average canonical correlation coefficient obtained using different view generation techniques, i.e., disjoint random sampling, uniform slice, clustering and partially joint random generation. Partially joint random view generation can apparently increase the chance of finding views with better domain transfer ability on the one hand, and to overcome the limitation ensemble techniques when the number of classifiers (equal to number of views in our case) is small on the other hand. Please note that for a more objective evaluation and comparison, each experiment was executed 10 times independently.

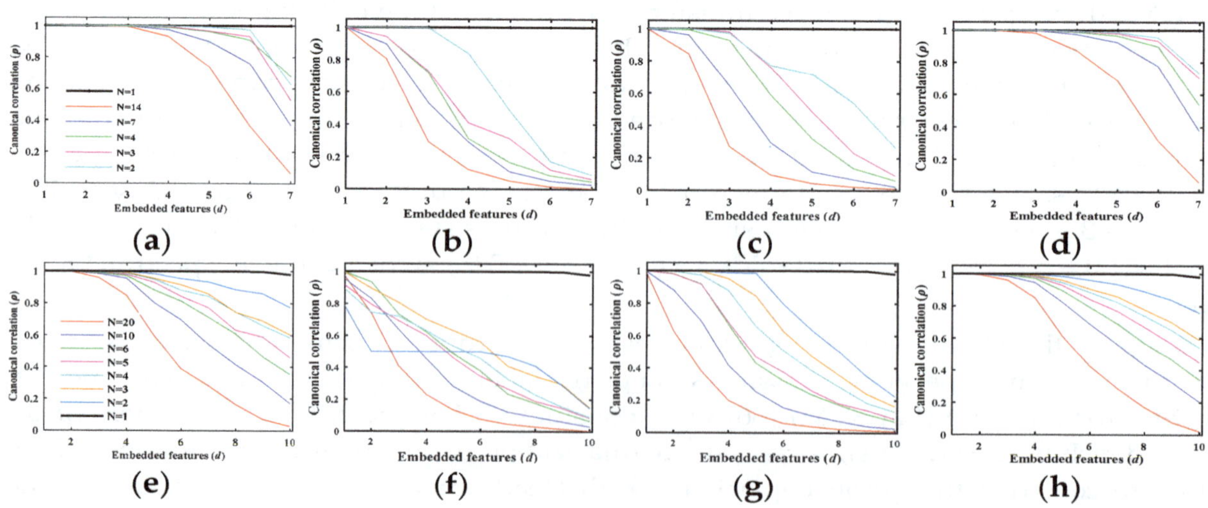

Figure 4. Average canonical correlation coefficient versus embedded features for: ROSIS (**a**–**d**); and Indian Pines (**e**–**h**) data sets using different view generation techniques: disjoint random sampling (**a**,**e**); uniform slice (**b**,**f**); clustering (**c**,**g**); and partially joint random generation (**d**,**h**).

In Figure 4, we see that the embedded features with the highest canonical correlation coefficient are obtained by directly applying CCA without multi view generation (i.e., $n = 1$). However, single view CCA may still fail to balance potential mismatches across heterogeneous domains by overfitting, as demonstrated in the results reported in the following sections. Additionally, the decreasing trend of the canonical correlation coefficient with an increasing number of views is obvious because of the increasing mismatch between the source and target views. However, the decreasing rates of the canonical correlation coefficient for disjoint random and partially joint random generation techniques are lower than those from disjoint uniform slice and disjoint clustering view generations. Therefore, partially joint random and disjoint random view generation techniques have been selected for the following experiments.

5.2. Parameter Analysis for SMVCCAE

In Figure 5, we report the results of a sensitivity analysis of SMVCCAE that involves its critical parameters: the dimension of the target view $d_T^i = \frac{d_T}{N}$, the view generating strategies including disjoint random (DJR) and partially joint random (PJR) generation, as well as the ensemble approaches MJV and CCWV. Please note that the number of views for PJR based SMCCAE was set to 35, which is a number that will be discussed later in this paper.

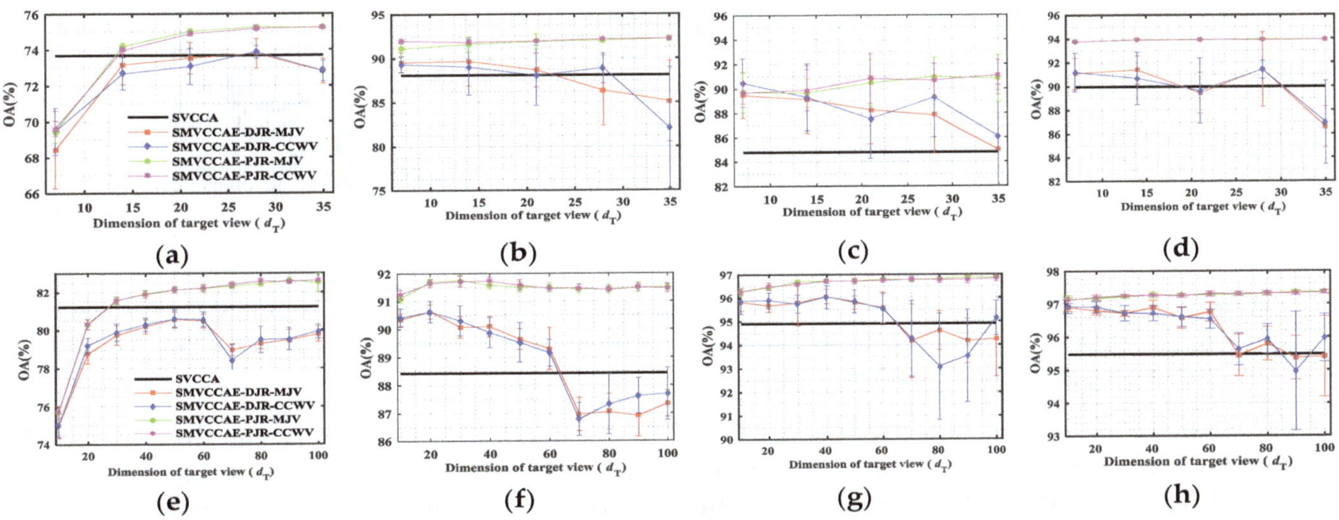

Figure 5. (**a–h**)Average OA values versus target view dimensionality for SMVCCAE with different fusion strategies using: spectral (**a**,**e**); spectral-OO (**b**,**f**); spectral-MPs (**c**,**g**); and spectral-OO-MPs (**d**,**h**) features on: ROSIS University (**a–d**); and Indian Pine datasets (**e–h**).

As illustrated in Figure 5 for the test data sets, the choice of PJR view generation with MJV and CCWV strategies allows the best overall accuracy values (OA curves in color green and pink). Concerning the dimensionality of the target views, they are different using different features. Specifically, for spectral features, the larger the dimensionality of the target views, the larger the OA values for PJR-based SMVCCAE because of the better domain transfer capacity with more ensemble classifiers. However, a dimensionality that is too large leads to too few view splits, i.e., a small number of ensemble elements, eventually resulting in a degraded performance. For example, when target view dimensionality is larger than four times the source view (7) dimensionality for ROSIS and larger than six times this value for Indian Pines, the OA value exhibits a decreasing trend (Figure 5a,e). Among the different types of features, (e.g., spectral and object-oriented features (labeled "spectral-OO"), spectral and morphological profile features (labeled "spectral-MPs"), and all of them together (labeled "spectral-OO-MPs"), the outcome is as expected, which is that the best results are obtained using spectral-OO-MPs. Interestingly, whereas the classification performances of the PJR-based approach are quite stable with respect to the dimensionality of the target views, the DJR-based results show a negative trend with an increasing number of target views. This finding is especially true when spatial (i.e., OO and morphological profiles) features are incorporated. This result can be explained by the trade-off between the diversity, OA and number of classifiers in an ensemble system. Specifically, the statistical diversity among spectral and spatial features tends to enhance the classification accuracy diversities more than using any view splitting strategy. As a result, the final classification performance could be limited or even degraded, especially when the number of classifiers is small.

Finally, in Figure 6, we focus on the computational complexity of the proposed approach by presenting OA, kappa statistics and CPU time values with respect to the number of views and the various fusion strategies.

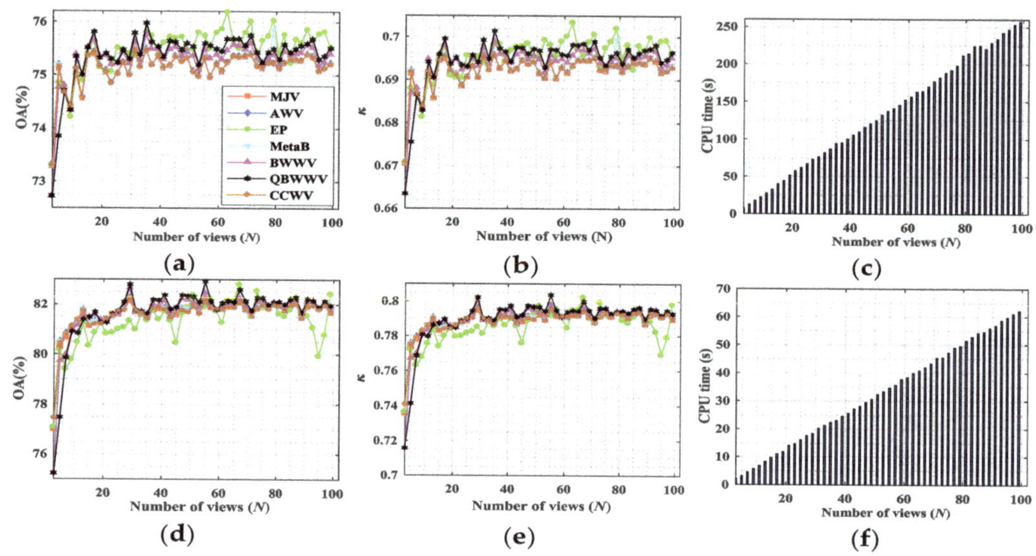

Figure 6. Average OA, Kappa (κ) and CPU time in seconds vs. the number of views for SMVCCA with PJR view generation and various fusion strategies applied to spectral features of ROSIS: University (**a**–**c**); and Indian Pines datasets (**d**–**f**).

According to Figure 6, the proposed CCWV fusion technique is effective as the other fusion techniques. Apparently, with regard to the improvements in the OA values (see Figure 6a,b,d,e), and the computational burden from the number of views (see Figure 6c,f), views between 30 and 40 produce the best tradeoff between computational burden and classification accuracy.

In summary, in a scenario in which low-dimensional and high-dimensional data sets require DA, a well-designed SMVCCAE requires us to set the dimensionality of each target view to three or four times the dimensionality of the source view, and to use a PJR view generation technique.

5.3. Validation of SMVCCAE

Figure 7 provides the SMVCCAE heterogeneous cross-domain classification maps with OA values for the ROSIS University dataset using spectral, spectral-OO, spectral-MPs and spectral-OO-MPs features. Compared with the maps produced by a single-view canonical correlation analysis (SVCCA) approach, the thematic maps obtained by SMVCCAE using the associated features are better, specifically with adequate delineations of the bitumen, gravel and bare soil areas (see the numbers in Table 4). These results experimentally verify our earlier assumptions that single view CCA could fail to balance potential mismatches across heterogeneous domains by overfitting. Additionally, the most accurate result is obtained with spectral-OO-MPs by SMVCCAE using the PJR view generation strategy, as shown by the results in Figure 7 and the numbers in bold in Table 4.

For the Indian Pines dataset, Figure 8 shows the thematic maps with OA values, whereas Table 5 reports the classification accuracies (Average accuracy (AA) and OA), and kappa statistics (k) with respect to various features. Once again, the thematic maps with larger OA values produced by SMVCCAE are better than the results produced by SVCCA, especially when the OO and MPs are incorporated. The numbers in bold in Table 5 show that the largest accuracies for various class types are obtained by the SMVCCAE with the PJR technique using spectral-OO-MPs features.

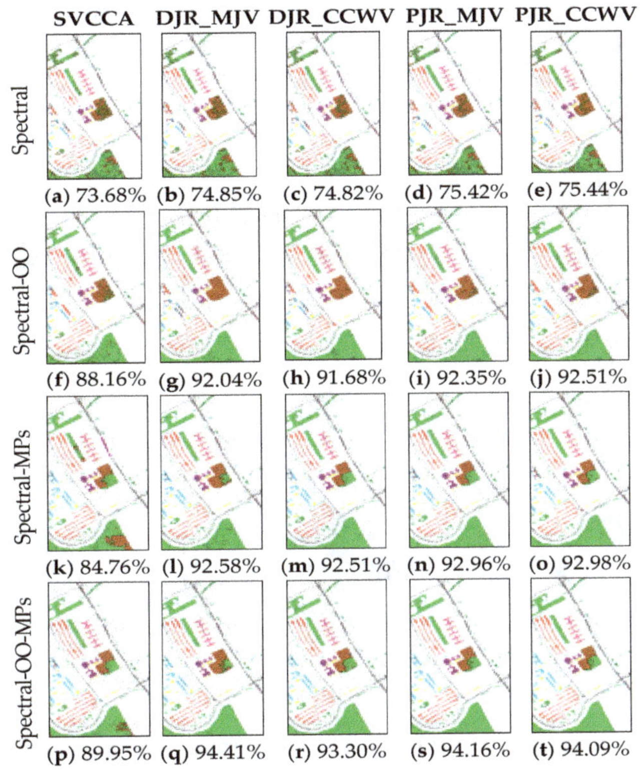

Figure 7. (**a–t**) Summary of the best classification maps with OA values for SMVCCAE with different fusion strategies using spectral, OO and MPs features of ROSIS University.

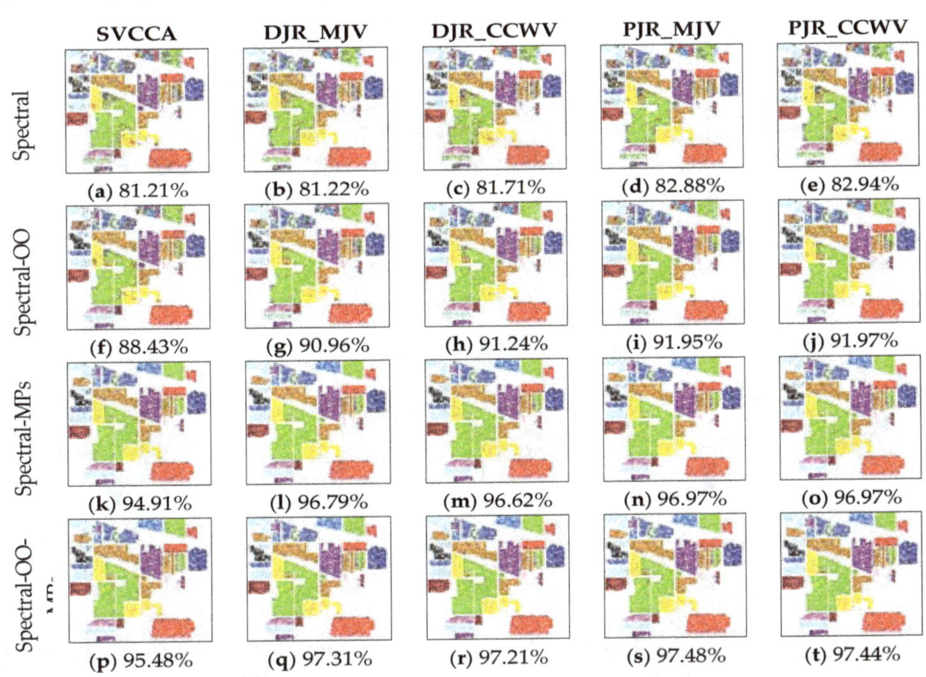

Figure 8. (**a–t**) Summary of the best classification maps with OA values for SMVCCAE with different fusion strategies using spectral, OO and MPs features of Indian Pines.

5.4. Parameter Analysis for the Semi-Supervised Version of the Algorithm

In Figures 9 and 10, we report the results of the sensitivity analysis for SSMVCCAE while considering the two critical parameters from the adopted SRKDA technique: (1) the regularization parameter δ; and (2) the number of neighbors NN used to construct the graph Laplacian L. The other parameters, such as the target view dimensionality, d^i_T and the number of total views N (i.e., the ensemble size), are set by default to $d^i_T = 4 \times d_s$ and $N = 35$, according to our previous experimental analysis for the supervised version of the same technique.

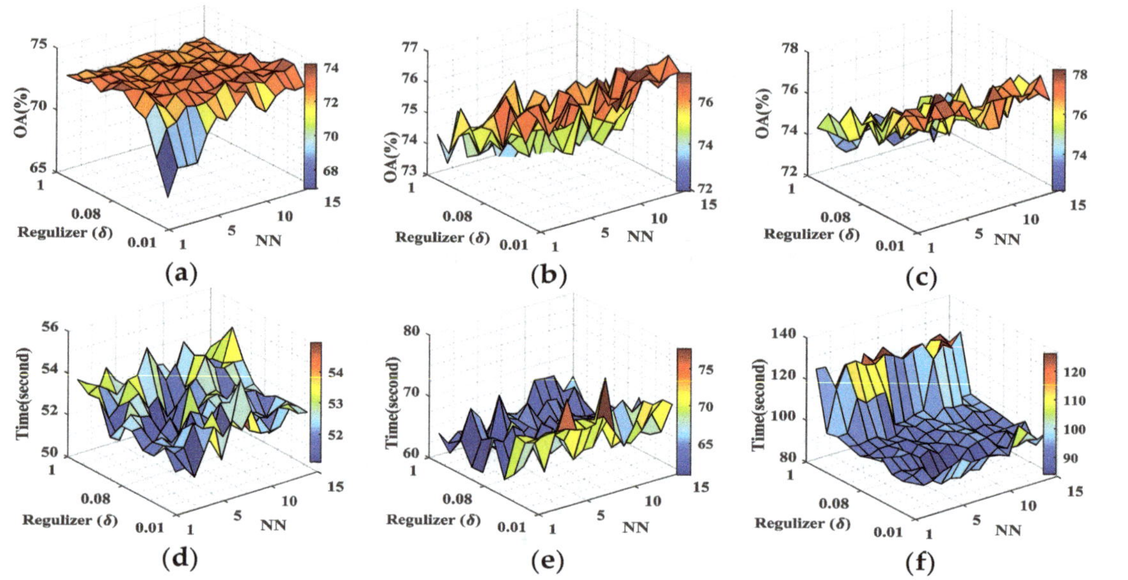

Figure 9. (**a–f**) OA values and CPU time (in seconds) versus the regularization parameter (δ) and nearest neighborhood size (NN) set of SSMVCCAE with DJR view generation strategy for ROSIS University using different sizes of labeled samples: 10 pixels/class (**a**,**d**); 50 pixels/class (**b**,**e**); and 100 pixels/class (**c**,**f**).

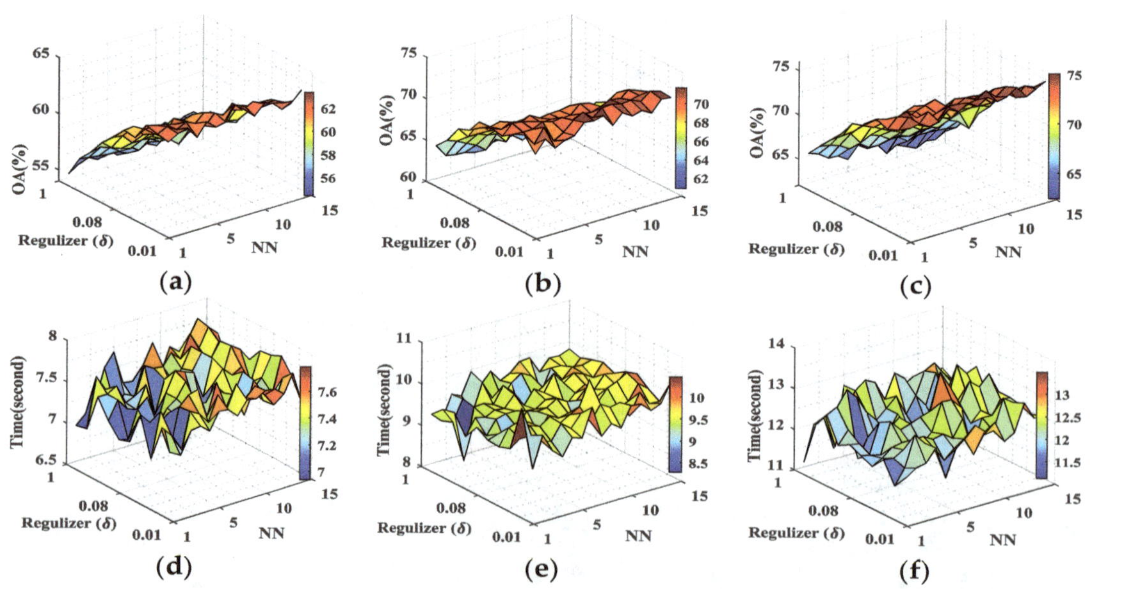

Figure 10. (**a–f**) OA values and CPU time (in seconds) versus the regularization parameter (δ) and nearest neighborhood size (NN) set of SSMVCCAE with the DJR view generation strategy for Indian Pine using different size of labeled samples: 10 pixels/class (**a**,**d**); 30 pixels/class (**b**,**e**); and 55 pixels/class (**c**,**f**).

Table 4. Classification accuracy values for the SVCCA and SMVCCAE ($d_T = 4 \times d_S$) methods for ROSIS University. Considered metrics: Overall accuracy (OA), Average accuracy (AA), Kappa statistic (Kappa).

Methods	Strategy	Features	Asphalt	Meadows	Gravel	Trees	Metal Sheets	Bare Soil	Bitumen	Bricks	Shadows	AA (%)	OA (%)	Kappa
SVCCA	~	F1	94.96	91.61	66.17	52.56	98.68	45.70	70.63	69.50	99.26	76.56	73.68	0.67
		F2	98.02	95.88	93.04	71.59	98.68	71.13	90.88	76.59	**100.00**	88.42	88.76	0.85
		F3	99.47	88.06	94.39	80.89	99.55	53.10	72.31	90.67	98.90	86.37	84.76	0.80
		F4	99.86	89.32	91.35	82.43	98.67	71.98	96.44	95.76	99.15	91.66	89.96	0.87
SMVCCAE	DJR_MJV	F1	94.72	93.38	73.72	51.80	99.40	45.96	73.73	68.49	99.62	77.87	73.80	0.68
		F2	95.35	97.59	77.93	61.68	99.61	82.74	81.94	68.45	99.90	85.02	86.37	0.82
		F3	99.36	90.58	95.88	74.91	99.37	72.69	94.79	85.28	99.21	90.23	88.16	0.85
		F4	99.49	92.05	96.76	79.38	99.77	88.84	96.45	86.15	99.26	93.13	91.44	0.89
	DJR_CCWV	F1	94.81	93.25	73.44	51.82	99.65	45.99	74.43	68.29	99.61	77.92	73.84	0.68
		F2	96.56	**98.56**	78.45	62.47	99.47	92.16	88.74	68.79	99.81	87.22	88.89	0.86
		F3	99.31	90.60	96.05	74.26	99.52	75.25	96.40	85.82	99.32	90.72	88.33	0.85
		F4	99.24	90.64	96.72	77.65	99.28	93.32	97.36	86.71	99.19	93.34	91.42	0.89
	PJR_MJV	F1	95.30	94.31	76.84	52.48	99.83	48.15	76.36	69.20	99.87	79.15	75.28	0.69
		F2	98.64	98.00	95.62	77.19	99.52	89.26	97.57	72.69	99.79	92.03	92.14	0.90
		F3	99.72	91.09	98.96	82.15	99.93	80.10	**99.62**	87.11	99.36	93.11	90.97	0.88
		F4	**99.89**	91.46	94.97	**84.63**	99.93	98.81	99.34	95.66	99.40	96.01	**93.97**	**0.92**
	PJR_CCWV	F1	95.26	94.33	77.45	52.44	99.83	47.96	76.56	69.06	99.86	79.19	75.20	0.69
		F2	98.55	98.02	95.55	77.33	99.54	89.40	97.57	72.62	99.79	92.04	92.16	0.90
		F3	99.71	90.90	**99.08**	82.14	**99.95**	77.94	99.56	86.99	99.36	92.85	90.69	0.88
		F4	**99.89**	91.36	95.16	84.40	**99.95**	**98.88**	99.35	**95.78**	99.38	**96.02**	93.92	**0.92**

F1: Spectral; F2: Spectral-OO; F3: Spectral-MPs; F4: Spectral-OO-MPs.

Table 5. Classification accuracy values (average) for the SVCCA and SMVCCAE methods for Indian Pines. Considered metrics: Overall accuracy (OA), Average accuracy (AA), Kappa statistic (Kappa).

Methods	**SVCCA**				**SMVCCAE**															
Strategy	~				**DJR_MJV**				**DJR_CCWV**				**PJR_MJV**				**PJR_CCWV**			
Features	**F1**	**F2**	**F3**	**F4**	**F1**	**F2**	**F3**	**F4**	**F1**	**F2**	**F3**	**F4**	**F1**	**F2**	**F3**	**F4**	**F1**	**F2**	**F3**	**F4**
Alfalfa	69.57	91.30	86.96	86.96	45.22	91.30	95.22	**95.65**	48.26	91.30	**95.65**	**95.65**	46.09	91.30	**95.65**	**95.65**	46.09	91.30	95.65	**95.65**
Corn-notill	81.42	87.42	90.67	93.92	78.04	87.86	90.45	91.92	77.91	87.05	90.41	91.70	80.83	89.12	91.06	92.78	80.82	89.38	91.14	**92.90**
Corn-mintill	70.71	88.71	96.86	96.57	64.70	84.54	97.00	98.16	65.34	83.29	97.31	97.91	65.87	86.54	98.03	99.20	65.43	86.93	97.97	**99.24**
Corn-notill	75.56	94.44	95.56	96.67	62.83	83.89	94.11	95.33	62.94	83.44	93.11	95.61	65.56	86.17	95.56	95.89	65.61	86.28	95.56	**95.95**
Grass-pasture	79.00	81.50	88.75	88.75	72.98	86.48	93.13	93.60	72.95	86.38	92.55	93.70	75.90	88.65	93.63	93.90	75.95	88.43	93.73	**94.00**
Grass-trees	91.00	92.17	99.00	99.00	96.03	96.80	99.70	99.69	95.68	96.73	99.67	99.64	96.58	97.33	**99.67**	**99.67**	96.50	97.49	**99.67**	**99.67**
Grass-pasture-mowed	85.71	85.71	85.71	85.71	91.43	**92.86**	**92.86**	**92.86**	**92.86**	**92.86**	**92.86**	**92.86**	**92.86**	**92.86**	**92.86**	**92.86**	**92.86**	**92.86**	**92.86**	**92.86**
Hay-windrowed	97.75	97.50	99.75	99.75	98.45	96.30	99.68	99.70	98.68	96.93	99.68	99.63	99.45	97.70	99.73	99.75	99.23	97.83	99.65	**99.75**
Oats	90.00	90.00	100.00	100.00	80.00	89.00	91.00	99.00	85.00	92.00	97.00	98.00	91.00	99.00	**100.**	**100.**	**89.00**	**98.00**	**100.**	**100.**
Soybean-notill	77.75	87.38	87.25	89.00	79.37	89.83	91.34	92.59	79.54	89.02	90.93	92.22	81.89	91.85	92.27	93.14	81.70	91.89	92.15	**92.92**
Soybean-mintill	78.55	87.45	95.77	95.64	76.94	91.20	97.35	98.44	77.20	91.22	97.54	98.16	78.52	92.82	98.16	98.54	78.74	92.88	98.25	**98.52**
Soybean-clean	71.60	82.80	93.80	93.80	73.42	81.86	95.20	96.36	73.90	82.72	95.18	95.94	77.90	86.10	97.14	97.16	77.50	86.16	96.90	**97.18**
Wheat	**99.33**	**99.33**	98.67	98.67	98.20	99.00	98.60	98.73	98.13	98.53	98.27	98.80	98.54	**99.33**	98.67	99.00	98.47	**99.33**	98.67	98.80
Woods	96.60	96.90	99.60	99.50	98.23	98.94	99.87	99.82	98.17	98.90	99.83	99.85	98.37	99.08	**99.91**	**99.91**	98.41	99.06	99.90	99.90
Buildings-grass-trees-drives	51.33	61.33	99.00	99.00	56.00	69.60	99.77	99.97	55.27	70.27	99.90	99.97	57.17	69.87	**100.**	**100.**	56.60	70.33	**100.**	**100.**
Stone-steel-towers	**100.**	**100.**	**100.**	**100.**	99.80	**100.**	**100.**	**100.**	99.60	**100.**	**100.**	**100.**	**100.**	**100.**	**100.**	**100.**	**100.**	**100.**	**100.**	**100.**
AA (%)	82.24	89.00	94.83	95.18	79.48	89.97	95.95	96.99	80.09	90.04	96.24	96.85	81.66	91.73	97.02	**97.34**	81.43	91.76	97.01	97.33
OA (%)	81.21	88.43	94.91	95.48	80.19	90.10	96.07	96.89	80.30	89.89	96.05	96.72	81.95	91.60	96.73	**97.27**	81.89	91.71	96.73	**97.27**
Kappa	0.78	0.87	0.94	0.95	0.77	0.89	0.95	0.96	0.77	0.88	0.95	0.96	0.79	0.90	0.96	**0.97**	0.79	0.90	0.96	**0.97**

F1: Spectral; F2: Spectral-OO; F3: Spectral-MPs; F4: Spectral-OO-MPs.

According to the results, the smaller the regularization parameter δ *is* and the larger the number of neighbors *NN*, the larger the OA values. Thus, $\delta = 0.01$ and $NN = 12$ were considered in all of the experiments. Computational complexity is primarily controlled by the labeled sample size (note the vertical axis in Figures 9d–f and 10d–f.

5.5. Validation of the Semi-Supervised MVCCAE

To validate the performances of the semi-supervised version of the proposed algorithm, comparisons with existing methods, specifically LLP and LapSVM, are presented for the ROSIS University data set, starting from a label set of increasing size.

Figure 11 shows the learning curves for SSMVCCAE, LLP, and LapSVM using different view generation and classifier ensemble strategies as a function of this size. Each point on the *x*-axis represents the size of the labeled samples (pixels) for each class type, while the *y*-axis represents the average overall classification accuracy. In Table 6, we report the average overall classification accuracies and kappa statistics (κ) over 10 independent runs, when a total of 100 labeled samples are considered for each class.

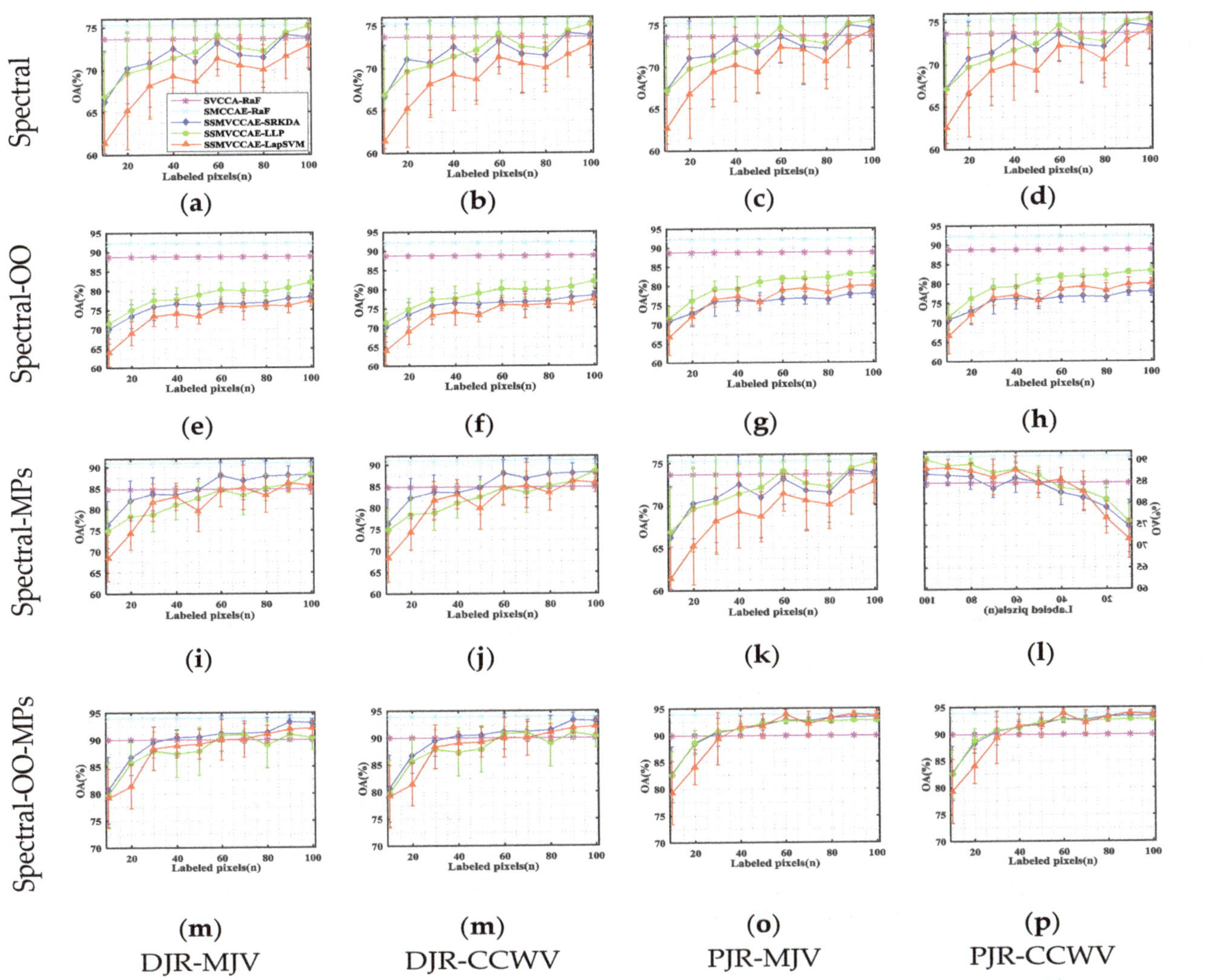

Figure 11. (**a–p**) Average OA values versus labeled pixels for SSMVCCAE with different view generation and fusion strategies for ROSIS University dataset.

According to the results in Figure 11 and Table 6, the proposed semi-supervised heterogeneous DA approach achieves comparable and sometimes better results in any case (see the learning curves in blue for SSMVCCAE-SRKDA vs. green for SSMVCCAE-LPP and red for SSMVCCAE-LapSVM in Figure 11). Moreover, larger OA values with faster convergence rates are shown by SSMVCCAE with PJR as opposed to DJR view generation, either by MJV fusion or by the CCWV fusion, especially using the spectral-OO-MPs features.

In Figure 12 and Table 7, the results of the same experiments are reported for the Indian Pines test set. Please note that because only a few samples are available for some classes in the Indian Pines case, class types that contain less than 70 pixels for training are not considered here. Even in this case, to obtain a more objective comparison and evaluation, each test is executed independently for 10 rounds.

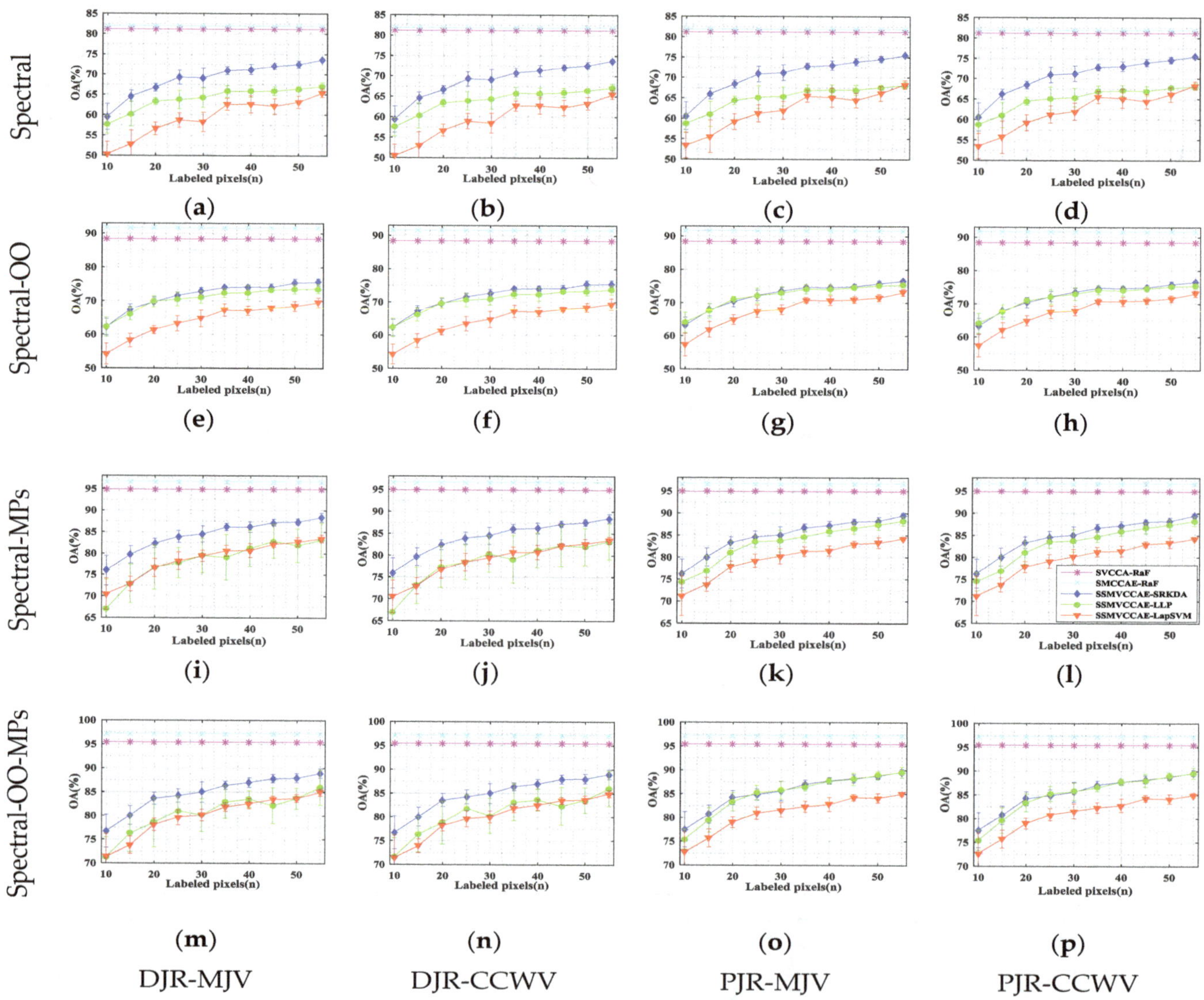

Figure 12. (**a**–**p**) Average OA values versus labeled pixels for SSMVCCAE with different view generation and fusion strategies on Indian Pines data.

Figure 12 shows that better classification results are obtained by the SSMVCCAE with SRKDA, not only using the original spectral features but also using spectral features that incorporate OO and MPs features (see the learning curves in blue vs. those in green and red). Moreover, the best classification results are obtained by SSMVCCAE-SRKDA with the PJR view generation technique, and when considering the spectral-OO-MPs stacked features (see the numbers in bold in Table 7).

Table 6. Average overall classification accuracies and kappa statistics (κ) for SSMVCCAE with different semi-supervised classifiers for the ROSIS University data. Total 100 labeled samples are available for each class over 10 independent runs.

Classifier	SSMVCCAE (SRKDA)								SSMVCCAE (LPP)								SSMVCCAE (LapSVM)							
View Generation	DJR				PJR				DJR				PJR				DJR				PJR			
Voting	MJV		CCWV		MJV		CCWV		MJV		CCWV		MJV		CCWV		MJV		CCWV		MJV		CCWV	
Statistics	OA	κ	OA	κ	OA	κ	OA	κ	OA	κ	OA	κ	OA	κ	OA	κ	OA	κ	OA	κ	OA	κ	OA	κ
Spectral	73.82	0.66	73.83	0.66	74.58	0.67	74.47	0.67	75.12	0.68	75.11	0.68	**75.45**	**0.68**	75.37	0.68	72.82	0.66	72.85	0.66	74.26	0.67	74.16	0.67
Spectral-OO	78.31	0.72	78.29	0.72	78.02	0.72	78.00	0.72	82.08	0.76	82.04	0.76	**83.44**	**0.78**	83.32	0.78	77.37	0.71	77.38	0.71	80.08	0.74	80.04	0.74
Spectral-MPs	88.23	0.85	88.24	0.85	86.73	0.82	86.89	0.83	88.44	0.85	88.45	0.85	90.16	0.87	**90.17**	**0.87**	85.65	0.82	85.72	0.82	88.14	0.85	88.12	0.85
Spectral-OO-MPs	93.17	0.91	93.14	0.91	93.47	0.91	93.46	0.91	90.36	0.87	90.31	0.87	92.78	0.90	92.77	0.90	92.08	0.90	92.10	0.90	**93.68**	**0.92**	93.67	0.92

Table 7. Average overall classification accuracies and kappa statistics (κ) for SSMVCCAE with different semi-supervised classifiers for the Indian Pines data. A total of 55 labeled samples are available for each class over 10 independent runs.

Classifier	SSMVCCAE (SRKDA)								SSMVCCAE (LPP)								SSMVCCAE (LapSVM)							
View Generation	DJR				PJR				DJR				PJR				DJR				PJR			
Voting	MJV		CCWV		MJV		CCWV		MJV		CCWV		MJV		CCWV		MJV		CCWV		MJV		CCWV	
Statistics	OA	κ	OA	κ	OA	κ	OA	κ	OA	κ	OA	κ	OA	κ	OA	κ	OA	κ	OA	κ	OA	κ	OA	κ
Spectral	73.67	0.70	73.81	0.70	75.49	0.72	**75.51**	**0.72**	67.05	0.62	67.13	0.62	68.18	0.64	68.17	0.64	65.43	0.61	65.48	0.61	68.34	0.64	68.38	0.64
Spectral-OO	75.64	0.72	75.53	0.72	76.61	0.73	**76.66**	**0.73**	73.69	0.70	73.75	0.70	75.42	0.72	75.47	0.72	69.59	0.65	69.54	0.65	73.20	0.69	73.18	0.69
Spectral-MPs	88.42	0.87	88.49	0.87	89.57	0.88	**89.54**	**0.88**	83.19	0.81	83.22	0.81	88.29	0.86	88.31	0.86	83.49	0.81	83.51	0.81	84.33	0.82	84.33	0.82
Spectral-OO-MPs	89.02	0.87	89.05	0.87	**89.66**	**0.88**	**89.66**	**0.88**	86.01	0.84	86.02	0.84	89.60	0.88	89.59	0.88	85.01	0.83	84.99	0.83	85.03	0.83	85.01	0.83

Finally, in Figures 13 and 14, the CPU time consumptions in seconds for the different implementations of the semi-supervised procedure are reported as a function of the labeled sample size for both Pavia and Indian Pines. According to the results, SSMVCCAE with SRKDA is only slightly more efficient than LapSVM for the ROSIS University data, but is much more efficient for the Indian Pines data. Moreover, the computational complexities of LapSVM and LLP increase linearly with the number of labeled samples, because they are more visible for the Indian Pines data, whereas the CPU time for SRKDA stays almost constant.

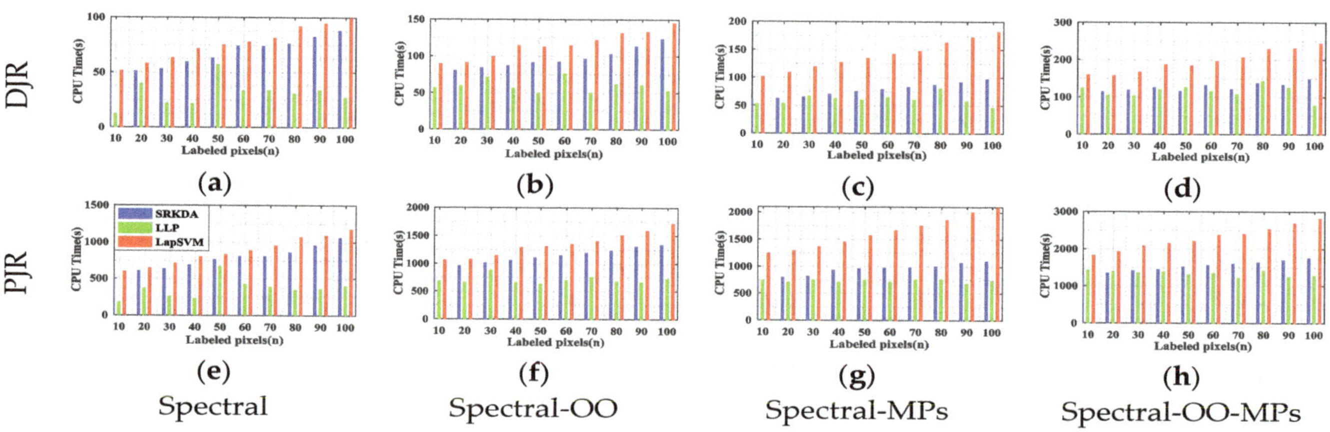

Figure 13. (**a**–**h**) CPU time consumption in seconds versus the size of the labeled samples for SSMVCCAE-SRKDA/-LLP/-LapSVM for the ROSIS University data.

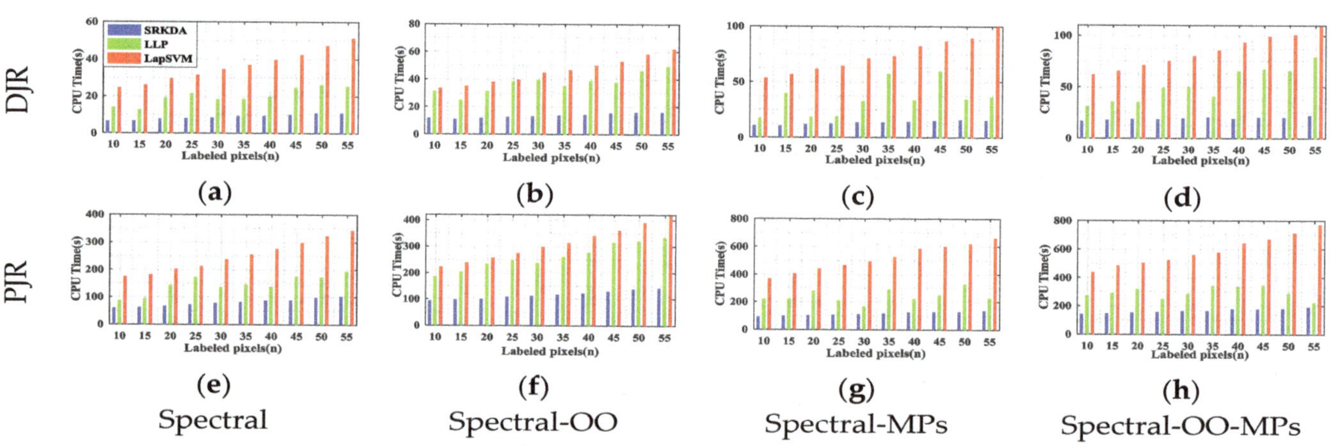

Figure 14. (**a**–**h**) CPU time versus the size of the labeled samples for SSMVCCAE-SRKDA/-LLP/-LapSVM for the Indian Pines data.

Summing the results presented in this section, it can be concluded that the novel proposed semi-supervised heterogeneous DA approach works properly and achieves satisfactory results better than the current state-of-the-art techniques when using a PJR view generation technique either with majority voting or with canonical correlation coefficient voting. A comparison of the results by SSMVCCAE with those by LLP and LapSVM shows that the performance of SRKDA is superior for both classification accuracy and computational efficiency. Finally, the computational burden caused by the sizes of the labeled samples and feature dimensionality is much smaller for SSMVCCAE with SRKDA, whereas it increases linearly with the sample size when using the other techniques.

6. Conclusions

In this paper, we have presented the implementation details, analyzed the parameter sensitivity, and proposed a comprehensive validation of two versions of an ensemble classifier that is suitable for heterogeneous DA and based on multiple view CCA. The main idea is to overcome the limitations of SVCCA by incorporating multi view CCA into EL. Superior results have been proven using two high dimensional (hyperspectral) images, the ROSIS Pavia University and the AVIRIS Indian Pine datasets, as high dimensional target domains, with synthetic low dimensional (multispectral) images as associated SDs. The best classification results were always obtained by jointly considering the original spectral features stacked with object-oriented features assigned to segmentation results, and the morphological profiles, which were subdivided into multiple views using the PJR view generation technique.

To further mitigate the marginal and/or conditional distribution gap between the source and the target domains, when few or even no labeled samples are available from the target domain, we propose a semi-supervised version of the same approach via training multiple speed-up SRKDA.

For new research directions, we are considering more complex problems, such as single SD vs. multiple TDs, as well as multiple SDs vs. multiple TDs supervised and semi-supervised adaptation techniques.

Acknowledgments: This work was partially supported by the Project funded by China Postdoctoral Science Foundation (2016M592872), the Xinjiang Uyghur Autonomous Region High Level Talents Introduction Project (Y648031) and the National Natural Science Foundation of China (No. 41601440, No. 41601354, and No. 41471098).

Author Contributions: Alim Samat developed the algorithms, executed all of the experiments, finished the original manuscript and the subsequent revisions, and provided part of the funding. Claudio Persello and Paolo Gamba offered valuable suggestions and comments, and carefully revised the original manuscript and its revisions. Jilili Abuduwaili provided part of the funding. Sicong Liu and Erzhu Li, contributed to revising of the manuscript.

References

1. Olofsson, P.; Foody, G.M.; Herold, M.; Stehman, S.V.; Woodcock, C.E.; Wulder, M.A. Good practices for estimating area and assessing accuracy of land change. *Remote Sens. Environ.* **2014**, *148*, 42–57. [CrossRef]
2. Curlander, J.C. Location of spaceborne SAR imagery. *IEEE Trans. Geosci. Remote Sens.* **1982**, *3*, 359–364. [CrossRef]
3. Bruzzone, L.; Cossu, R. A multiple-cascade-classifier system for a robust and partially unsupervised updating of land-cover maps. *IEEE Trans. Geosci. Remote Sens.* **2002**, *40*, 1984–1996. [CrossRef]
4. Torralba, A.; Efros, A.A. Unbiased Look at Dataset Bias. In Proceedings of the IEEE Conference on Computer Vision and Pattern Recognition, Colorado Springs, CO, USA, 20–25 June 2011; pp. 1521–1528.
5. Khosla, A.; Zhou, T.; Malisiewicz, T.; Efros, A.A.; Torralba, A. Undoing the Damage of Dataset Bias. In Proceedings of the IEEE Conference on Computer Vision and Pattern Recognition, Providence, RI, USA, 20–25 June 2012; Springer: Berlin/Heidelberg, Germany, 2012; pp. 158–171.
6. Schott, J.R.; Salvaggio, C.; Volchok, W.J. Radiometric scene normalization using pseudoinvariant features. *Remote Sens. Environ.* **1988**, *26*, 1–16. [CrossRef]
7. Woodcock, C.E.; Macomber, S.A.; Pax-Lenney, M.; Cohen, W.B. Monitoring large areas for forest change using Landsat: Generalization across space, time and Landsat sensors. *Remote Sens. Environ.* **2001**, *78*, 194–203. [CrossRef]
8. Olthof, I.; Butson, C.; Fraser, R. Signature extension through space for northern landcover classification: A comparison of radiometric correction methods. *Remote Sens. Environ.* **2005**, *95*, 290–302. [CrossRef]
9. Rakwatin, P.; Takeuchi, W.; Yasuoka, Y. Stripe noise reduction in MODIS data by combining histogram matching with facet filter. *IEEE Trans. Geosci. Remote Sens.* **2007**, *45*, 1844–1856. [CrossRef]
10. Inamdar, S.; Bovolo, F.; Bruzzone, L.; Chaudhuri, S. Multidimensional probability density function matching for preprocessing of multitemporal remote sensing images. *IEEE Trans. Geosci. Remote Sens.* **2008**, *46*, 1243–1252. [CrossRef]
11. Bruzzone, L.; Marconcini, M. Domain adaptation problems: A DASVM classification technique and a circular validation strategy. *IEEE Trans. Pattern Anal. Mach. Intell.* **2010**, *32*, 770–787. [CrossRef] [PubMed]
12. Pan, S.J.; Yang, Q. A survey on transfer learning. *IEEE Trans. Knowl. Data Eng.* **2010**, *22*, 1345–1359. [CrossRef]

13. Banerjee, B.; Bovolo, F.; Bhattacharya, A.; Bruzzone, L.; Chaudhuri, S.; Buddhiraju, K.M. A Novel Graph-Matching-Based Approach for Domain Adaptation in Classification of Remote Sensing Image Pair. *IEEE Trans. Geosci. Remote Sens.* **2015**, *53*, 4045–4062. [CrossRef]
14. Matasci, G.; Volpi, M.; Kanevski, M.; Bruzzone, L.; Tuia, D. Semisupervised Transfer Component Analysis for Domain Adaptation in Remote Sensing Image Classification. *IEEE Trans. Geosci. Remote Sens.* **2015**, *53*, 3550–3564. [CrossRef]
15. Tuia, D.; Persello, C.; Bruzzone, L. Domain Adaptation for the Classification of Remote Sensing Data: An Overview of Recent Advances. *IEEE Geosci. Remote Sens. Mag.* **2016**, *4*, 41–57. [CrossRef]
16. Samat, A.; Gamba, P.; Abuduwaili, J.; Liu, S.; Miao, Z. Geodesic Flow Kernel Support Vector Machine for Hyperspectral Image Classification by Unsupervised Subspace Feature Transfer. *Remote Sens.* **2016**, *8*, 234. [CrossRef]
17. Daumé, H., III; Kumar, A.; Saha, A. Frustratingly Easy Semi-Supervised Domain Adaptation. In Proceedings of the 2010 Workshop on Domain Adaptation Natural Language Processing, Uppsala, Sweden, 15 July 2010; pp. 53–59.
18. Li, W.; Duan, L.; Xu, D.; Tsang, I.W. Learning with augmented features for supervised and semi-supervised heterogeneous domain adaptation. *IEEE Trans. Pattern Anal. Mach. Intell.* **2014**, *36*, 1134–1148. [CrossRef] [PubMed]
19. Patel, V.M.; Gopalan, R.; Li, R.; Chellappa, R. Visual domain adaptation: A survey of recent advances. *IEEE Sign. Process. Mag.* **2015**, *32*, 53–69. [CrossRef]
20. Gao, J.; Fan, W.; Jiang, J.; Han, J. Knowledge Transfer via Multiple Model Local Structure Mapping. In Proceedings of the 14th ACM SIGKDD International Conference on Knowledge Discovery and Data Mining, Las Vegas, NV, USA, 24–27 August 2008; pp. 283–291.
21. Jiang, J.; Zhai, C. Instance Weighting for Domain Adaptation in NLP. In Proceedings of the ACL, Prague, Czech Republic, 23–30 June 2007; Volume 7, pp. 264–271.
22. Sugiyama, M.; Nakajima, S.; Kashima, H.; Buenau, P.V.; Kawanabe, M. Direct Importance Estimation with Model Selection and Its Application to Covariate Shift Adaptation. In *Advances in Neural Information Processing Systems*; Springer: Vancouver, BC, Canada, 2008; pp. 1433–1440.
23. Persello, C.; Bruzzone, L. Active learning for domain adaptation in the supervised classification of remote sensing images. *IEEE Trans. Geosci. Remote Sens.* **2012**, *50*, 4468–4483. [CrossRef]
24. Bruzzone, L.; Persello, C. A novel approach to the selection of spatially invariant features for the classification of hyperspectral images with improved generalization capability. *IEEE Trans. Geosci. Remote Sens.* **2009**, *47*, 3180–3191. [CrossRef]
25. Pan, S.J.; Kwok, J.T.; Yang, Q. Transfer Learning via Dimensionality Reduction. In Proceedings of the AAAI, 8, Stanford, CA, USA, 26–28 March 2008; pp. 677–682.
26. Long, M.; Wang, J.; Ding, G.; Pan, S.J.; Yu, P.S. Adaptation regularization: A general framework for transfer learning. *IEEE Trans. Knowl. Data Eng.* **2014**, *26*, 1076–1089. [CrossRef]
27. Pan, S.J.; Tsang, I.W.; Kwok, J.T.; Yang, Q. Domain adaptation via transfer component analysis. *IEEE Trans. Neural Netw.* **2011**, *22*, 199–210. [CrossRef] [PubMed]
28. Mihalkova, L.; Huynh, T.; Mooney, R.J. Mapping and Revising Markov Logic Networks for Transfer Learning. In Proceedings of the AAAI, 7, Vancouver, BC, Canada, 22–26 July 2007; pp. 608–614.
29. Yeh, Y.R.; Huang, C.H.; Wang, Y.C.F. Heterogeneous domain adaptation and classification by exploiting the correlation subspace. *IEEE Trans. Image Proc.* **2014**, *23*, 2009–2018.
30. Gopalan, R.; Li, R.; Chellappa, R. Domain Adaptation for Object Recognition: An Unsupervised Approach. In Proceedings of the IEEE International Conference on Computer Vision, Barcelona, Spain, 6–13 November 2011; pp. 999–1006.
31. Gopalan, R.; Li, R.; Chellappa, R. Unsupervised adaptation across domain shifts by generating intermediate data representations. *IEEE Trans. Pattern Anal. Mach. Intell.* **2014**, *6*, 2288–2302. [CrossRef] [PubMed]
32. Duan, L.; Xu, D.; Tsang, I.W. Domain adaptation from multiple sources: A domain-dependent regularization approach. *IEEE Trans. Neural Netw. Learn. Syst.* **2012**, *23*, 504–518. [CrossRef] [PubMed]
33. Wang, W.; Zhou, Z.H. A New Analysis of Co-Training. In Proceedings of the 27th International Conference on Machine Learning, (ICML-10) 2010, Haifa, Israel, 21–24 June 2010; pp. 1135–1142.
34. Di, W.; Crawford, M.M. View generation for multiview maximum disagreement based active learning for hyperspectral image classification. *IEEE Trans. Geosci. Remote Sens.* **2012**, *50*, 1942–1954. [CrossRef]

35. Sun, S. A survey of multi-view machine learning. *Neural Comput. Appl.* **2013**, *23*, 2031–2038. [CrossRef]
36. Kuncheva, L.I.; Rodríguez, J.J.; Plumpton, C.O.; Linden, D.E.; Johnston, S.J. Random subspace ensembles for fMRI classification. *IEEE Trans. Med. Imaging* **2010**, *29*, 531–542. [CrossRef] [PubMed]
37. Samat, A.; Du, P.; Liu, S.; Li, J. E2LMs: Ensemble extreme learning machines for hyperspectral image classification. *IEEE J. Sel. Top. Appl. Earth Obs. Remote Sens.* **2014**, *7*, 1060–1069. [CrossRef]
38. Hady, M.F.A.; Schwenker, F. Semi-Supervised Learning. In *Handbook on Neural Information Processing*; Springer: Berlin/Heidelberg, Germany, 2013; pp. 215–239.
39. Kulis, B.; Saenko, K.; Darrell, T. What you Saw Is not What you Get: Domain Adaptation Using Asymmetric Kernel Transforms. In Proceedings of the IEEE International Conference on Computer Vision and Pattern Recognition (CVPR), Colorado Springs, CO, USA, 20–25 June 2011; pp. 1785–1792.
40. Kumar, A.; Saha, A.; Daume, H. Co-regularization based semi-supervised domain adaptation. In *Advances in Neural Information Processing Systems 23 (NIPS 2010)*; Springer: Vancouver, BC, Canada, 2010; pp. 478–486.
41. Cai, D.; He, X.; Han, J. Speed up kernel discriminant analysis. *VLDB J. Int. J. Very Large Data Bases* **2011**, *20*, 21–33. [CrossRef]
42. Dai, W.; Chen, Y.; Xue, G.R.; Yang, Q.; Yu, Y. Translated learning: Transfer learning across different feature spaces. In *Advances in Neural Information Processing Systems*; Springer: Vancouver, BC, Canada, 2008; pp. 353–360.
43. Yang, Q.; Chen, Y.; Xue, G.R.; Dai, W.; Yu, Y. Heterogeneous Transfer Learning for Image Clustering via the Social Web. In Proceedings of the Joint Conference 47th Annual Meeting of the ACL and the 4th International Joint Conference Natural Language Process, AFNLP, Singapore, 2–7 August 2009; Volume 1, pp. 1–9.
44. Zhu, Y.; Chen, Y.; Lu, Z.; Pan, S.J.; Xue, G.R.; Yu, Y.; Yang, Q. Heterogeneous Transfer Learning for Image Classification. In Proceedings of the AAAI, San Francisco, CA, USA, 7–11 August 2011.
45. Evangelopoulos, N.; Zhang, X.; Prybutok, V.R. Latent semantic analysis: Five methodological recommendations. *Eur. J. Inf. Syst.* **2012**, *21*, 70–86. [CrossRef]
46. Hong, L. A Tutorial on Probabilistic Latent Semantic Analysis. Available online: https://arxiv.org/pdf/1212.3900.pdf (accessed on 21 December 2012).
47. Koltchinskii, V. Rademacher penalties and structural risk minimization. *IEEE Trans. Inf. Theory* **2001**, *47*, 1902–1914. [CrossRef]
48. Gong, B.; Shi, Y.; Sha, F.; Grauman, K. Geodesic Flow Kernel for Unsupervised Domain Adaptation. In Proceedings of the 2012 IEEE Conference Computer Vision and Pattern Recognition (CVPR), Providence, RI, USA, 16–21 June 2012; pp. 2066–2073.
49. Shi, X.; Liu, Q.; Fan, W.; Yu, P.S.; Zhu, R. Transfer Learning on Heterogenous Feature Spaces via Spectral Transformation. In Proceedings of the IEEE 10th International Conference on Data Mining, Sydney, Australia, 13–17 December 2010; pp. 1049–1054.
50. Zhou, J.T.; Tsang, I.W.; Pan, S.J.; Tan, M. Heterogeneous Domain Adaptation for Multiple Classes. In Proceedings of the AISTATS, Reykjavik, Iceland, 22–25 April 2014; pp. 1095–1103.
51. Hardoon, D.R.; Szedmak, S.; Shawe-Taylor, J. Canonical correlation analysis: An overview with application to learning methods. *Neural Comput.* **2004**, *16*, 2639–2664. [CrossRef] [PubMed]
52. Rokach, L. *Pattern Classification Using Ensemble Methods*; World Scientific Publishing Company: Singapore, 2010; Volume 75.
53. Xu, C.; Tao, D.; Xu, C. Multi-View Learning with Incomplete Views. *IEEE Trans. Image Proc.* **2015**, *24*, 5812–5825. [CrossRef] [PubMed]
54. Cai, D.; He, X.; Han, J. Spectral Regression: A Unified Subspace Learning Framework for Content-Based Image Retrieval. In Proceedings of the 15th International Conference on Multimedia, Augsburg, Germany, 24–29 September 2007; pp. 403–412.
55. Baudat, G.; Anouar, F. Generalized discriminant analysis using a kernel approach. *Neural Comput.* **2000**, *12*, 2385–2404. [CrossRef] [PubMed]
56. Vassilevich, D.V. Heat kernel expansion: User's manual. *Phys. Rep.* **2003**, *388*, 279–360. [CrossRef]
57. Vapnik, V.N. An overview of statistical learning theory. *IEEE Trans. Neural Netw.* **1999**, *10*, 988–999. [CrossRef] [PubMed]
58. Cai, D.; He, X.; Han, J. Document clustering using locality preserving indexing. *IEEE Trans. Knowl. Data Eng.* **2005**, *17*, 1624–1637. [CrossRef]

59. Subramanya, A.; Talukdar, P.P. Graph-based semi-supervised learning. *Synth. Lect. Arti. Intell. Mach. Learn.* **2014**, *8*, 1–125. [CrossRef]
60. Camps-Valls, G.; Marsheva, T.V.B.; Zhou, D. Semi-supervised graph-based hyperspectral image classification. *IEEE Trans. Geosci. Remote Sens.* **2007**, *45*, 3044–3054. [CrossRef]
61. Breiman, L. Random forests. *Mach. Learn.* **2001**, *45*, 5–32. [CrossRef]
62. Du, P.; Samat, A.; Waske, B.; Liu, S.; Li, Z. Random Forest and Rotation Forest for fully polarized SAR image classification using polarimetric and spatial features. *ISPRS J. Photogramm. Remote Sens.* **2015**, *105*, 38–53. [CrossRef]
63. Benediktsson, J.A.; Palmason, J.A.; Sveinsson, J.R. Classification of hyperspectral data from urban areas based on extended morphological profiles. *IEEE Trans. Geosci. Remote Sens.* **2005**, *43*, 480–491. [CrossRef]
64. Fauvel, M.; Benediktsson, J.A.; Chanussot, J.; Sveinsson, J.R. Spectral and spatial classification of hyperspectral data using SVMs and morphological profiles. *IEEE Trans. Geosci. Remote Sens.* **2008**, *46*, 3804–3814. [CrossRef]
65. Kobayashi, T.; Watanabe, K.; Otsu, N. Logistic label propagation. *Pattern Recognit. Lett.* **2012**, *33*, 580–588. [CrossRef]
66. Melacci, S.; Belkin, M. Laplacian support vector machines trained in the primal. *J. Mach. Learn. Res.* **2011**, *12*, 1149–1184.

5

Deep Salient Feature Based Anti-Noise Transfer Network for Scene Classification of Remote Sensing Imagery

Xi Gong [1,2], Zhong Xie [1,2], Yuanyuan Liu [1,*], Xuguo Shi [1] and Zhuo Zheng [1]

1 Department of Information Engineering, China University of Geosciences, Wuhan 430075, China; gongxi_cug@126.com (X.G.); xiezhong@cug.edu.cn (Z.X.); shixg@cug.edu.cn (X.S.); zhuozheng_2017@163.com (Z.Z.)

2 National Engineering Research Center of Geographic Information System, Wuhan 430075, China

* Correspondence: liuyy@cug.edu.cn.

Abstract: Remote sensing (RS) scene classification is important for RS imagery semantic interpretation. Although tremendous strides have been made in RS scene classification, one of the remaining open challenges is recognizing RS scenes in low quality variance (e.g., various scales and noises). This paper proposes a deep salient feature based anti-noise transfer network (DSFATN) method that effectively enhances and explores the high-level features for RS scene classification in different scales and noise conditions. In DSFATN, a novel discriminative deep salient feature (DSF) is introduced by saliency-guided DSF extraction, which conducts a patch-based visual saliency (PBVS) algorithm using "visual attention" mechanisms to guide pre-trained CNNs for producing the discriminative high-level features. Then, an anti-noise network is proposed to learn and enhance the robust and anti-noise structure information of RS scene by directly propagating the label information to fully-connected layers. A joint loss is used to minimize the anti-noise network by integrating anti-noise constraint and a softmax classification loss. The proposed network architecture can be easily trained with a limited amount of training data. The experiments conducted on three different scale RS scene datasets show that the DSFATN method has achieved excellent performance and great robustness in different scales and noise conditions. It obtains classification accuracy of 98.25%, 98.46%, and 98.80%, respectively, on the UC Merced Land Use Dataset (UCM), the Google image dataset of SIRI-WHU, and the SAT-6 dataset, advancing the state-of-the-art substantially.

Keywords: scene classification; saliency detection; deep salient feature; anti-noise transfer network; DSFATN

---

1. Introduction

Many RS images have been accumulated due to the rapid development of Remote Sensing (RS) sensors and imaging techniques. The interpretation of such huge amount of RS imagery is a challenging task of significant sense for disaster monitoring, urban planning, traffic controlling and so on [1–5]. RS scene classification, which aims at automatically classifying extracted sub-regions of the scenes into a set of semantic categories, is an effective method for RS image interpreting [6,7]. However, the complex spatial arrangement and the variety of surface objects in RS scenes make the classification quite challenging, especially for scenes in low quality (e.g., various scales and noises), since their within-class differences are more indistinct and between-class similarity are more distinct. How to automatically recognize and represent the RS scene from these different scale and quality RS image data effectively has become a critical task. To deal with such a challenge, this paper proposes a deep salient feature based anti-noise transfer network (DSFATN) approach that effectively enhances and explores the high-layer features for RS scene classification in different scales and noise conditions with great efficiency and robustness.

Many attempts have been made for RS scene classification. Among various previous approach, the bag-of-visual-words (BoVW) based models have drawn much attention for their good performance [1,8–10]. The BoVW based models encode local invariant features of an image and represent the image as a histogram of visual word occurrences. However, the BoVW based models utilize a collection of local features, which may not fully exploit the spatial layouts information thus result in information loss [11]. To solve the problem, the spatial pyramid matching kernel (SPMK) [12] introduced the spatial layout to form improved local features. Even though SPMK shows inspiring results, it only considers the absolute spatial arrangement of visual words. Thus, the improved version of SPMK, spatial co-occurrence kernel (SCK) [1], and its pyramidal version spatial pyramid co-occurrence kernel (SPCK) [13], were proposed to capture both absolute and relative spatial arrangements. Other alternative models, e.g., latent Dirichlet allocation (LDA) model [14–16] and the probabilistic latent semantic analysis (pLSA) model [17,18], represent the image scene as a finite random mixture of topics and obtain competitive performance. In general, these approaches have made some achievements in RS scene classification but demand prior knowledge in handcrafted feature extraction, which is still opening challenging task in scene classification.

Recently, deep learning (DL) methods have achieved dramatic improvements and state-of–the-art performance in many fields (e.g., image recognition [19], object detection [20,21], and image synthesis [22]) due to automatic high-level feature representations from images and powerful ability of abstraction. DL methods also draw much attention in RS image classification [23,24]. For example, Lu et al. [25] proposed a discriminative representation for high spatial resolution remote sensing image by utilizing a shallow weighted deconvolution network and spatial pyramid model (SPM), and classified the representation vector by support vector machine (SVM). Chen et al. [26] utilized the single-layer restricted Boltzmann machine (RBM) and multilayer deep belief network (DBN) based model to learn the shallow and deep features of hyperspectral data, the learnt features can be used in logistic regression to achieve the hyperspectral data classification. As one of the most popular DL approaches, convolutional neural networks (CNNs) show incomparable superiority on several benchmark datasets such as Imagenet [27], and have been widely used in the recognition, detection tasks and obtained impressive results [28–30]. However, training a powerful CNN is complicated since many labeled training samples and techniques are needed, while the available labeled RS scene datasets are not comparable to any natural scene dataset. For example, compared with the dataset ImageNet containing 15 million labeled images in 22,000 classes, the most famous and widely used UC Merced Land Use (UCM) [1] RS scene dataset only contains 21 classes and 2100 label images.

To address the data limitation, an effective strategy is data augmentation. It generates more training image samples by adding rotated, flipped versions and random cropped, stretched patches of the training images [31,32], or patches sampled by some optimized strategy [11,33]. Another effective strategy is transfer learning based on a pre-trained CNN model. Castelluccio et al. [34] fine-tuned the pre-trained CNNs on the UCM dataset. The best result reached 97.10% when fine-tuning the GoogLeNet [35] while training a GoogLeNet from scratch just reached 91.2%. Penatti et al. [36] and Hu et al. [37] investigated the deep features extracted from different pre-trained CNNs for RS scene representation and classification, and proved the effectiveness and superiority of the features from the 1st full-connected layer of CNNs. The features extracted from pre-trained CNNs also have some invariance to small-scale deformations, larger-scale and so on [38,39]. Compared with training a new CNN, transfer learning methods are faster and the classification results are much promising without large amount of training data. It is known that most of the pre-trained CNNs have been trained in dataset with large number of natural images such as ImageNet. In natural image scenes, the objects are almost centrally focused, and the center pixels have more influence on the image semantic labels [11], while, in RS image scenes, the surface objects are usually distributed randomly, and the central parts may not relate closely with the semantic label. Hence, due to the objects distributions difference between natural scenes and RS scenes, the pre-trained CNNs based on transfer learning method is applicable for a limit amount of training date but lacks robustness to low quality variance (e.g., various scales and noises) in RS scene classification.

To address the challenging task, we propose a deep salient feature based anti-noise transfer network (DSFATN) for classification of RS scenes with different scales and various noises. Our method aims at improving both feature representation of RS scene and classification accuracy. In DSFATN, a novel deep salient feature (DSF) and an anti-noise transfer network are introduced to suppress the influences of different scales and noise variances. The saliency-guided DSF extraction conducts a patch-based visual saliency (PBVS) algorithm to guide pre-trained CNNs for producing the discriminative high-level DSF. It compensates the affect caused by objects distribution difference between natural scenes and RS scenes, thus makes the DSF extracted exactly from the most relevant, informative and representative patches of the RS scene related to its category. The anti-noise transfer network is trained to learn and enhance the robust and anti-noise structure information of RS scene by minimizing a joint loss. DSFATN performs excellent with RS scenes in different scales and qualities, even with noise.

The major contributions of this paper are as follows:

- We propose a novel DSF representation using "visual attention" mechanisms. DSF can achieve discriminative high-level feature representation learnt from pre-trained CNN for the RS scenes.
- An anti-noise transfer network is improved to learn and enhance the robust and anti-noise structure information of RS scene, where a joint loss is used to minimize the network by considering anti-noise constraint and softmax classification loss. The simple architecture of the anti-noise transfer network makes it easier to be trained with the limited availability of training data.
- The proposed DSFATN is evaluated on several public RS scene classification benchmarks. The significant performance demonstrated our method is of great robustness and efficiency in various scales, occlusions, and noise conditions and advanced the state-of-the-arts methods.

This paper is organized as follows. In Section 2, we illustrate the proposed DSFATN method in detail. In Section 3, we introduce the experimental data and protocol, provide the performance of the proposed DSFATN and discuss the influence of serval factors. Section 4 concludes the paper with a summary of our method.

2. The Proposed DSFATN Method

2.1. Framework of DSFATN

DSFATN consists of two main steps, as shown in Figure 1.

1. **Saliency-guided DSF extraction**: To achieve discriminative high-level feature representation for RS scenes, we introduce saliency-guided DSF extraction. Instead of using the whole RS scene for feature extraction, saliency-guided DSF extraction produces a novel DSF representation based on saliency-guided RS scene patches using "visual attention" mechanisms. First, we conduct an improved patch-based visual saliency (PBVS) method to detect salient region and sample multi-scales salient patches in an image. Next, the multi-scales salient patches are fed to a pre-trained CNN model to extract the DSF. The saliency-guided DSF extraction ensures the most informative and representative parts are definitely centrally focused in the salient patches. Compared with randomly or densely sampling methods, the saliency-guided sampling is also more targeted and effective. The different scales of the salient patches also help to improve the scale invariance of DSF in the anti-noise transfer network training process.
2. **Anti-noise transfer network based classification**: To suppress the influences of various scales and noises of RS scenes, an anti-noise transfer network is trained as the classifier successively. It introduces an anti-noise layer to tackle with DSFs extracted from RS scene patches in low quality

even with various noises. Except for the anti-noise layer, the anti-noise transfer network only has a fully-connected (FC) layer and a softmax layer, which is a simple CNN architecture and can be trained easily. Different from the traditional CNN model, we optimize a new objective function to train the anti-noise transfer network by imposing an anti-noise constraint, which enforces the training samples before and after adding noises to share the similar features. Meanwhile, for anti-noise transfer network learning, the input scenes contain origin scenes and scenes with various noises, such as: (1) salt and pepper noise; (2) partial occlusions; and (3) their mixed noise. The whole framework works perfectly on three different scale RS scene datasets and even outperforms the state-of-the-art methods.

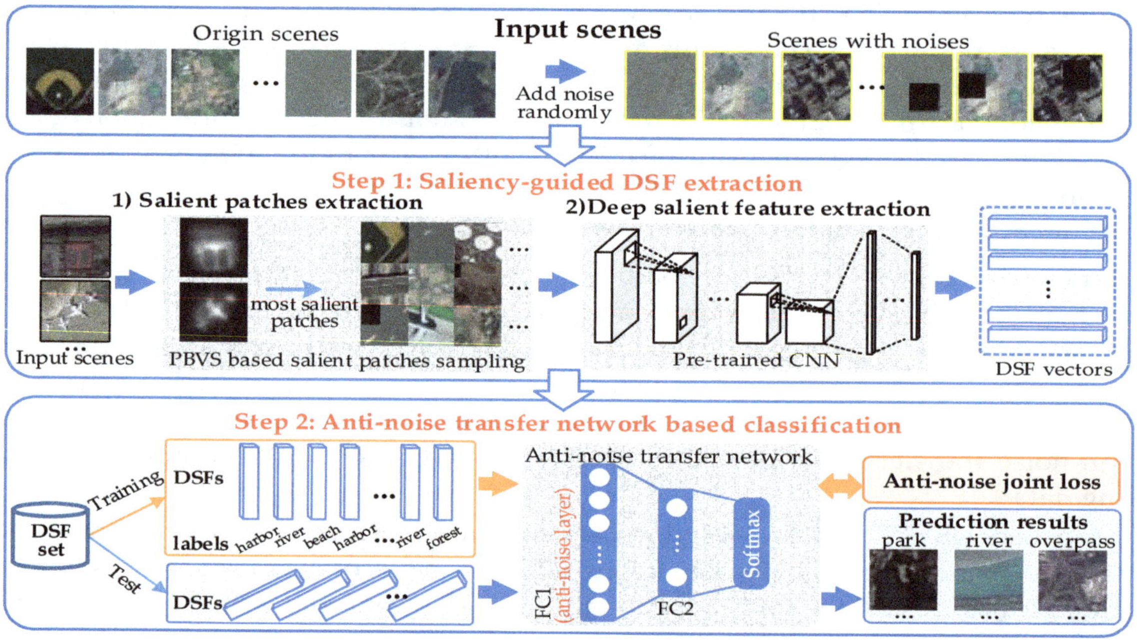

Figure 1. The framework of deep salient feature based anti-noise transfer network (DSFATN) contains two main steps: saliency-guided deep salient feature (DSF) extraction and anti-noise transfer network based classification. The saliency-guided DSF extraction conducts a patch-based visual saliency (PBVS) to guide pre-trained convolutional neural networks (CNNs) for producing the discriminative high-level DSF for remote sensing (RS) scene with different scale and various noises. Then, the anti-noise transfer network is trained to learn and enhance the robust and anti-noise structure information of RS scene by minimizing a joint loss. For anti-noise learning, the input scenes include origin scenes and scenes with various noises (e.g., salt and pepper, occlusions and mixtures).

2.2. *Saliency-Guided DSF Extraction*

The saliency-guided DSF extraction provides the effective and discriminative high-level features from the most relevant scene patches using "visual attention" mechanisms. This extraction is inspired by the human visual system which interprets complex scenes in real time to get most relevant features of the scenes and reduce the complexity of scene analysis [40]. It also can be divided into two steps (Figure 1): (1) salient patch extraction; and (2) DSF extraction. The first step provides the scene patches sampled from the salient regions of input RS scenes. Inspired by graph-based visual saliency (GBVS) [41,42] method, we introduce a patch-based visual saliency (PBVS) algorithm to support the salient patch extractor. The second step is mainly accomplished by a pre-trained CNN, i.e., VGG-19 [19], where the 4096-dimensional activations of the first FC layer are used as the final DSFs.

2.2.1. Salient Patch Extraction

We improved the PBVS method for salient patch extraction. Different from traditional GBVS algorithm which can only detect the salient region from an image, our PBVS can provide multi-scales salient patches of the image. PBVS can be organized into the following procedures: (1) salient region detection; and (2) salient patch extraction. Figure 2 shows the flowchart of the PBVS based salient patch extraction. The details are described in the following section.

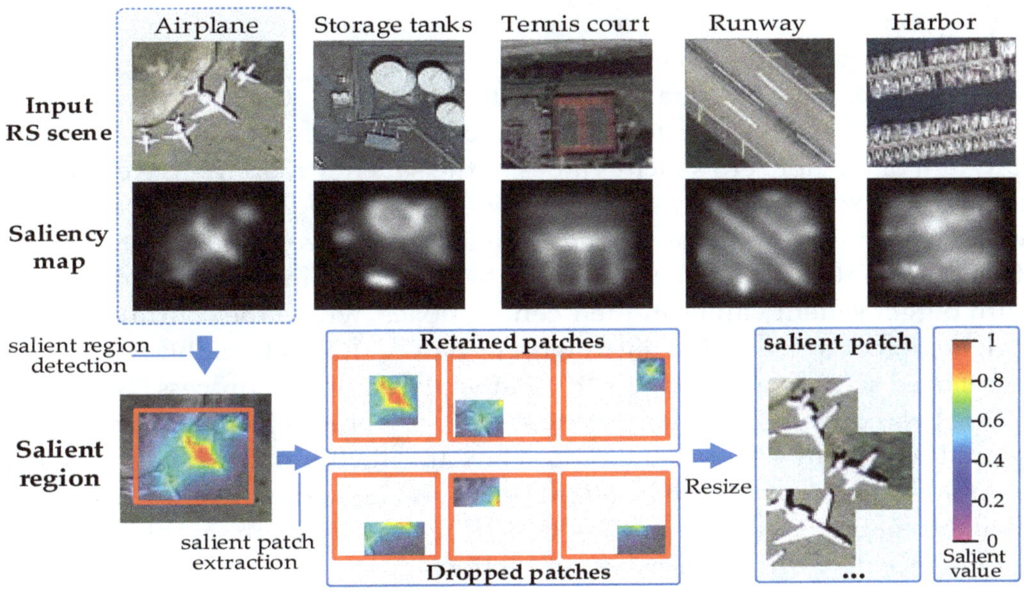

Figure 2. The flowchart of PBVS based salient patch extraction. The brightness in the saliency map indicates the salient level of the corresponding parts in the input RS scenes: brighter in saliency map, more salient in RS scene. The overlay of RS scene and saliency map make the salient level reflected in the input RS scene, the bigger salient value corresponds higher salient level. The red rectangle is the salient region of the scene.

(1) **Salient region detection**. Given a set of n scenes $\mathrm{S} = \{s_1, s_2, \cdots, s_n\}$. For expository simplicity, suppose arbitrarily RS scene $s \in \mathrm{S}$ is a square image of size $n \times n$. At first step, PBVS extracts feature vectors at locations over s to form the feature map of $M^s_{\mathrm{Fea}} : n \times n \to \mathbb{R}$, $M^s_{\mathrm{Fea}}(i,j)(1 \leq i \leq n, 1 \leq \mathrm{j} \leq n)$ is the value of locations (i,j) in M^s_{Fea}. The dissimilarity between $M^s_{\mathrm{Fea}}(i,j)$ and $M^s_{\mathrm{Fea}}(p,q)$ is defined as

$$d((i,j)||(p,q)) := \left| \log \frac{M^s_{\mathrm{Fea}}(i,j)}{M^s_{\mathrm{Fea}}(p,q)} \right| \tag{1}$$

Then, the activation map M^s_{Act} of s needs to be formed. By connecting every node of the feature map M^s_{Fea}, the fully connected directed graph $graph_{\mathrm{Act}}$ is obtained. The directed edge from node (i,j) to node (p,q) of $graph_{\mathrm{Act}}$ is assigned a weight, as shown in Equation (2). σ is a free parameter that is set to approximately 1/10 to 1/5 of the map width because it has been proven the results were not very sensitive to perturbations around these values. Then, the $graph_{\mathrm{Act}}$ is treated as a Markov chain to compute the equilibrium distribution namely get the activation map M^s_{Act}. More details can be found in [41].

$$w_{\mathrm{Act}}((i,j)||(p,q)) := d((i,j)||(p,q)) \cdot \exp\left(-\frac{(i-p)^2 + (j-q)^2}{2\sigma^2} \right) \tag{2}$$

Then, activation map M^s_{Act} will be normalized to get the normalization map M^s_{Nor}. Similar to the process of forming M^s_{Act}, another graph $graph_{\mathrm{Nor}}$ can be constructed based on activation map M^s_{Act}, but the weight assigned to the edges is defined as Equation (3). Again, a Markov chain on $graph_{\mathrm{Nor}}$ is obtained to help obtain the normalization map namely the final saliency map M^s_{Sal}. If multiple activations were generated, these maps will be combined into one saliency map M^s_{Sal} after normalization.

$$w_{\text{Nor}}((i,j)||(p,q)) := M^{I}_{\text{Act}}(p,q) \cdot \exp\left(-\frac{(i-p)^2+(j-q)^2}{2\sigma^2}\right) \tag{3}$$

(2) **Salient patch extraction**. The Salient patch extraction provides multi-scales salient patches from the salient region. As shown in Figure 2, if an object is salient in the image, the corresponding location of its saliency map is high-lighted with bigger salient values. In an image, the salient values of its saliency map range from [0, 1], where 1 indicates the current location in the corresponding RS scene is the most salient, and 0 corresponds to the most non-salient. By finding the minimum bounding rectangle (MBR) [43] of the nonzero salient values in the saliency map M^{s}_{Sal}, we primarily determine a salient region r_s of RS scene s. Then, α patches will be sampled from r_s by an iterative sampling procedure, where α is the threshold of patches' number. The size of the patch can be scaled as the random rate from 30% to 80% of the salient region. The iterative sampling procedure prefers to sample the patches with bigger salient values in their central boxes, where the central box is defined as the central rectangle region of the sampled patch with its half width and height. In this work, we regard [0.8, 1] as the preferred salient value range γ to conduct the sampling process. Algorithm 1 shows the iterative sampling procedure for RS image scene s. At each iteration, a patch is randomly sampled in the salient region. If its salient values in the central box are all within the preferred salient value range γ, this patch should be considered as the salient patch and be kept, otherwise it should be dropped. The iteration will be continued until α patches with different scales are sampled. In our work, we set $\alpha = 9$, and the influence of α will be discussed in Section 3.5.4.

Algorithm 1. The iterative sampling procedure.

Input: Salient region r_s of RS image scene s
Output: $\mathbf{P} = \{p_1, p_2, \cdots, p_\alpha\}$

Initialization:
    set salient patch set $\text{P} = \{\varnothing\}$
    set salient patches' number $n_{\text{patch}} = 0$
Iterations:
while $(n < \alpha)$
    randomly sampled a patch p_{tmp} in $\boldsymbol{r_s}$
if (each salient value $v \in \gamma$ in central box)
    put p_{tmp} to P and note p_{tmp} as $\text{p}_{n_{\text{patch}}+1}$ in P
    $n_{\text{patch}} = n_{\text{patch}} + 1$
Return $\text{P} = \{p_1, p_2, \cdots, p_\alpha\}$

2.2.2. DSF Extraction

After selecting the training patches, we employed the VGG-19 architecture [19] (Figure 3) pre-trained with the ImageNet dataset to derive DSF representation. Additionally, we have compared different pre-trained CNN models in the Experimental Section and showed that VGG-19 performed the best. VGG-19 is one of the very deep CNN models proposed by Simonyan et al. [19]. Hu et al. [37] compared the performance of the activation vectors from different layers of the model, and found the activation vectors from the 1st FC layer are more capable to represent the image feature. Hence, the 4096-dimensional activation vector from the 1st FC layer of VGG-19 is adopted for deep salient feature representation in the case.

The pre-trained VGG-19 model includes 16 convolutional layers, five maxpool layers and three FC layers. When the multi-scales salient patches are fed to VGG-19 and preprocessed to the size of 224×224, the DSF can be extracted on the 1st FC layer. Supposing a set of n scenes $S = \{s_1, s_2, \cdots, s_n\}$, the t-th DSF vector can be described as:

$$\boldsymbol{d}_t = f(h_j(\phi_k(s_t), \alpha))(\, k \in \{0,\ 1,\ 2,\ 3\},\ j \leq \alpha\,), \tag{4}$$

where $\phi_k(s_t)$ returns scene s_t added with the k-th kind of noises (see Figure 1) and $k = 0$ means none noise is added. PBVS function $h_j(\cdot)$ returns the j-th salient patch of the corresponding scene. α is the threshold of sampled salient patches, as described in Section 2.2.1. f defines the deep feature extraction from the 1-st FC layer from VGG-19.

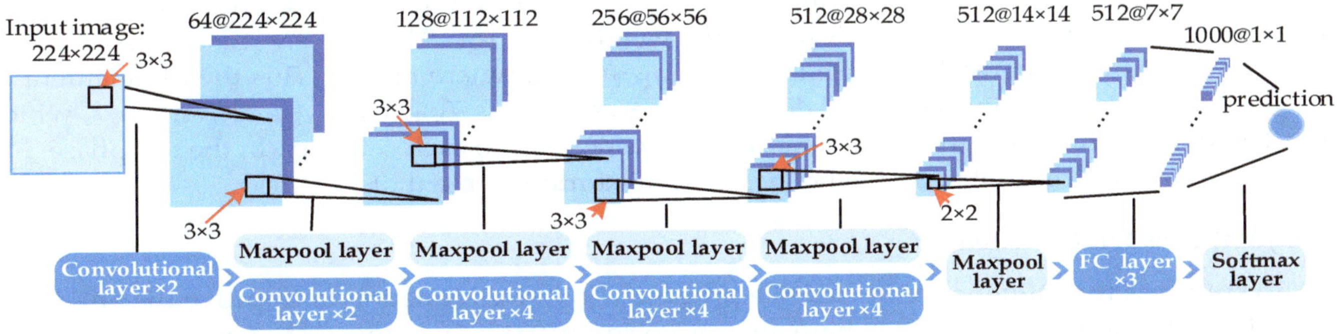

Figure 3. The architecture of the very deep CNN with 19 layers (VGG-19)

2.3. *Anti-Noise Transfer Network Based Classification*

Different from the traditional CNN models, an anti-noise transfer network is introduced to deal with the 4096-dimensional DSF vectors, as shown in Figure 4. The anti-noise transfer network is designed with simple architecture that can be trained easily with limited availability of the training data. It works well for DSFs of different RS scenes even with lower quality due to the anti-noise layer. The anti-noise layer imposes an anti-noise constraint to enforce the training samples before and after adding noises to share the similar output features. Thus, it can produce more robust and discriminative scene features to make the classification easier. Combining the anti-noise constraint to the softmax classification loss function, the anti-noise transfer network is learned by minimizing a joint loss, which is very different from the training of the traditional CNN models. The architecture and loss function of the anti-noise transfer network is described in detail below.

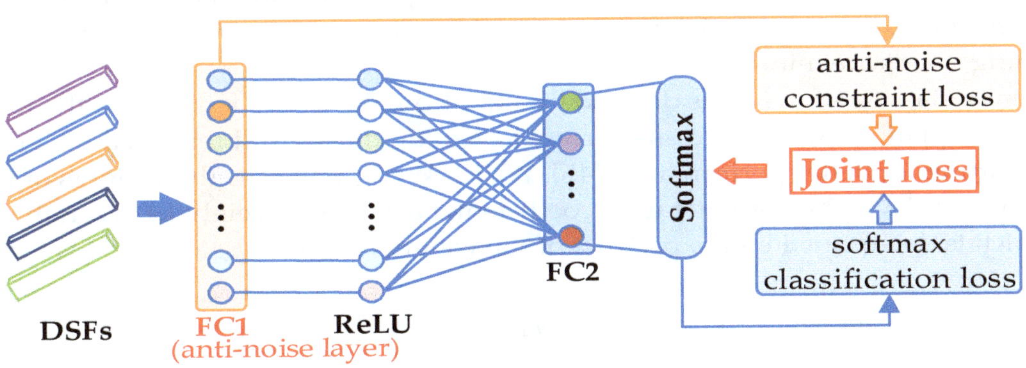

Figure 4. The anti-noise transfer network.

2.3.1. DSF Based Anti-Noise Transfer Network Architecture

As Figure 4 demonstrates, the anti-noise transfer network consists of two FC layers named FC1 and FC2 and a softmax layer, where rectified linear units (ReLU) [44] function is adopted to activate the output of FC1. FC1 and FC2 generate 4096-dimensional and N-dimensional vectors, respectively. N is the category number of the dataset. FC2 transfers the output vector of FC1 into N-dimensional vector thus it can be processed by softmax to produce the final classification results. The 4096-dimensional input DSF vector d_t will be fed to anti-noise layer FC1 and activated by ReLU as:

$$\boldsymbol{o}_{FC1}(\boldsymbol{d}_t) = \sigma(\mathbf{W}_{FC1}\boldsymbol{d}_t + \boldsymbol{b}_{FC1}), \tag{5}$$

where $\sigma(x) = (0, x)$ is the ReLU function, $\boldsymbol{b}_{FC1}$ is the bias. Since the output of FC1 is 4096-dimensional, the weights $\mathbf{W}_{FC1} \in \boldsymbol{R}^{4096\times4096}$. Analogously, $\boldsymbol{o}_{FC1}(\boldsymbol{d}_t)$ will be processed by FC2 and the last softmax layer as follows:

$$\boldsymbol{o}_{FC2}(\boldsymbol{d}_t) = \varphi(\mathbf{W}_{FC2}\boldsymbol{d}_t + \boldsymbol{b}_{FC2}), \tag{6}$$

where $\varphi(x) = e^x / \sum e^x$ is the softmax function, $\boldsymbol{b}_{FC2}$ is the bias. $\boldsymbol{o}_{FC2}(\boldsymbol{d}_t)$ is N-dimensional, N equals the category number of scene categories, thus the weights $\mathbf{W}_{FC2} \in \boldsymbol{R}^{4096\times N}$. $\boldsymbol{o}_{FC2}(\boldsymbol{d}_t)$ is also the final output of the transfer network T_{net}. Setting $y_i = \boldsymbol{o}_{FC2}(\boldsymbol{d}_t, i)$, where $\boldsymbol{o}_{FC2}(\boldsymbol{d}_t, i)$ is the i-th element of $\boldsymbol{o}_{FC2}(\boldsymbol{d}_t)$, the final prediction vector of $\boldsymbol{d}_t$ can be represented as $T_{net}(\boldsymbol{d}_t) = \{y_1, y_2, \ldots, y_N\}$, which indicates the probabilities of the corresponding DSF $\boldsymbol{d}_t$ belongs to each category. In the test phase, i-th category is the prediction label of $\boldsymbol{d}_t$ when y_i is the maximum element of $T_{net}(\boldsymbol{d}_t)$.

2.3.2. Joint Loss Function Learning

To suppress the influence of noises, we propose a joint loss function to improve the anti-noise capability of the transfer network, where an anti-noise constraint is imposed to enforce the training samples before and after adding noise to share similar features. More specifically, for each training RS scene s_t and its corresponding scene with the l-th noise $\phi_l(s_t)(l \in \{1, 2, 3\})$, their DSFs $\boldsymbol{d}_t^0$ and $\boldsymbol{d}_t^l$ are enforced to generate similar output features in the transfer network by the anti-noise layer FC1. To achieve this goal, the novel joint loss function is proposed to learn parameters. Given the training RS scene set S_{tr}=$\{s_t, \phi_l(s_t) | s_t \in S\}$, their DSF set can be obtained as $D_{tr} = \left\{\boldsymbol{d}_t \middle| \boldsymbol{d}_t \in D^0 \cup D^l\right\}$, where D^0 is the DSF set of origin scenes (e.g., s_t) and D^l is the DSF set of corresponding scenes with the l-th ($l \in \{1,2,3\}$) noise (e.g., $\phi_2(s_t)$). Y_{tr} is the true label set of D_{tr}. The joint loss value L can be computed by:

$$L = loss(D_{tr}, Y_{tr}) + dis\left(D^0, D^l\right), \tag{7}$$

where the first term $loss(D_{tr}, Y_{tr})$ is the softmax classification loss function and the second term $dis\left(D^0, D^l\right)$ is the anti-noise constraint. The joint loss L is feedback for backpropagation update. Stochastic Gradient Descent (SGD) approach is employed here to solve the optimization problem, which is a widely used method for neural work training. By minimizing the joint loss value L, both the softmax classification loss and the distance between features extracted from training samples before and after adding noises are minimized.

The softmax classification loss is defined by Equation (8), where $y_{\boldsymbol{d}_t} \in Y_{tr}$ is the true label of $\boldsymbol{d}_t$, the first term of Equation (8) is the cross-entropy loss of $\boldsymbol{d}_t$, the second term is the L2 regularization to avoid over-fitting for better performance [45], $\mathbf{W}_i = \{\mathbf{W}_{FC1}, \mathbf{W}_{FC2}\}$ is the weights of the anti-noise transfer network, and λ is the regularization coefficient, balance the weight between the two terms to be added, which is determined by the product of the weights decay.

$$loss(D_{tr}, Y_{tr}) = -\sum_{\boldsymbol{d}_t \in D_{tr}} y_{\boldsymbol{d}_t} \log(\boldsymbol{o}_{FC2}(\boldsymbol{d}_t)) + \frac{\lambda}{2}\|\mathbf{W}_i\|^2, \tag{8}$$

The anti-noise constraint is proposed to enforce the training DSFs before and after adding noises to share the similar output features extracted by FC1, which introduced as the anti-noise layer in the transfer network. We define the constraint term by measuring the distance between DSFs before and after adding noises as:

$$dis\left(D^0, D^l\right) = \frac{1}{M}\sum_{\boldsymbol{d}_t^0 \in D^0} \|\boldsymbol{o}_{FC2}\left(\boldsymbol{d}_t^0\right) - \boldsymbol{o}_{FC2}\left(\boldsymbol{d}_t^l\right)\|^2, \tag{9}$$

where $d_t^0 \in D^0$ and $d_t^l \in D^l$ are extracted from one RS scene before and after adding the l-th ($l \in \{1, 2, 3\}$) noises. M is the number of D^0, namely half of the joint number of the training samples.

By incorporating Equations (10) and (11) into Equation (9), the joint loss value L is defined as:

$$L = -\sum_{d_t \in D_{tr}} y_{d_t} \log(\boldsymbol{o}_{FC2}(\boldsymbol{d}_t)) + \frac{\lambda}{2}\|\mathbf{W}_i\|^2 + \frac{1}{M}\sum_{d_t^0 \in D^0} \|\boldsymbol{o}_{FC2}\left(\boldsymbol{d}_t^0\right) - \boldsymbol{o}_{FC2}\left(\boldsymbol{d}_t^l\right)\|^2 \quad (10)$$

3. Experiments and Analysis

3.1. Dataset and Experimental Protocol

Three different scale datasets are utilized; their specific categories are shown in Figure 5.

1. UC Merced Land Use Dataset [1] (UCM) is collected from the large aerial orthoimagery of USGS National Map Urban Area Imagery collection. There are 100 images for each of 21 classes. Each image measures 256 × 256 pixels, with a 1-ft spatial resolution.
2. The Google image dataset designed by RS_IDEA Group in Wuhan University (SIRI-WHU) [10] is acquired from Google Earth (Google Inc., Mountain View, CA, USA) and mainly covers urban areas in China. It contains 12 scene categories. Each class consists of 200 images with a size of 200 × 200 pixels and a spatial resolution of 2 m.
3. SAT-6 dataset [46] is extracted from the National Agriculture Imagery Program and consists of a total of 405,000 image patches of size 28 × 28 and covering six classes. We choose 200 images from each class for our experiments.

Figure 5. The categories sequences of the UC Merced Land Use Dataset (UCM), The Google image dataset designed by RS_IDEA Group in Wuhan University (SIRI-WHU) and the SAT-6 dataset: the numbers before the category names will be used to represent the corresponding categories in the experiments.

All experiments are implemented with a 4.0 GHz Intel Core i7-6700K CPU, and two 8 GB GeFore GTX 1080 GPUs. We carried out experiments with five-fold cross-validation protocol on each RS scene dataset. The training set contained 80% of the RS scenes for each class, and the remaining scenes were used for testing. The numbers of training and test images of each RS scene dataset are listed in Table 1. Moreover, in this paper, three kinds of noise-adding strategies are applied: (1) salt and pepper noise with fixed noise density 0.1; (2) partial occlusion at random position that covers 20–30% of the image; and (3) their mixed noise. In the mixed noise strategy, origin scenes, scenes with salt and pepper noise and scenes with partial occlusion account 1/3 of the total scenes, respectively. Although much fewer images are utilized in this work than benchmark datasets such as ImageNet, DSFATN performs in different scales and noise conditions with great efficiency and robustness.

Table 1. Training and test images' numbers of the three RS scene datasets.

	UCM	SIRI-WHU	SAT-6
Training	1680	1920	960
Test	420	480	240
Total	2100	2400	1200

We mainly analyze the performance of DSFATN by the following aspects: (a) the effectiveness and applicability of DSFATN on the three different datasets; (b) the representation ability of DSF; (c) the robustness of the model by the anti-noise layer learning; and (d) the influence factors including patches' number and pre-training models. Comparisons with the state-of-the-arts also demonstrate the superiority of our method.

3.2. Performance on Different Datasets

RS scenes from the three datasets employed in our experiments have a tremendous difference in image resolution and size. The UCM and SIRI-WHU datasets can provide different high-resolution RS scene images with proper image size, while the RS scenes from SAT-6 are really blurry with a quite small size. The diversity of the datasets can test DSFATN to the utmost.

(1) **UCM dataset.** We compared DSFATN with the state-of-the-arts such as the second extended spatial pyramid co-occurrence kernel (SPCK++) [13], pyramid-of-spatial-relations (PSR) [47], saliency-guided unsupervised feature learning (SG+UFL) [33] on the UCM dataset as shown in Table 2. Although most CNN methods can obtain results higher than 90%, especially the fine-tuning on GoogLeNet [34] get the second highest accuracy in the table, it is still 1.15% lower than the result of DSFATN. The CNN (including six convolutional layers and two FC layers) derived from [48] performs badly with the limited amount of data, while DSFATN deals with it well and obtains the highest accuracy, topping the accuracy of random forest (RF) [49] by almost 55%.

Table 2. Accuracy comparison of state-of-the-art methods and DSFATN on UCM dataset.

Rank	Methods	Accuracy (%)
1	RF [49]	44.77
2	CNN(6conv+2fc)	76.40
3	SPCK++ [13]	77.38
4	LDA [15]	81.92 ± 1.12
5	SG + UFL [33]	82.72 ± 1.18
6	PSR [47]	89.10
7	OverFeat [36]	90.91 ± 1.19
8	Caffe-Net [36]	93.42 ± 1.00
9	GoogLeNet [34]	97.10
10	**DSFATN**	**2**

Figure 6 displays the confusion matrix of DSFATN on the UCM dataset. Most scenes can be classified into the right category, especially, the 6th class chaparral whose accuracy equals 1. While the 20th class storage tanks, as the lowest accuracy owner, are mistaken for several other classes, particularly the 5th category buildings and 14th category mobile home park, which is reasonable since some storage tanks are located on the roofs of buildings. The accuracy of storage tanks is higher than 96%, and the whole classification accuracy of DSFATN on the UCM dataset is quite satisfactory.

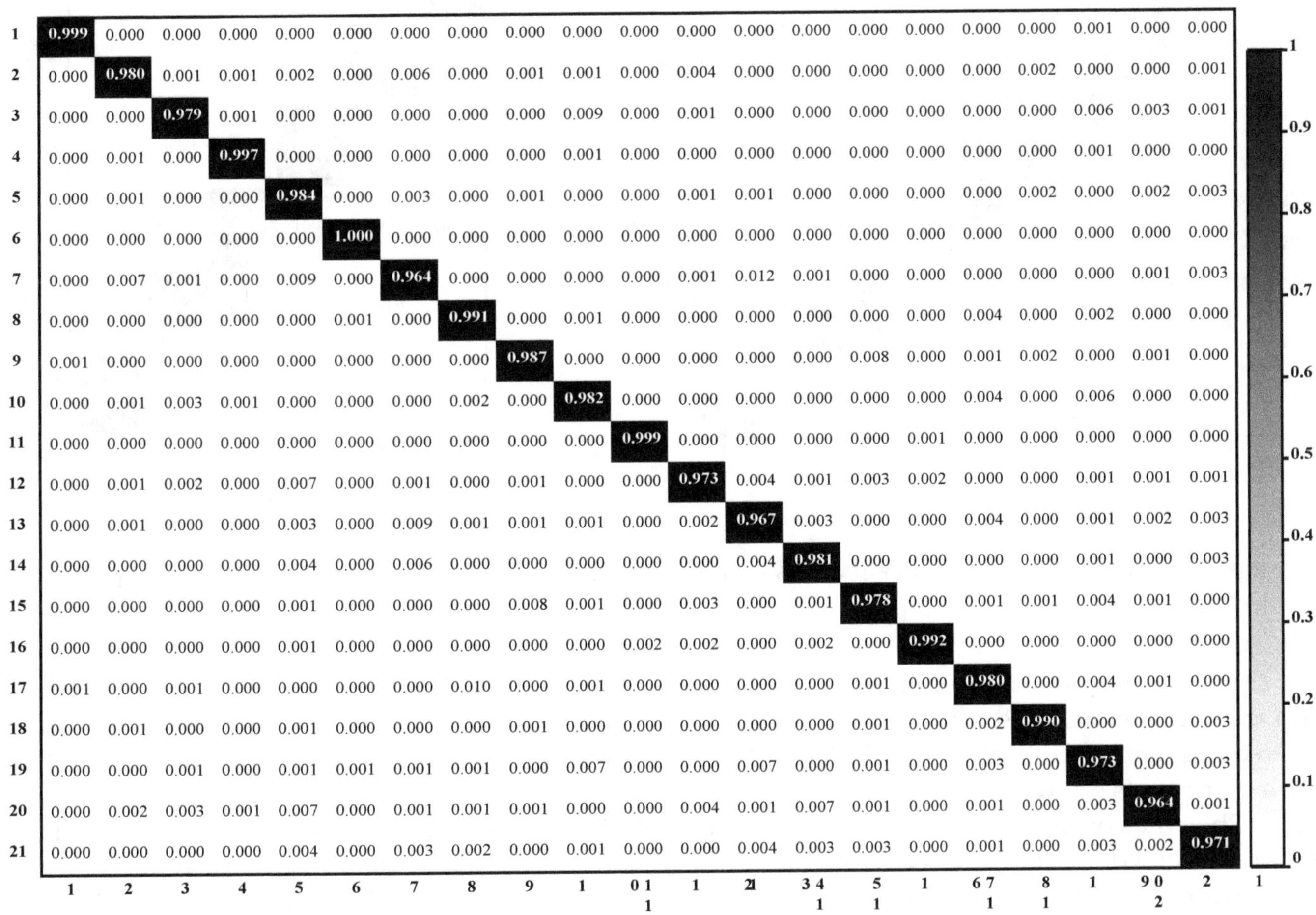

Figure 6. Confusion matrix of DSFATN on the UCM dataset: the horizontal and vertical axes represent the predict labels and true labels respectively. All categories obtain accuracy higher than 0.96.

(2) **SIRI-WHU dataset**. Table 3 shows the results of DSFATN and several compared methods such Spatial Pyramid Matching (SPM) [12] on the SIRI-WHU dataset. Similar to the results on the UCM dataset, DSFATN obtains a high classification result of over 98%. RF and CNN(6conv+2fc) obtain the higher results than the UCM dataset because the SIRI-WHU dataset has fewer categories and more images in each category. It is obvious that DSFATN outperforms the other methods. Moreover, Figure 7a is the confusion matrix of DSFATN on the SIRI-WHU dataset. The accuracy of each category is higher than 97%. The worst misclassification probability is resulted by the 3rd class harbor: 0.9% of the harbor scenes are mistaken for 12th class water. The reason is that these two classes both consist of ship and water. For the same reason, the majority of the confusion occurs among categories that have the same component parts. For example, both the 2nd class commercial and 10th class residential consist of buildings and roads, while both the 9th class pond and the 11th class river are mainly made up of water. All categories achieve accuracies of over 97%.

Table 3. Classification results on the SIRI-WHU dataset.

Methods	RF [49]	LDA [15]	CNN(6conv+2fc)	SPM [12]	DSFATN
Accuracy (%)	0	60.32 ± 1.20	78.20	77.69 ± 1.01	**98.46**

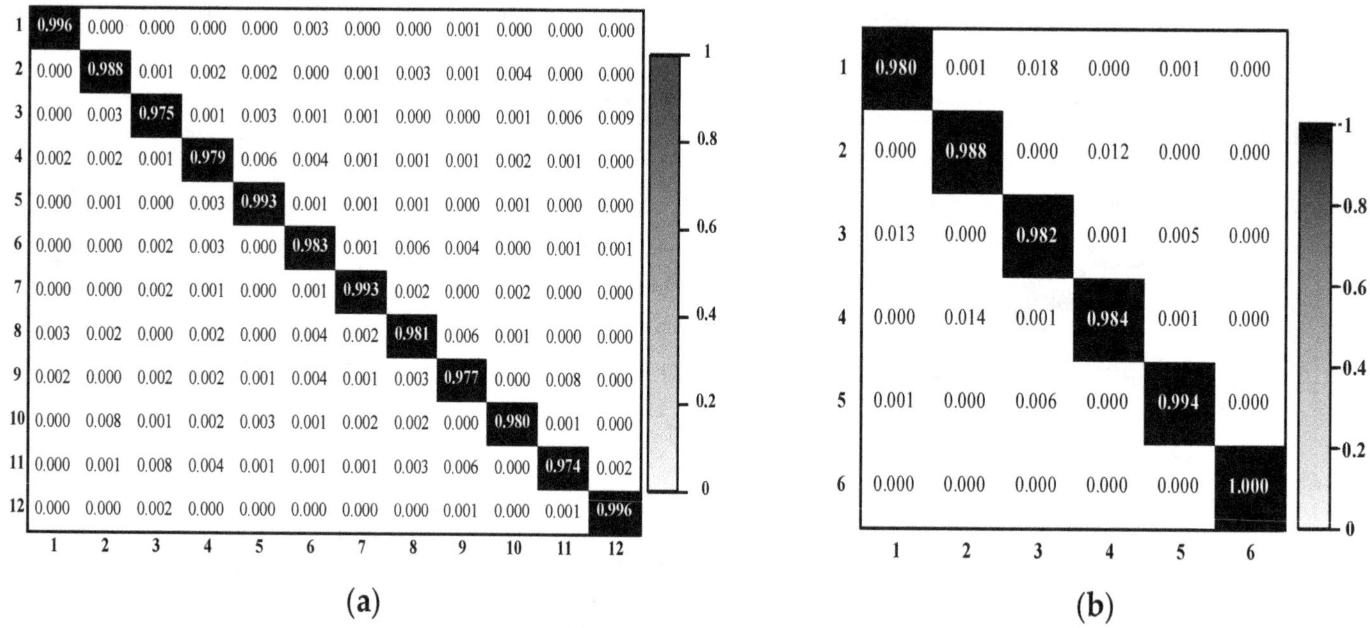

Figure 7. Confusion matrix of DSFATN on: (**a**) the SIRI-WHU dataset; and (**b**) the SAT-6 dataset. The horizontal and vertical axes represent the predict labels and true labels respectively.

(3) **SAT-6 dataset.** Note that the image scenes in the SAT-6 dataset are already salient patches with the dimension of 28 × 28 from RS imageries. Even though the image resolution and scale in the SAT-6 dataset are identically low, DSFATN obtains the average accuracy of 98.80%, as shown in Table 4. The experiments show the impressive representation ability of DSFATN for small scale image scenes. Figure 7b is the confusion matrix of DSFATN on the SAT-6 dataset. Compared with results of the UCM and SIRI-WHU datasets, the misclassification probabilities of the SAT-6 dataset are much higher due to the high similarity between the scenes in smaller scale. The majority of the confusion occurs between the 1st class barren-land and the 3rd class grassland, and the 2nd class buildings and the 4th class roads, because these two pairs of categories have similar color and texture distribution, e.g., the former pair of categories both consist of green grass and brown earth.

Table 4. Classification results on SIRI-WHU dataset.

Methods	RF [49]	CNN(6conv+2fc)	DeepSat [46]	DSFATN
Accuracy (%)	89.29	92.67	93.92	**91.96**

3.3. *Representative Ability Comparison of Different Features*

In this section, to demonstrate the discriminative ability of DSF, we compared the DSF with several different features including histogram of oriented gradients (HOG) [50], scale invariant feature transform (SIFT) [51], and local binary patterns (LBP) [52], as shown in Table 5. For CNN(6conv+2fc), we extract its activations from the 1st FC layer as the representation features. After obtaining these features, we simply implement scene classification by training a linear support vector machine (SVM) classifier with each kind of features.

Table 5. Classification results on three datasets with different features.

Features	UCM		SIRI-WHU		SAT-6	
	Accuracy (%)	Kappa	Accuracy (%)	Kappa	Accuracy (%)	Kappa
Raw image	33.10	0.3361	35.83	0.3469	87.08	0.8116
HOG [50]	52.14	0.4975	44.79	0.3977	57.92	0.4950
SIFT [51]	58.33	0.5625	53.96	0.4977	45.00	0.3400
LBP [52]	31.43	0.2800	46.25	0.4136	77.08	0.7250
CNN(6conv+2fc)	63.10	0.6424	60.42	0.5523	94.58	0.9188
DSF	98.07	0.9801	88.96	0.8766	96.25	0.9437

As Table 5 shows, no matter the accuracy or kappa coefficient, DSF obtained much higher results than other features on the three datasets. The high kappa values indicate the almost perfect coherence of DSF. On the UCM and SIRI-WHU datasets, the classification results of raw images are worse than the classification results of low-level features (e.g., HOG and LBP), and as expected both are worse than the classification results of high-level features extracted from CNNs including the CNN(6conv+2fc) feature and DSF. The raw images of SAT-6 perform much better than those low-level features owing to the characteristics of the SAT-6 dataset. The distinctive colors of the raw image in the SAT-6 dataset help a lot in the raw image classification but does not help in the low-level features extraction. Instead, the small image size and blurry image quality of SAT-6 image scenes make the low-level features extracted from raw images more unrepresentative. However, the features extracted from CNNs are discriminated, both the CNN(6conv+2fc) feature and DSF obtain accuracies over 90%. Especially the CNN(6conv+2fc) feature, although it does not perform well on the more complex UCM and SIRI-WHU datasets, it works quite well on the SAT-6 dataset due to the fewer categories and small RS scene image size of the SAT-6 dataset. The DSF performs more efficient and robust than the others in all three datasets.

Moreover, we embed the high-dimensional features to 2-D space by t-SNE [53], thus to visualize and compare the features extracted from these datasets. As shown in Figure 8, subfigures from top to bottom are the 2-D feature visualization images of HOG, LBP, SIFT, CNN(6conv+2fc) feature and DSF in order, and from left to right are the 2-D feature visualization images of the UCM, SIRI-WHU and SAT-6 datasets respectively. Each color in the images represents a category in the corresponding dataset. Obviously, the 2-D features of HOG, SIFT, and LBP are distributed disordered and only form very few clusters. In contrast, the 2-D features of DSF form clusters separated much clearly. Moreover, the 2-D features of CNN(6conv+2fc) also form more clusters than HOG, SIFT and LBP since the high-level features that contain more abstract semantic information than the low-level features. Notice that CNN(6conv+2fc) feature performs very well in the SAT-6 dataset, obtaining a high result of 94.58%, which very close to the result 96.25% obtained by DSF; this is also reflected in the 2-D feature visualization images, both kinds of features can form the main six clusters. Barren-land class, grassland class and trees class are very close to each other and have some overlap, since in the small scale and resolution SAT-6 dataset, these three categories all consist of soils and vegetation with different vegetation cover rate. The grassland has a middle vegetation plant cover rate; therefore, its features locate between features of barrenland and trees. The buildings class and the roads class have similar situations because the roof of the buildings and the roads are both mainly made up of cement concrete. Particularly, the water class is not similar to the five other categories, and the overlap between the grassland class and the water class in the CNN (6conv+2fc) situation turns out to be unreasonable. While the DSFATN discriminates the difference since the pink area that represents the water features locates far away from the five other categories. Moreover, compared with DSFATN, the CNN(6conv+2fc) feature generates more points that do not locate in the clusters they belong to. In general, DSF learns to be more discriminative.

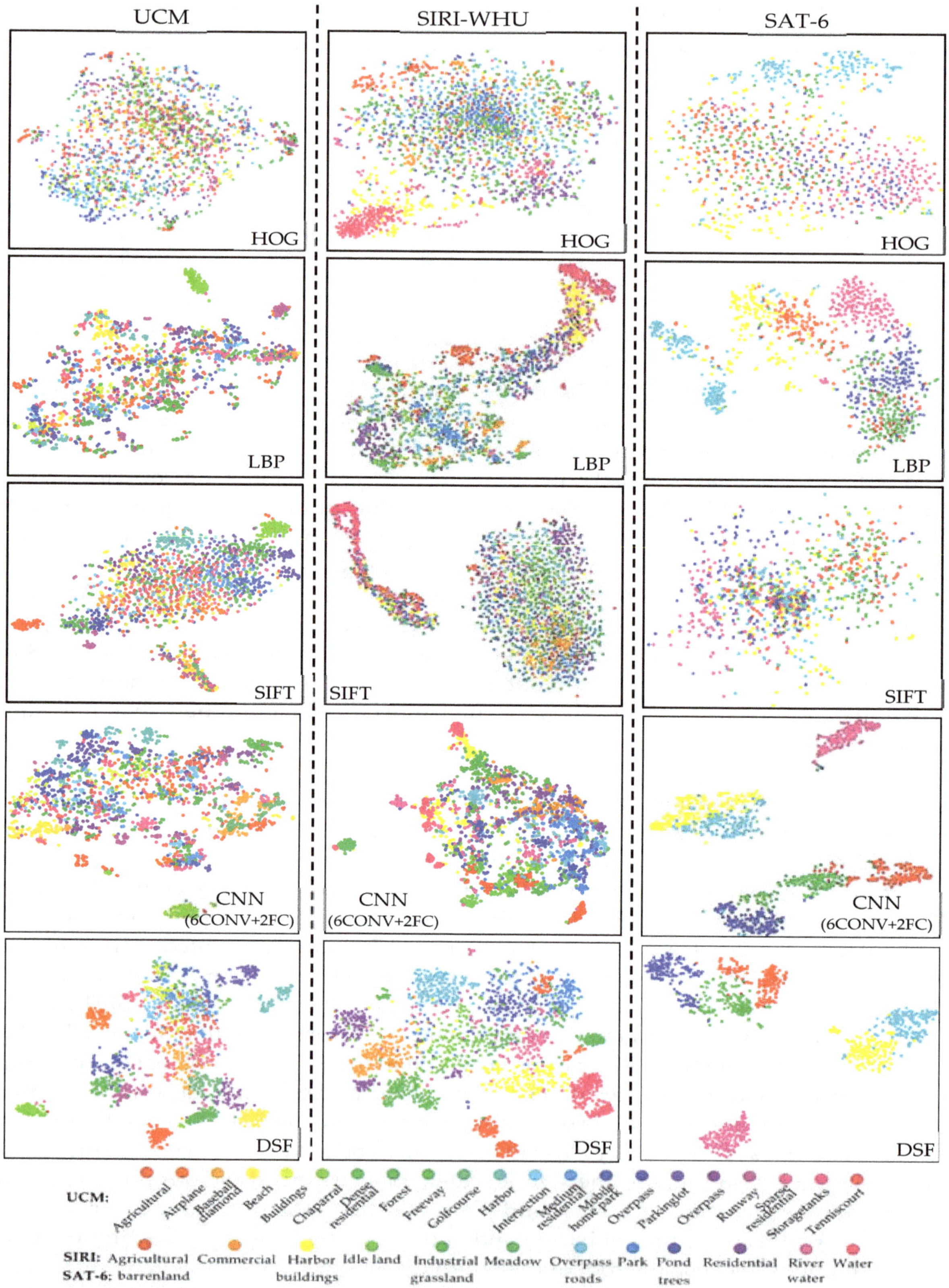

Figure 8. The comparison of different features on the three datasets by per-class two-dimensional feature visualization. From left to right: the UCM dataset, the SIRI-WHU dataset and the SAT-6 dataset. From top to bottom: histogram of oriented gradients (HOG), local binary patterns (LBP), scale invariant feature transform (SIFT), CNN(6conv+2fc) feature, and DSF. It is obvious that DSF (the last row) has more clearly separated clusters.

3.4. Evaluation of Image Distortion

In this section, we validate the robustness of DSFATN for two kinds image distortion conditions: (1) images with noises; and (2) images in different scales. In Section 3.2, we have already proven DSFATN worked well on the SAT-6 dataset which contains RS scenes in small scale and low resolution, thus, in this section, we perform the anti-noise tests on the UCM and SIRI-WHU datasets.

3.4.1. Evaluation of Noises

To validate the anti-noise ability of DSFATN, we compared DSFATN with several different methods under three kinds of noise. Particularly, to prove the indispensability and effectiveness of multi-scales salient patches and anti-noise layer, two variant models derived from DSFATN are introduced. Table 6 lists their difference with the proposed DSFATN. TN-1 refers to DSFATN without multi-scales salient patches sampling and anti-noise layer training. TN-2 refers to DSFATN without multi-scales salient patches sampling but with anti-noise layer training. The absence of anti-noise layer training is simply achieved by learning the joint loss without the anti-noise constraint.

Table 6. Difference between DSFATN and its variant compared models.

Model	Multi-Scales Salient Patch Sampling	Anti-Noise Layer Training
TN-1	×	×
TN-2	×	√
DSFATN	√	√

Table 7 compares the models on the UCM dataset and the SIRI-WHU dataset. Obviously, RF and CNN(6conv+2fc) have a very weak anti-noise property for obtaining accuracies less than 50% with all three kinds of noises, while the classification results of DSFATN are all above 95%. In particular, the result difference between TN-2 and DSFATN almost reaches 10%, which indicates the great importance of saliency patches sampling. Analogously, TN-1 has much worse results compared with TN-2, where the result difference even reaches 47.66% on the SIRI-WHU dataset with salt and pepper noise. The averaged result difference between TN-1 and TN-2 on the UCM and SIRI-WHU datasets with the three kinds of noises reaches 19.92%, which shows the effectiveness of anti-noise layer. As expected, on both the UCM and SIRI-WHU datasets, the results with the three kinds of noises rank in the order: TN-1 < TN-2 < DSFATN. This reflects the important role played by salient patches and the anti-noise layer.

Table 7. Classification results on the UCM dataset and SIRI-WHU dataset with three kinds of noises.

Model	Classification Accuracy (%)					
	UCM			SIRI-WHU		
	Salt and Pepper Noise	Partial Occlusion	Mixed Noise	Salt and Pepper Noise	Partial Occlusion	Mixed Noise
RF		0.0	0.0	0.0	0.0	0.0
CNN(6conv+2fc)	1.60	0.32.00	0.380	0.460	0.5520	0.66240
TN-1	-0.2	-0.04	-0.05	-0.06	-0.07	-0.08
TN-2	-0.4	88.76	88.33	83.83	52.1	84.79
DSFATN	**0.0**	**0.0**	**0.0**	**0.0**	**0.0**	**0.0**

Figure 9 shows the per-class accuracies of TN-1, TN-2, and DSFATN on the UCM and SIRI-WHU datasets with the three kinds of noises. Similar to the trend of the whole results, in most cases, the accuracies are in the following order: DSFATN > TN-2 > TN-1. It is interesting to find that TN-1 and TN-2 perform well in several classes with accuracies over 90%, which even equal or exceed the results of DSFATN, such as the 1st class agricultural, the 11th class harbor, the 16th class overpass, the 18th class runway of the UCM dataset and the 12th class water of the SIRI-WHU dataset. These scenes including duplicate texture information (e.g., ships, water, roads, and roads) make saliency detection confusing. Moreover, irrelevant objects (e.g., a ship in the water class scene) occasionally appearing misled the saliency detection results. Nevertheless, TN-1 and TN-2 behave poor in the other categories, and the corresponding accuracies dropped in different degrees. In general, on both the UCM and SIRI-WHU datasets, TN-2 obtains mediocre performance, better than TN-1 and worse than DSFATN, while TN-1 obtains quite uneven results—most of its results are below 80%. In sharp contrast to this is the stable performance of the DSFATN, which ensures most results are higher than 90%.

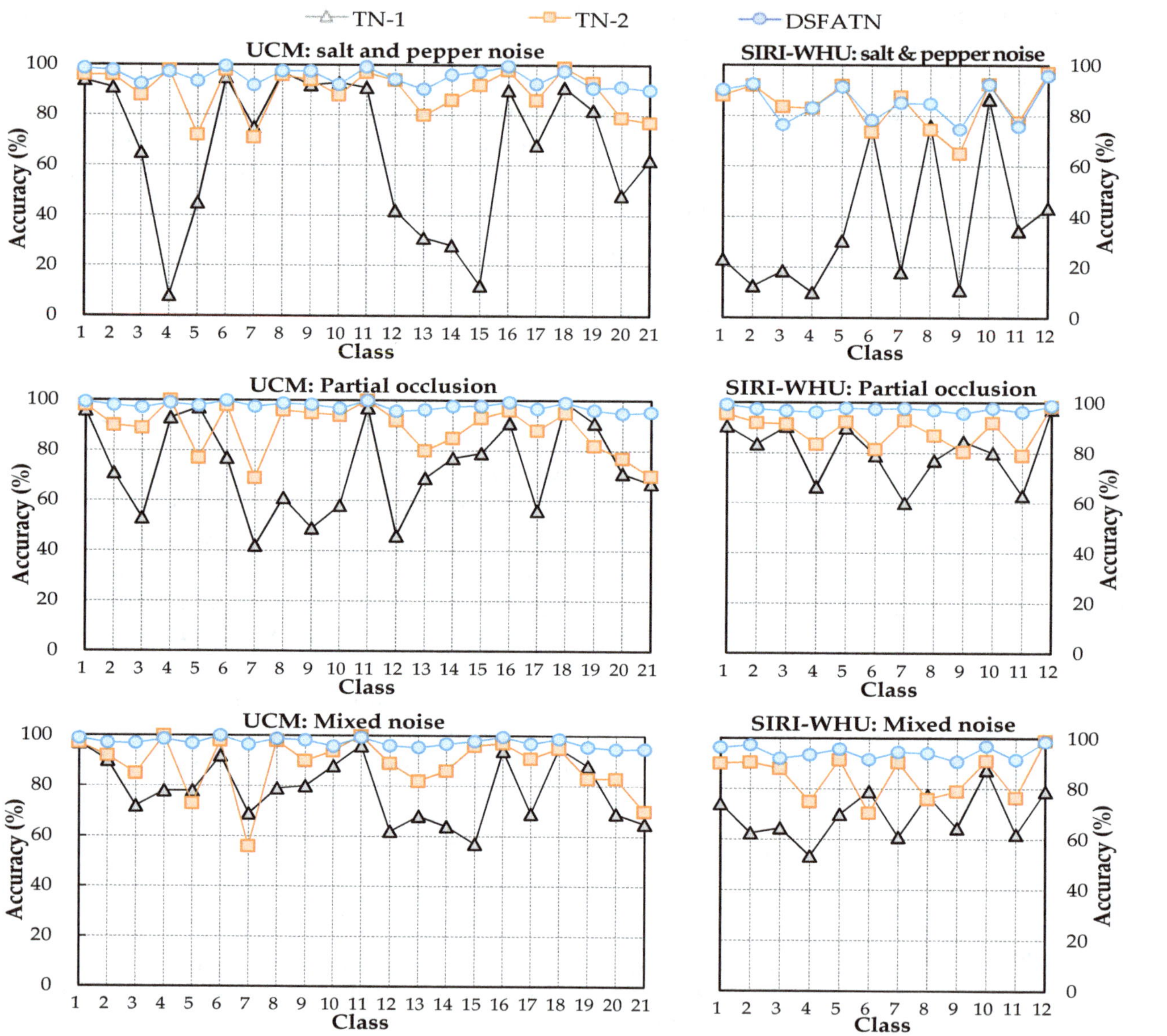

Figure 9. The per-class accuracy comparisons on: the UCM dataset with salt and pepper noise (**top left**); the SIRI-WHU dataset with salt and pepper noise (**top right**); the UCM dataset with partial occlusion (**middle left**); the SIRI-WHU dataset with partial occlusion (**middle right**); the UCM dataset with mixed noise (**bottom left**); and the SIRI-WHU dataset with mixed noise (**bottom right**). In most cases, the accuracies rank in the following order: DSFATN > TN-2 > TN-1.

To further confirm the ability of the anti-noise layer FC1 in the anti-noise transfer network, we compare the FC1 layer's output feature $O_{FC1}(\mathbf{DSF})$ with different features under the three kinds of noises (see Table 8), similar to in Section 3.3. Compared with the accuracies in Table 5, the results of corresponding features in Table 8 have declined in different degrees due to the influence of the noises.

CNN(6con+2fc) feature and DSF show some superiority compared with the low-level features in this anti-noise experiments, obtaining accuracies even higher than the results obtained by the low-level features extracted from origin RS scene images without any noise (see Table 5). However, it is not robust enough to represent the images with noises. The last row of Table 8 shows the accuracies obtained by features extracted from the FC1 layer; most of them are higher than 0.90, and all the results are significantly enhanced compared to the results classified by DSF. The great difference between DSF and $O_{FC1}(\mathbf{DSF})$ indicates introducing the FC1 layer to the anti-noise transfer network is indeed very important.

Table 8. Anti-noise analysis on the UCM and SIRI-WHU datasets with different features.

Features	Classification Accuracy (%)					
	UCM			SIRI-WHU		
	Salt and Pepper Noise	Partial Occlusion	Mixed Noise	Salt and Pepper Noise	Partial Occlusion	Mixed Noise
HOG [50]	41.19	34.76	25.47	40.21	40.63	31.46
SIFT [51]	62.62	41.43	44.05	51.25	46.46	46.46
LBP [52]	25.00	18.10	10.48	37.92	39.38	25.42
CNN(6conv+2fc)	56.19	38.57	47.62	52.92	65.63	53.96
DSF	89.76	83.10	82.62	79.58	87.29	83.54
$O_{FC1}(\mathbf{DSF})$	**96.61**	**98.04**	**97.70**	**86.39**	**97.52**	**94.56**

3.4.2. Evaluation of Multiple Scales

To evaluate the impact of image scale, we resized the RS scene images from the UCM and SIRI-WHU datasets to five different scales, i.e., a quarter of original image size (height and width dimensions), half of original image size, three quarters of original image size, original image size and one and a quarter size.

We compared DSFATN with CNN(6conv+2fc) and the TN-2 model. As the results in Tables 9 and 10 show, DSFTAN performs the best on the UCM and SIRI-WHU datasets at all five scales. Almost all the accuracies are over 98%, and the results of DSFATN are quite stable for obtaining the lowest STD value. The results of CNN(6conv+2fc) are very unstable with the image scale variances. Particularly, TN-2, which does not conduct multi-scales patches sampling, compared with DSTATN, also obtained high accuracies around 90% on the two datasets. However, the STD values of TN-2 are much higher. Moreover, the UCM, SIRI-WHU and SAT-6 datasets also have different scales and resolutions, especially the SAT-6 dataset. Our method demonstrated robustness across the three datasets.

Table 9. Classification results on the UCM dataset with five kinds of scales.

Models	25%	50%	75%	100%	125%	STD
CNN(6conv+2fc)	76.60	80.00	77.80	76.40	80.00	1.58
TN-2	92.20	91.40	91.60	92.60	91.20	0.52
DSFATN	97.87	98.53	98.46	98.25	98.22	0.23

Table 10. Classification results on the SIRI-WHU dataset with five kinds of scales.

Models	25%	50%	75%	100%	125%	STD
CNN(6conv+2fc)	77.00	77.40	77.00	78.20	75.40	0.91
TN-2	87.60	89.20	89.60	90.00	90.60	1.01
DSFATN	98.30	98.39	98.73	98.46	98.92	0.23

3.5. The Analysis of Influence Factors

In this section, we analyze several influence factors in DSFATN: (a) the threshold of salient patches' number α; (b) the regularization coefficient λ; (c) the pre-trained CNN models; and (d) the noise level. For simplicity and equity, all comparison experiments were conducted on the UCM dataset.

3.5.1. Influence of Salient Patches' Number α

Figure 10 shows the influence of salient patches' number α. Time consumption refers to the time for obtaining all the DSF of the RS image scenes utilized in the corresponding experiments. $\alpha = 0$ means salient regions are not detected thus the DSF are directly extracted from the origin image scenes. As α increases from 0 to 9, the time consumption increases slowly while the accuracy rises sharply. When $\alpha \in [9, 36]$, the accuracy keeps a flat level of growth while the time consumption steepens. Only when $\alpha = 9$, a high classification accuracy can be gained without much time consumption. Hence, $\alpha = 9$ is selected in our method.

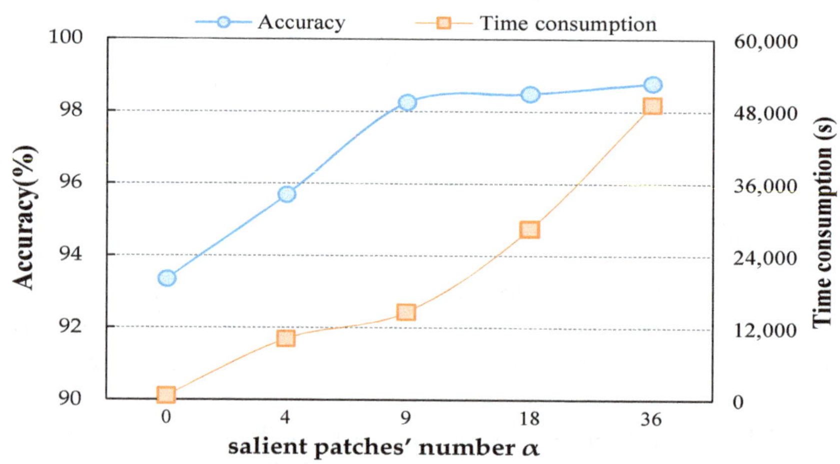

Figure 10. The influence of salient patches' number α in DSFATN on UCM dataset.

3.5.2. Influence of the Regularization Coefficient λ

Figure 11 shows the classification results of DSFATN in different regularization coefficient λ. When λ is very small (i.e., $\lambda \in \left[1 \times 10^{-7}, 1 \times 10^{-1}\right]$), DSFATN performs quite good, and the accuracy levels out at around 98%. When λ is assigned bigger values (i.e., $\lambda > 1 \times 10^{-1}$), the accuracy declines fast. When $\lambda = 1 \times 10^{-4}$, DSFATN achieves the highest result.

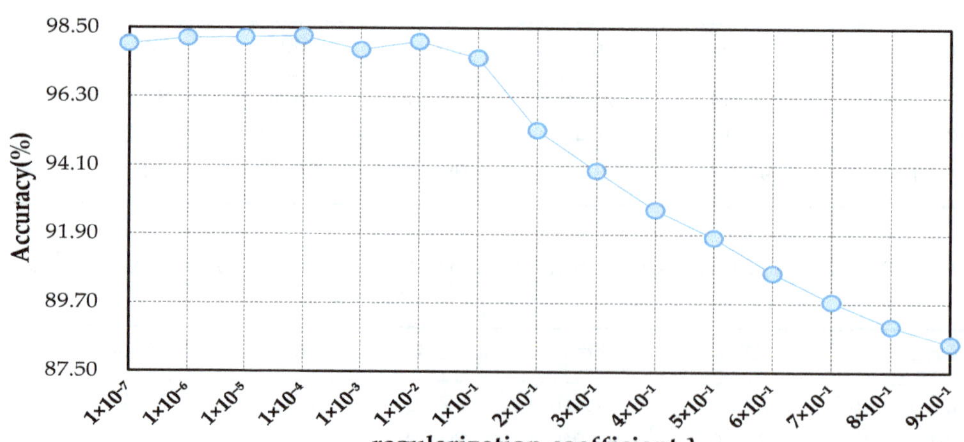

Figure 11. The influence of the regularization coefficient λ.

3.5.3. Influence of Pre-Trained CNNs

We changed the pre-trained CNN in DSFATN from VGG-19 to several other kinds of CNNs, while keeping the rest of the structure of DSFATN unchanged. Table 11 presents the classification results of different pre-trained CNNs. Note that, for the pre-trained CNN models which contain three FC layers (Rows 1–7), we extracted the features from the 1st FC layer as the feature presentation. For the other per-trained CNNs (Rows 8–15), we regarded the output of the layer that generate one-dimensional vectors (e.g., logits layer in InceptionV3) as the representation. All extracted representations have the same anti-noise transfer network architecture but are trained separately. As shown in Table 11, compared with VGG-19, most pre-trained CNNs can achieve comparable results over 96% (e.g., Rows 1–6 and 11–12). Although the inception models perform well too, they are not so competitive with other models for deep feature extraction, since they are not deep enough compared with Resnetv1_50 and Resnetv1_101. The fully connected layers, which appear in each traditional CNNs (e.g., Rows 1–7), play a great role for deep feature extraction. Nevertheless, the results of inceptions are still higher than 90%. One can see that our DSFATN with VGG-19 outperforms the others.

Table 11. Result comparison with different pre-trained CNNs.

No.	Pre-Trained CNNs	Classification Accuracy (%)
1	Alexnet [48]	96.85
2	Caffenet [54]	97.35
3	VGG -F [55]	97.54
4	VGG -M [55]	97.57
5	VGG -S [55]	97.12
6	VGG-16 [19]	97.91
7	VGG-19 [19]	1.4
8	Inceptionv1 [35]	91.25
9	Inceptionv2 [56]	90.54
10	Inceptionv3 [57]	91.82
11	Resnetv1_50 [58]	97.89
12	Resnetv1_101 [58]	97.94

3.5.4. Influence of Noise Levels

We investigate the robustness sensitivity of DSFATN at five different levels of noises. Table 12 shows the parameters of the salt and pepper noise and partial occlusion at these five levels of noise conditions; the mixed noise is still the mixture of the former two kinds noise and original image scenes with the same proportion. Note that Level 2 noise condition has been adopted as the setting in the preceding experiment part (see Section 3). Figure 12 demonstrates these five noise levels of an example tennis court scene. Obviously, when the noise level becomes higher, the scenes with salt and pepper noise are blurrier with more noise pixels, and the scenes with partial occlusion are covered with larger black region. At Levels 3–5, the tennis court cannot be seen in the image scene.

Table 12. The parameters of the salt and pepper noise and partial occlusion at five levels.

Level	Salt and Pepper Noise Density	Partial Occlusion Covering Scale
1	0.05	10–20%
2	0.1	20–30%
3	0.15	30–40%
4	0.2	40–50%
5	0.25	50–60%

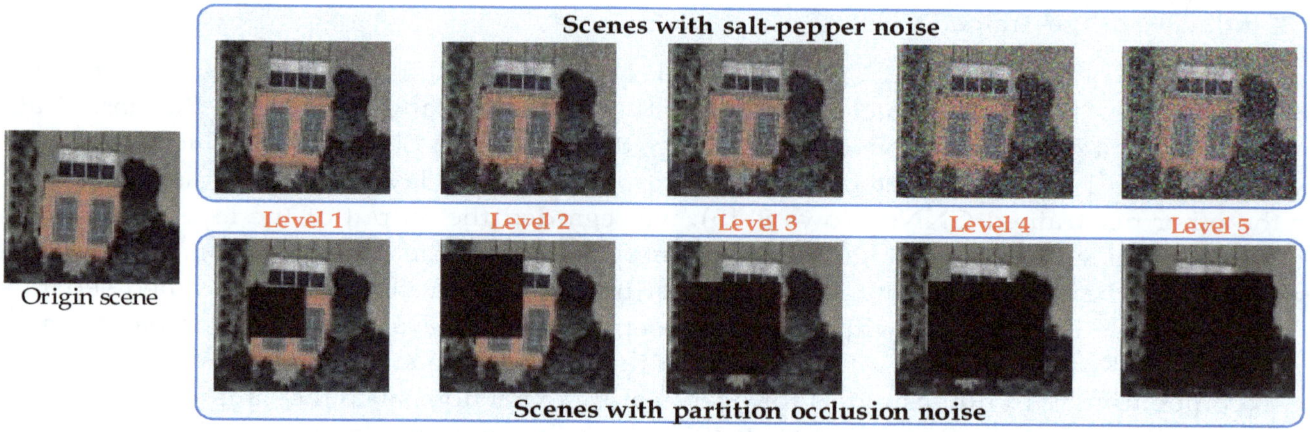

Figure 12. The example scenes of an example tennis court scene at five levels of noise conditions.

The average accuracies of DSFATN at these five noise levels are shown in Figure 13. As expected, the higher the noise level is, the lower the classification accuracy is. In salt and pepper noise condition, the salt and pepper noise with higher noise level brings more noise pixels to the scene, which makes the performance of saliency detection degenerate. In partial occlusion condition, the higher partial occlusion level leads to the semantic information loss. When the noise level is higher than 2, the results of salt and pepper noise and partial collusion conditions declines more sharply than the result of mixed noise condition. The origin scenes in the mixed noise condition, which supplement the information loss caused by the salt and pepper noise and partial occlusion to some extent. In general, although the accuracies have a declining trend, all results are higher than 80%, even for partial occlusion covering almost half scale of the scenes. The results are also higher than the accuracies obtained by some traditional methods in the origin scenes without any noises (see Table 2).

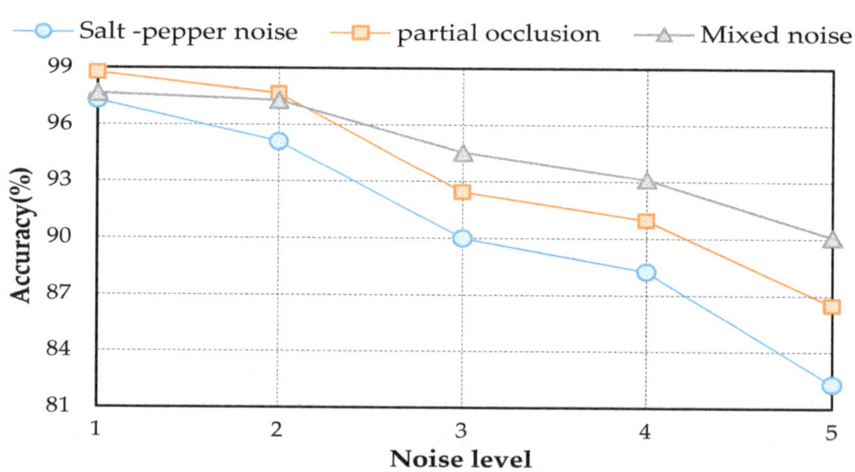

Figure 13. The results of DSFATN at five levels of noise conditions: the classification accuracies decrease when the noise level increases.

4. Conclusions

This paper proposes a deep salient feature based anti-noise transfer network (DSFATN) method for RS scene classification with different scales and various noises. In DSFATN, the saliency-guided DSF extraction extracts the discriminative high-level DSF from the most relevant, informative and representative patches of the RS scene sampled by the Patch-Based Visual Saliency (PBVS) method. The VGG-19 is selected as the pre-trained CNN to extract DSF among various candidate CNNs for its better performance. DSF achieves discriminative high-level feature representation learned from pre-trained VGG-19 for the RS scenes. Meanwhile, an anti-noise transfer network is introduced to learn and enhance the robust and anti-noise structure information of RS scene by directly propagating the label information to fully-connected layers. By minimizing the joint loss concerning anti-noise constraint and softmax classification loss simultaneously, the anti-noise transfer network can be trained easily with limited amount of data and without accuracy loss. DSFATN performs excellent with RS scenes in different quality, even with noise.

The results on three different scale datasets with limited data are encouraging: the classification results are all above 98%, which outperforms the results of state-of-the-art methods. DSFATN also obtains satisfactory results under various noises. For example, the results on the widespread UCM with noises are higher than 95%, which is even better than the best results of some state-of-the-art methods on UCM without noise. The remarkable results indicate the effectiveness and wide applicability of DSFATN and prove the robustness of DSFATN.

However, the strong anti-noise property of DSFATN is dependent on different datasets; for example, under salt and pepper noise, the accuracy of DSFATN reaches 95.12% on the UCM dataset while it dropped to 84.98% on the SIRI-WHU dataset. In the future, we will conduct an end-to-end multi-scale and multi-channel network to jointly extract more adaptive representation for RS scene with limited availability of training data for complex scene understanding.

Acknowledgments: This study was financially supported by the National Natural Science Foundation of China (No. 41671400, No. 41701446 and No. 61602429); National key R & D program of China (No. 2017YFC0602204); Hubei Natural Science Foundation of China (2015CFA012).

Author Contributions: Yuanyuan Liu and Zhong Xie conceived and designed the experiments. Xi Gong and Zhuo Zheng performed the experiments. Yuanyuan Liu and Xi Gong contributed to the analysis. The manuscript was written by Xi Gong with contributions from Yuanyuan Liu, Xuguo Shi.

References

1. Yang, Y.; Newsam, S. Bag-of-visual-words and spatial extensions for land-use classification. In Proceedings of the 18th SIGSPATIAL International Conference on Advances in Geographic Information Systems, San Jose, CA, USA, 3–5 November 2010; pp. 270–279.
2. Cheriyadat, A.M. Unsupervised feature learning for aerial scene classification. *IEEE Trans. Geosci. Remote Sens.* **2014**, *52*, 439–451. [CrossRef]
3. Yang, W.; Yin, X.; Xia, G.S. Learning high-level features for satellite image classification with limited labeled samples. *IEEE Trans. Geosci. Remote Sens.* **2015**, *53*, 4472–4482. [CrossRef]
4. Shao, W.; Yang, W.; Xia, G.S. Extreme value theory-based calibration for the fusion of multiple features in high-resolution satellite scene classification. *Int. J. Remote Sens.* **2013**, *34*, 8588–8602. [CrossRef]
5. Wang, Q.; Meng, Z.; Li, X. Locality Adaptive Discriminant Analysis for Spectral–Spatial Classification of Hyperspectral Images. *IEEE Geosci. Remote Sens. Lett.* **2017**, *14*, 2077–2081. [CrossRef]
6. Bosch, A.; Muñoz, X.; Martí, R. Which is the best way to organize/classify images by content? *Image Vis. Comput.* **2007**, *25*, 778–791. [CrossRef]
7. Zhong, Y.; Zhu, Q.; Zhang, L. Scene classification based on the multifeature fusion probabilistic topic model for high spatial resolution remote sensing imagery. *IEEE Trans. Geosci. Remote Sens.* **2015**, *53*, 6207–6222. [CrossRef]
8. Zhou, L.; Zhou, Z.; Hu, D. Scene classification using a multi-resolution bag-of-features model. *Pattern Recognit.* **2013**, *46*, 424–433. [CrossRef]
9. Zhao, B.; Zhong, Y.; Zhang, L.; Huang, B. The fisher kernel coding framework for high spatial resolution scene classification. *Remote Sens.* **2016**, *8*, 157. [CrossRef]
10. Zhao, B.; Zhong, Y.; Xia, G.-S.; Zhang, L. Dirichlet-derived multiple topic scene classification model fusing heterogeneous features for high spatial resolution remote sensing imagery. *IEEE Trans. Geosci. Remote Sens.* **2016**, *54*, 2108–2123. [CrossRef]
11. Zhong, Y.; Fei, F.; Zhang, L. Large patch convolutional neural networks for the scene classification of high spatial resolution imagery. *J. Appl. Remote Sens.* **2016**, *10*, 025006. [CrossRef]

12. Lazebnik, S.; Schmid, C.; Ponce, J. Beyond bags of features: Spatial pyramid matching for recognizing natural scene categories. In Proceedings of the IEEE Conference on Computer Vision and Pattern Recognition, New York, NY, USA, 17–22 June 2006; Volume 2, pp. 2169–2178.
13. Yang, Y.; Newsam, S. Spatial pyramid co-occurrence for image classification. In Proceedings of the IEEE International Conference on Computer Vision, Barcelona, Spain, 6–13 November 2011; pp. 1465–1472.
14. Blei, D.M.; Ng, A.Y.; Jordan, M.I. Latent dirichlet allocation. *J. Mach. Learn. Res.* **2003**, *3*, 993–1022.
15. Lienou, M.; Maitre, H.; Datcu, M. Semantic annotation of satellite images using latent dirichlet allocation. *IEEE Geosci. Remote Sens. Lett.* **2010**, *7*, 28–32. [CrossRef]
16. Vaduva, C.; Gavat, I.; Datcu, M. Latent dirichlet allocation for spatial analysis of satellite images. *IEEE Trans. Geosci. Remote Sens.* **2013**, *51*, 2770–2786. [CrossRef]
17. Bosch, A.; Zisserman, A.; Muñoz, X. Scene classification using a hybrid generative/discriminative approach. *IEEE Trans. Pattern Anal. Mach. Intell.* **2008**, *30*, 712–727. [CrossRef] [PubMed]
18. Cheng, G.; Guo, L.; Zhao, T.; Han, J.; Li, H.; Fang, J. Automatic landslide detection from remote-sensing imagery using a scene classification method based on BoVW and pLSA. *Int. J. Remote Sens.* **2013**, *34*, 45–59. [CrossRef]
19. Simonyan, K.; Zisserman, A. Very deep convolutional networks for large-scale image recognition. In Proceedings of the International Conference on Learning Representations, San Diego, CA, USA, 7–9 May 2015.
20. Zou, Z.; Shi, Z. Ship detection in spaceborne optical image with SVD networks. *IEEE Trans. Geosci. Remote Sens.* **2016**, *54*, 5832–5845. [CrossRef]
21. Gao, J.; Wang, Q.; Yuan, Y. Embedding structured contour and location prior in siamesed fully convolutional networks for road detection. In Proceedings of the IEEE International Conference on Robotics and Automation, Singapore, 29 May–3 June 2017; pp. 219–224.
22. Li, C.; Wand, M. Combining markov random fields and convolutional neural networks for image synthesis. *arXiv*, **2016**.
23. Zhang, L.; Zhang, L.; Du, B. Deep Learning for Remote Sensing Data: A Technical Tutorial on the State of the Art. *IEEE Geosci. Remote Sens. Mag.* **2016**, *4*, 22–40. [CrossRef]
24. Zhu, X.X.; Tuia, D.; Mou, L.; Xia, G.S.; Zhang, L.; Xu, F. Deep Learning in Remote Sensing: A Comprehensive Review and List of Resources. *IEEE Geosci. Remote Sens. Mag.* **2017**, *5*, 8–36. [CrossRef]
25. Lu, X.; Zheng, X.; Yuan, Y. Remote sensing scene classification by unsupervised representation learning. *IEEE Trans. Geosci. Remote Sens.* **2017**, *55*, 5148–5157. [CrossRef]
26. Chen, Y.; Zhao, X.; Jia, X. Spectral–spatial classification of hyperspectral data based on deep belief network. *IEEE J. Sel. Top. Appl. Earth Obs. Remote Sens.* **2015**, *8*, 2381–2392. [CrossRef]
27. Deng, J.; Dong, W.; Socher, R.; Li, L.J.; Li, K.; Fei-Fei, L. Imagenet: A large-scale hierarchical image database. In Proceedings of the IEEE Conference on Computer Vision and Pattern Recognition, Miami, FL, USA, 20–25 June 2009; pp. 248–255.
28. Li, H.; Fu, K.; Yan, M.; Sun, X.; Sun, H.; Diao, W. Vehicle detection in remote sensing images using denoising-based convolutional neural networks. *Remote Sens. Lett.* **2017**, *8*, 262–270. [CrossRef]
29. Yang, Y.; Zhuang, Y.; Bi, F.; Shi, H.; Xie, Y. M-FCN: Effective fully convolutional network-based airplane detection framework. *IEEE Geosci. Remote Sens. Lett.* **2017**, *14*, 1293–1297. [CrossRef]
30. Chen, X.; Xiang, S.; Liu, C.L.; Pan, C.H. Vehicle detection in satellite images by hybrid deep convolutional neural networks. *IEEE Geosci. Remote Sens. Lett.* **2017**, *11*, 1797–1801. [CrossRef]
31. Simard, P.Y.; Steinkraus, D.; Platt, J.C. Best practices for convolutional neural networks applied to visual document analysis. In Proceedings of the Seventh International Conference on Document Analysis and Recognition, Edinburgh, UK, 3–6 August 2003; pp. 958–962.
32. Dieleman, S.; Willett, K.W.; Dambre, J. Rotation-invariant convolutional neural networks for galaxy morphology prediction. *Mon. Not. R. Astron. Soc.* **2015**, *450*, 1441–1459. [CrossRef]
33. Zhang, F.; Du, B.; Zhang, L. Saliency-guided unsupervised feature learning for scene classification. *IEEE Trans. Geosci. Remote Sens.* **2015**, *53*, 2175–2184. [CrossRef]
34. Castelluccio, M.; Poggi, G.; Sansone, C.; Verdoliva, L. Land use classification in remote sensing images by convolutional neural networks. *arXiv*, 2015.

35. Szegedy, C.; Liu, W.; Jia, Y.; Sermanet, P.; Reed, S.; Anguelov, D. Going deeper with convolutions. In Proceedings of the IEEE Conference on Computer Vision and Pattern Recognition, Boston, MA, USA, 11–12 June 2015; pp. 1–9.
36. Penatti, O.A.; Nogueira, K.; dos Santos, J.A. Do deep features generalize from everyday objects to remote sensing and aerial scenes domains? In Proceedings of the IEEE Conference on Computer Vision and Pattern Recognition Workshops, Boston, MA, USA, 12 June 2015; pp. 44–51.
37. Hu, F.; Xia, G.S.; Hu, J.; Zhang, L. Transferring deep convolutional neural networks for the scene classification of high-resolution remote sensing imagery. *Remote Sens.* **2015**, *7*, 14680–14707. [CrossRef]
38. Gong, Y.; Wang, L.; Guo, R.; Lazebnik, S. Multi-scale orderless pooling of deep convolutional activation features. In *The European Conference on Computer Vision (ECCV)*; Springer: Cham, Swizerland, 2014; pp. 392–407.
39. Lee, H.; Grosse, R.; Ranganath, R.; Ng, A.Y. Convolutional deep belief networks for scalable unsupervised learning of hierarchical representations. In Proceedings of the 26th annual international conference on machine learning, Montreal, QC, Canada, 14–18 June 2009; pp. 609–616.
40. Itti, L.; Koch, C.; Niebur, E. A model of saliency-based visual attention for rapid scene analysis. *IEEE Trans. Pattern Anal. Mach. Intell.* **1998**, *20*, 1254–1259. [CrossRef]
41. Harel, J.; Koch, C.; Perona, P. Graph-based visual saliency. In Proceedings of the Advances in Neural Information Processing Systems, Vancouver, BC, Canada, 3 December 2007; pp. 545–552.
42. Harel, J. A Saliency Implementation in MATLAB. Available online: http://www.vision.caltech.edu/~harel/share/gbvs.php (accessed on 10 January 2018).
43. Caldwell, D.R. Unlocking the mysteries of the bounding box. *Coord. Online J. Map Geogr. Round Tab. Am. Libr. Assoc.* **2005**. Available online: http://www.stonybrook.edu/libmap/coordinates/seriesa/no2/a2.pdf (accessed on 10 January 2018).
44. Glorot, X.; Bordes, A.; Bengio, Y. Deep sparse rectifier neural networks. In Proceedings of the Fourteenth International Conference on Artificial Intelligence and Statistics, Lauderdale, FL, USA, 11–13 April 2011; pp. 315–323.
45. Peng, X.; Lu, C.; Yi, Z.; Tang, H. Connections between Nuclear Norm and Frobenius-Norm-Based Representations. *IEEE Trans. Neural Netw. Learn. Syst.* **2016**. [CrossRef] [PubMed]
46. Basu, S.; Ganguly, S.; Mukhopadhyay, S.; Dibiano, R.; Karki, M.; Nemani, R. Deepsat: A learning framework for satellite imagery. In Proceedings of the 23rd ACM SIGSPATIAL International Conference on Advances in Geographic Information Systems, Seattle, WA, USA, 3–6 November 2015; p. 37.
47. Chen, S.; Tian, Y. Pyramid of spatial relatons for scene-level land use classification. *IEEE Trans. Geosci. Remote Sens.* **2015**, *53*, 1947–1957. [CrossRef]
48. Krizhevsky, A.; Sutskever, I.; Hinton, G.E. ImageNet classification with deep convolutional neural networks. In Proceedings of the Twenty-Sixth International Conference on Neural Information Processing Systems, Lake Tahoe, NY, USA, 3–8 December 2012; pp. 1097–1105.
49. Breiman, L. Random forests. *Mach. Learn.* **2001**, *45*, 5–32. [CrossRef]
50. Dalal, N.; Triggs, B. Histograms of oriented gradients for human detection. In Proceedings of the IEEE Computer Society Conference on Computer Vision and Pattern Recognition, San Diego, CA, USA, 20–26 June 2005; pp. 886–893. [CrossRef]
51. Lowe, D.G. Distinctive image features from scale-invariant keypoints. *Int. J. Comput. Vis.* **2004**, *60*, 91–110. [CrossRef]
52. Ojala, T.; Pietikainen, M.; Maenpaa, T. Multiresolution gray-scale and rotation invariant texture classification with local binary patterns. *IEEE Trans. Pattern Anal. Mach. Intell.* **2002**, *24*, 971–987. [CrossRef]
53. Van der Maaten, L.; Hinton, G. Visualizing data using t-SNE. *J. Mach. Learn. Res.* **2008**, *9*, 2579–2605.
54. Jia, Y.; Shelhamer, E.; Donahue, J.; Karayev, S.; Long, J.; Girshick, R.; Guadarrama, S.; Darrell, T. Caffe: Convolutional architecture for fast feature embedding. In Proceedings of the ACM International Conference on Multimedia, Orlando, FL, USA, 3–7 November 2014.
55. Chatfield, K.; Simonyan, K.; Vedaldi, A.; Zisserman, A. Return of the devil in the details: Delving deep into convolutional nets. In Proceedings of the British Machine Vision Conference, Nottingham, UK, 1–5 September 2014.

56. Ioffe, S.; Szegedy, C. Batch normalization: Accelerating deep network training by reducing internal covariate shift. In Proceedings of the International Conference on Machine Learning, Lille, France, 6–11 July 2015; pp. 448–456.
57. Szegedy, C.; Vanhoucke, V.; Ioffe, S.; Shlens, J.; Wojna, Z. Rethinking the inception architecture for computer vision. In Proceedings of the IEEE Conference on Computer Vision and Pattern Recognition, Las Vegas, NV, USA, 26 June–1 July 2016; pp. 2818–2826.
58. He, K.; Zhang, X.; Ren, S.; Sun, J. Deep residual learning for image recognition. In Proceedings of the IEEE conference on computer vision and pattern recognition, Las Vegas, NV, USA, 26 June–1 July 2016; pp. 770–778.

6

Automatic Counting of Large Mammals from Very High Resolution Panchromatic Satellite Imagery

Yifei Xue [1,*], Tiejun Wang [1,*] and Andrew K. Skidmore [1,2]

1 Faculty of Geo-Information Science and Earth Observation (ITC), University of Twente, P.O. Box 217, 7500 AE Enschede, The Netherlands; a.k.skidmore@utwente.nl
2 Department of Environmental Science, Macquarie University, Sydney, NSW 2109, Australia
* Correspondence: y.xue@utwente.nl (Y.X.); t.wang@utwente.nl (T.W.).

Academic Editors: Qi Wang, Nicolas H. Younan and Carlos López-Martínez

Abstract: Estimating animal populations by direct counting is an essential component of wildlife conservation and management. However, conventional approaches (i.e., ground survey and aerial survey) have intrinsic constraints. Advances in image data capture and processing provide new opportunities for using applied remote sensing to count animals. Previous studies have demonstrated the feasibility of using very high resolution multispectral satellite images for animal detection, but to date, the practicality of detecting animals from space using panchromatic imagery has not been proven. This study demonstrates that it is possible to detect and count large mammals (e.g., wildebeests and zebras) from a single, very high resolution GeoEye-1 panchromatic image in open savanna. A novel semi-supervised object-based method that combines a wavelet algorithm and a fuzzy neural network was developed. To discern large mammals from their surroundings and discriminate between animals and non-targets, we used the wavelet technique to highlight potential objects. To make full use of geometric attributes, we carefully trained the classifier, using the adaptive-network-based fuzzy inference system. Our proposed method (with an accuracy index of 0.79) significantly outperformed the traditional threshold-based method (with an accuracy index of 0.58) detecting large mammals in open savanna.

Keywords: GeoEye-1; wavelet transform; fuzzy neural network; remote sensing; conservation

---

1. Introduction

Global biodiversity loss is a pressing environmental issue [1]. Populations of a number of wild animals have been reduced by half over the past four decades [2,3]. Counting wild animals to determine population size is an essential element of wildlife conservation and environmental management [4]. However, accurate population estimation using ground-based methods remains challenging, requiring considerable investment in resources and time [5]. Aerial surveys have been used as an alternative approach to detect large mammal populations and generate statistical estimates of their abundance in open areas [6]. In developed countries, wildlife such as caribou, elk, deer and moose have been monitored using aerial surveys [7–9]. For developing nations, where scores of endangered and threatened fauna are found, such an alternative is not always feasible due to limitations in access, technology, aircraft availability and skilled human resources [10,11]. It is therefore desirable to develop alternative approaches for conducting wildlife population counts in such regions.

Advances in satellite technology have provided new avenues in remote sensing for environmental applications, including the remote counting and mapping of animal populations. Lower spatial resolution satellite images have proven inadequate to detect and count individual animals [12], but the availability of commercial satellite images with a spatial resolution of one meter or less (e.g., IKONOS,

QuickBird, GeoEye and WorldView) has made such an undertaking more feasible [13]. As a result, studies have been undertaken utilizing satellite remote sensing data to detect animals. For example, Fretwell et al. [14] successfully estimated the abundance of penguins from fecal staining of ice by using a combination of medium resolution (15–30 m) Landsat-7 ETM+ and very high resolution (0.6–2.5 m) QuickBird satellite images, but they did not attempt to count individual birds. Stapleton et al. [15] used different very high resolution (VHR) satellite images (i.e., QuickBird, WorldView-1 and WorldView-2) to track the distribution and abundance of polar bears. Although their findings demonstrated the potential of remote sensing applications for wildlife detection and monitoring, they also revealed the need for more automated detection processes to expedite analysis. Yang et al. [16] explored mammal detection in open savanna country from VHR (0.5–2 m) GeoEye-1 satellite images, using a hybrid image classification approach. Through a two-step process of pixel-based and object-based image classification, they were able to demonstrate the feasibility of automated detection and counting of large wild animals in vast open spaces. However, the method they proposed requires the input by an expert of a number of parameters, and therefore this method remains subjective and labor-intensive. Fretwell et al. [17] compared a number of classification techniques endeavoring to automatically detect whale-like objects. They found that a simple thresholding technique of the panchromatic and coastal band delivered the best results. Neither Stapleton et al. [15] nor Fretwell et al. [17] made full use of the multispectral band, while the panchromatic band played an important role in their research. To our knowledge, there has been no substantial exploration of the feasibility of using a single panchromatic (black and white) band for wildlife detection. The typical panchromatic band data obtained from airborne platforms have a much wider spectral range than is utilized by multispectral bands (red, green, blue) [18], and also have a higher radiometric resolution (number of bits per pixel). Moreover, panchromatic satellite images have a higher spatial resolution than multispectral images [19].

Object counting can also be achieved with computer vision techniques, such as local feature-based subspace clustering algorithms [20,21] and global feature-based saliency detection approaches [22–25]. The conventional clustering method, such as the K-means clustering algorithm, has been used to extract local features, but its performance relies on finding "similar" records in the training data and could therefore be highly influenced by noise [21]. Data in a specific category can also be well-represented by low-dimensional subspace where noise can be reduced [26]. To achieve a good result by eliminating the influence of errors (e.g., noise, outliers), Peng et al. [20] proposed a graph-oriented learning method, which applied the L2-Graph for subspace learning and subspace clustering, for facial recognition and moving-vehicle detection [26]. However, studies on subspace clustering mainly concentrate on high-dimensional data clustering, such as facial recognition and motion image segmentation. Saliency detection is a well-researched problem in computer vision. It aims at indicating the saliency likelihood of each pixel by generating bounding boxes, binary foreground and background segmentation, or saliency maps [27]. The aforementioned methods have proven to be useful for multi-level features with multi-band images, but are difficult to apply to a single-band image where the object consists of few pixels.

Aerial photographs have been used for bird censuses since the 1980s, counting image points falling below an established threshold [28,29]. Bajzak and Piatt [29] studied the greater snow goose, contrasting its white plumage against the surrounding mud flats by size and tonal class. Similarly, a panchromatic image can use thresholding as a simple image segmentation method that divides an image into objects and background [30–32]. It works well when targets contrast sharply with their background. However, thresholding methods have their limitations: (1) targets cannot be separated from ground elements with similar brightness values; (2) gray value thresholding does not make full use of geometric information; and (3) threshold values are defined manually and depend heavily on the user's expertise.

Animal detection using remote sensing then predominantly switched to a two-step process [33]: (1) highlighting suspected targets; and then (2) classifying them, using geometric information. Groom et al. [33] proposed a scheme using geometric feature (object-size) filters to count birds against a monochromatic background. As targets were visually small and dim, they were not easily discerned against their background [34]. Using filters and image processing techniques, targets embedded in the scene could be visualized and detected [35–38]. However, the performance of such filters remains dependent on the brightness contrast between the target and background [34]. Several studies have employed wavelet-based techniques to address this concern [39–41]. The discernibility of targets from the background may vary at different scales, which can be problematic for object detection [19,42]. Wavelet analysis can transform signals into multiple resolutions, using an adaptive window [43], and thereby latently detect targets in cluttered backgrounds.

After highlighting the targets, the major challenge becomes how to make full use of geometric features to help separate a target from its surroundings. Spectral characteristics, cluster size, shape and other spatial features have been used in rule sets for image segmentation [44]. McNeill et al. [45] analyzed potential regions using shapes, by rejecting those with a compactness greater than a specified threshold value. Descamps et al. [46] counted large birds by fitting suspected objects (birds) into bright ellipses surrounded by a darker background. Expert knowledge can also play a critical role in image classification [16,47,48]. For example, Yang et al. [16] developed a specific rule set using expert knowledge to remove misclassified objects generated by object-based analysis. In another study, Wang et al. [47] proposed a hybrid neural network and expert system to quantify understory bamboo from satellite imagery, and they concluded that integration of a neural network and expert system appeared to be more efficient than when using either a neural network or an expert system alone. However, these methods rely on experts' subjective experience and knowledge, which can be challenging for practical applications.

An alternative approach to using an expert system is machine learning: a data analysis technique that automates model building through algorithms that iteratively learn from a given dataset. Though different classifiers based on machine learning generate varying levels of accuracy for different datasets [49], the most recent machine-learning techniques have a proven ability to solve complex problems [50]. For example, convolutional neural networks (CNNs) [51] have emerged as state-of-the-art models for image classification and object detection [52–57]. Local connections, shared weight, pooling and multiple layers are four architectural factors that make CNNs excel in processing natural signals [58]. However, the human involvement level is high when tailoring the CNN algorithm to a specific task [59], and large data sets are required for training purposes to ensure a high quality output [60]. Another major limitation of CNNs is their intrinsic black-box nature: their internal workings are hidden and not easily understood [61], so the models they generate are unexplainable [62]. The fuzzy neural network (FNN) is an alternative model that incorporates both the explicit knowledge representation of an fuzzy inference system (FIS) and the learning ability of an artificial neural network [63,64].The McCulloch–Pitts model [65] was one of the earliest applications to use fuzzy sets with a neural network concept. Since the 1990s, Takagi and others have developed a solid foundation for the fuzzy neural network [66]. In 1993, Jang proposed the adaptive-network-based fuzzy inference system (ANFIS) [67]. This algorithm has been widely employed in applied mathematics [68–71], and, unlike traditional expert systems, does not require a high level of expert knowledge when developing decision rules.

This study aims to detect and count large mammals in open spaces from a single, VHR GeoEye-1 panchromatic image, using a novel semi-supervised object-based scheme that combines a wavelet algorithm and a fuzzy neural network.

2. Materials and Methods

2.1. Study Area and Animal Species

The study area is located in the Maasai Mara National Reserve (also known as Maasai Mara or the Mara), a large game reserve in the Great Rift Valley in the southern part of Kenya (Figure 1).

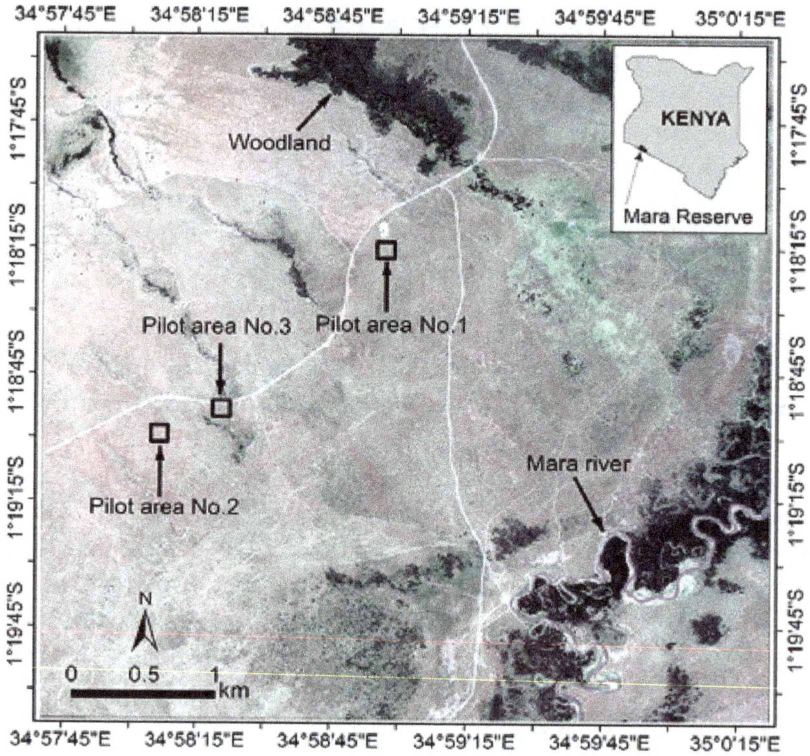

Figure 1. Location of the Maasai Mara National Reserve in Kenya and the three pilot study areas on a natural color composite of a GeoEye-1 image, acquired on 11 August 2009.

The reserve's topography is mainly open savanna (grassland) with clusters of acacia trees along the southeastern area of the park [72]. The reserve not only protects the habitat of resident species, but also preserves a critical part of the route used by wildebeests and zebras during the great migration that traverses the Maasai Mara via the Serengeti National Park. The wildebeest is the dominant species of the Maasai Mara, and herd sizes can range from a few individuals to many thousands [73]. Serengeti wildebeests migrate seasonally, and are seen intermittently in the Mara between August and November [74]. The sheer numbers of animals that congregate during migration make the wildebeest an ideal candidate species to map through the use of satellite technology.

2.2. Satellite Images

We acquired two GeoEye-1 satellite images of part of the Maasai Mara National Reserve through the DigitalGlobe Foundation (www.digitalglobefoundation.org/), each covering an area of 25 km^2. Both images are cloud free, and include one panchromatic (0.5 m) and four multispectral (2 m) bands. The image captured on 11 August 2009 depicts large numbers of animals. The other image, without any large animals present, was captured on 10 August 2013. To address our research objective, we carefully selected three small pilot study areas from the first image, each covering an area of 120 × 120 m (Figure 2). These pilot study areas were chosen to represent different levels of complexity regarding three criteria: (a) complexity of the landscape; (b) abundance of animals; and (c) feasibility and reliability of the visual interpretation of target animals. Pilot area No. 1 represents low complexity, with a few dozen animals viewed against a uniform background; Pilot area No. 2 represents moderate complexity, with more than one hundred animals viewed against a slightly less uniform background; and Pilot area No. 3 represents high complexity, with several hundred animals viewed against a non-uniform background.

Figure 2. The panchromatic band of the GeoEye-1 image taken on 11 August 2009, showing large mammals in the Maasai Mara National Reserve. Pilot area No. 1 represents low complexity regarding animal numbers and uniformity of background; Pilot area No. 2 represents moderate complexity; and Pilot area No. 3 represents high complexity. The rectangle visible in the top-left corner of Pilot area No. 3 is a white vehicle.

2.3. *Visual Interpretation to Establish Ground Truth for Large Animals Discerned on GeoEye-1 Imagery*

Ground truth is required to calibrate the model, as well as validate the classification result. Using the panchromatic band of the GeoEye-1 image, large mammals (e.g., wildebeests and zebras) are visualized as 3–4 pixels long and 1–2 pixels wide [16]. Due to their similarity in size, large animals can be confused with small ground features such as bushes and termite mounds [75]. To facilitate the visual interpretation of target animals and avoid the problem of subjectivity, we used one pan-sharpened GeoEye-1 image with, and one without, the presence of large animals (Figure 3). We invited two experienced wildlife researchers from Africa as independent visual interpreters. Together we visually compared the two separate temporal images of the three pilot study locations at multiple scales under the ArcGIS 10.3.1 environment (ESRI Inc., Redlands, CA, USA). After the observers had discussed their interpretation results, especially regarding uncertain objects, and had agreed which identified objects were indeed large mammals, their knowledge was recorded as confirmed animal ground truth points. In total, we identified 50, 128 and 426 large mammals in the pilot study areas 1, 2 and 3, respectively.

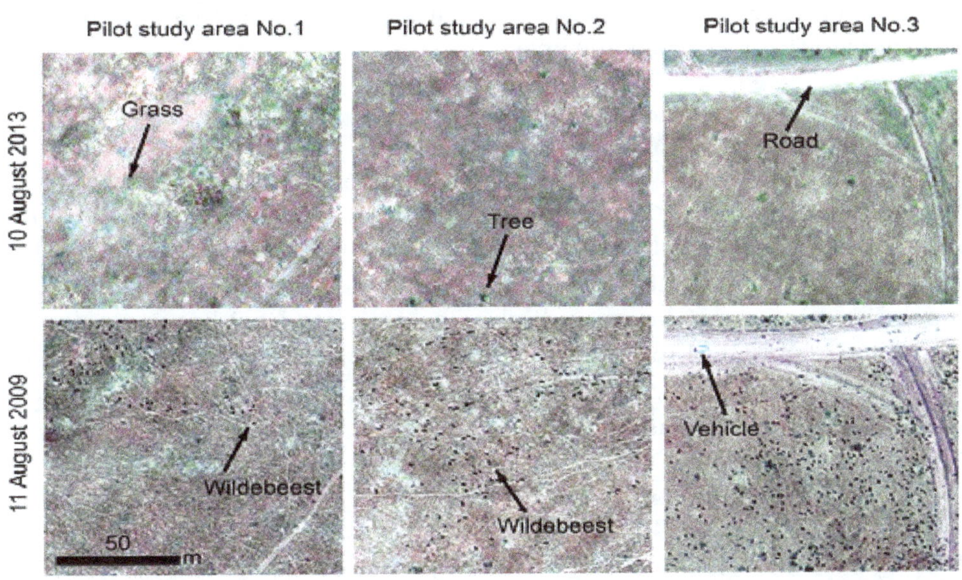

Figure 3. Visual interpretation of target animals by comparing two pan-sharpened GeoEye-1 images (0.5 m): one acquired 10 August 2013, without large animals present (top), and one acquired 11 August 2009, with large animals (bottom). The three pilot study areas represent the complexity of the landscape and the abundance of animals appearing in these images, from left to right: low, moderate and high.

2.4. *Semi-Automatic Animal Detection Algorithm*

Large mammals were identified by a series of multistage, semiautomatic techniques in VHR panchromatic satellite images. Our proposed scheme includes four principal steps (Figure 4): image preprocessing, preclassification, reclassification and accuracy assessment. Visual interpretation was incorporated for the purpose of reclassification and accuracy assessment.

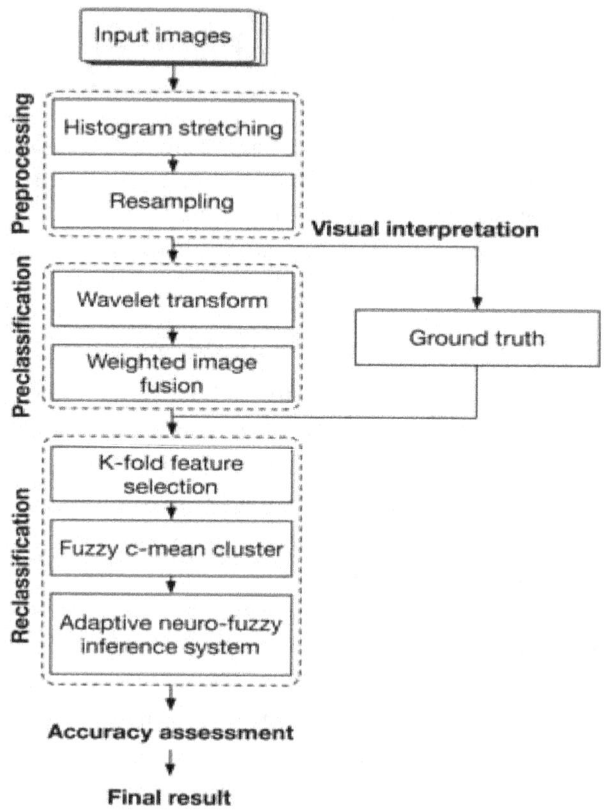

Figure 4. Workflow of the proposed method for counting large mammals from a single, very high resolution panchromatic GeoEye-1 satellite image.

2.4.1. Image Preprocessing

To highlight large mammals in the panchromatic imagery, we applied a histogram stretch in ENVI 5.2 (Exelis Visual Information Solutions, Inc., Boulder, CO, USA). Due to the limited resolution of the panchromatic band of VHR satellite images, an individual animal is represented as a cluster of pixels consisting of no more than eight pixels. To fully use their geometric information, we resampled the original image. Bicubic interpolation, which uses weighted arithmetic means, was chosen, as it maintains the quality of detailed information through antialiasing [76]. The image was carefully resized to eight times the original size, taking the wavelet decomposition performance into account, as well as memory and computation time, using

$$\mathrm{I} = \left[a_{i,j}\right]_{m\times n} \tag{1}$$

where the original image $\left[a_{i,j}\right]_{m\times n}$ is a matrix with m rows and n columns. We describe the resampled image as

$$\mathrm{I}\prime = f\left(\lambda\left[a_{i,j}\right]_{m\times n}\right) \tag{2}$$

where $\mathrm{I}\prime$ is the new image, λ represents the diagonal matrix of the resized scale and f is the bicubic interpolation function.

2.4.2. Wavelet-Based Preclassification

Based on the generally accepted methodology of image decomposition and reconstruction, we used the wavelet-based method when highlighting suspected large mammals, to enhance their contrast against the immediate surroundings and to suppress irrelevant background [77,78]. Wavelet transform (WT) is based on the theory of Short-Time Fourier Transform (STFT) [79]. The WT differs from STFT in that it replaces infinite triangle function bases with finite decay wavelet bases. The finite decay wavelet bases, which are stretched (or squeezed) and translated from the mother wavelet, have an average value of 0 [80]. The WT of a continuous signal is defined as

$$T(a,b) = w(a) \int_{-\infty}^{\infty} x(t)\psi^*(\frac{t-b}{a})dt \quad (3)$$

where a is scale, b is the position parameter, $w(a)$ is a weighting function and $\psi^*(\frac{t-b}{a})$ is the wavelet base [81]. If the wavelet base sufficiently corresponds to an input signal, the WT coefficient at this position is high [82]. The optimal mother wavelet and parameters were selected by comparing the performance of mainstream wavelet families regarding maintaining geometry features of suspected targets in our experimental imagery. A Haar wavelet (or db1 wavelet) was selected as it is not continuous and is therefore able to detect signals containing a sudden transition [83].

The image was transformed into a series of sub-images: A1 (low-frequency image), H1 (high-frequency image in the horizontal direction) and V1 (high-frequency image in the vertical direction); and then the same procedure was applied to the low frequency image (A1). Such a method permits multiresolution processing in both directions. After three transformation iterations, nine sub-images were generated, containing details as well as background. To highlight suspected targets and suppress background information, a weighted fusion algorithm was used. We then calculated the mean-square error (MSE) [84] between sub-images (resized to the original) and the original image. Sub-images containing more high-frequency information yielded higher MSE values. The weight of each sub-image should be

$$\omega_i = \frac{\sigma_i^2}{\sum_{j=1}^{n} \sigma_j^2} (i,j = 1,2,\ldots,n) \quad (4)$$

where i,j are the serial numbers of the current image, $\sigma_{i(j)}$ is the MSE of the current sub-image, and n is the total number of calculated sub-images. The weighted fusion algorithm creates a high signal-to-noise ratio (SNR) image. We then used Ostu's method [85] in MATLAB (The Mathworks Inc., Natick, MA, USA), to discriminate between each suspected animal blob and the background.

2.4.3. Selecting Geometric Features

The next concern was how to identify which suspected large mammals were true large mammals. This entailed deciding which geometric features to use, typically length and area. We also considered gray value (hue) pixels. We used cross-validation (a model assessment technique) to verify the performance of classifiers [86]. This basically involves grouping raw data: one group is used as training set and the other for validation. K-fold cross-validation (K-CV) is a commonly used validation technique in object detection [86,87]. We divided the data into ten groups, and used each group once as the training dataset while the other nine groups acted as the validation dataset. We determined the most suitable combination for this experiment by calculating the average value of the training errors and checking errors using the dataset mentioned above at situations of different feature combinations. After employing the K-fold cross-validation multiple times, we decided a combination of feature area, major axis length, minor axis length and bounding box area was most suitable for this experiment.

2.4.4. ANFIS-Based Reclassification

A total of 100 blobs (or unknown objects) were randomly selected from the database to train the final model. The distribution of training data was comparable to the distribution of the whole dataset.

Before we trained these data using ANFIS, a number of rules was decided upon. The Fuzzy C-Mean (FCM, or Fuzzy ISODATA), which was originally designed by Dunn [88], is a well-accepted clustering algorithm ideally suited to solving a natural problem [89,90]. As shown in Figure 5, this algorithm generated 10 cluster centers (corresponding to 10 membership functions for each variable). To limit the number of feature fields, we used expert knowledge to eliminate redundant classes. Finally, we input the 100 randomly selected blobs to train ANFIS in MATLAB. With the function ′genfis2′, we built an initial fuzzy inference system (FIS) structure. We then loaded the initial FIS structure into the function ′anfis′ to train the ANFIS and develop the model. A hybrid method, including least-squares and backpropagation gradient descent, was applied to optimise the model. ANFIS model evaluation was conducted according to the 'evalfis' function. Required parameters for the 'anfis' function, including training error goal, initial training step size, step size decrease rate and step size increase rate, were set to default values (0, 0.01, 0.9, 1.1), which were proven to be adequate for most situations [91]. In order to avoid overfitting, we set the epoch number to 75 by considering both training error and checking error (see Appendix A). The adaptive tuning stops when the least-squares error is less than the training error goal, or has reached the epoch number. By loading all the datasets containing feature values into the model, all suspected blobs were classified by the inference system into targets and non-targets.

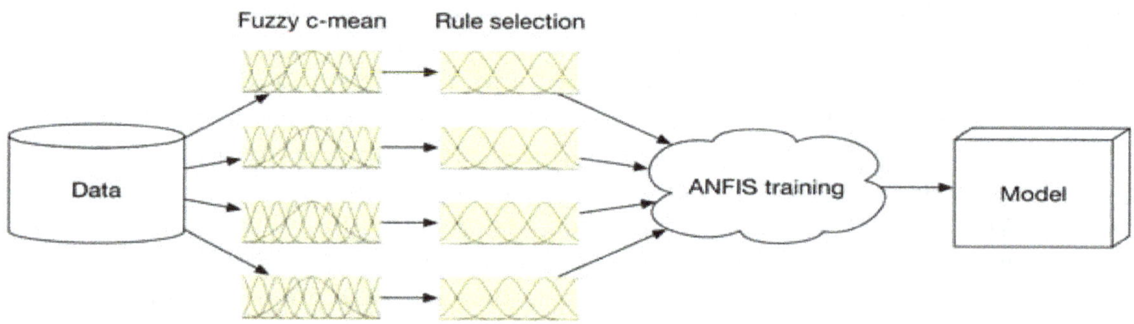

Figure 5. Flow diagram of the adaptive-network-based fuzzy inference system (ANFIS) based reclassification system.

2.5. Accuracy Assessment

We assessed the accuracy of the classification results by comparing the number of large mammals detected by the computer model with the ground truthing, and then calculated the omission error and commission error [92]. Detection accuracy (DA), which is the most commonly used metric, is highly inversely correlated ($DA + omission\ error = 1$) [93]. The values for both the omission error and the commission error are always between 0 and 1. The closer their values are to 0, the better the result.

The accuracy index (AI), which was devised by Pouliot et al. [94], was computed as:

$$\mathrm{AI} = \frac{\mathrm{N} - \mathrm{TP} - \mathrm{FN}}{\mathrm{N}} \quad (5)$$

where TP (true positive) denotes the number of targets occurring in both the ground truth and our processing result; FN (false negative) denotes the number of targets that do appear in the ground truth, but not in our processing result; FP (false positive) denotes the number of targets occurring in our processing result, but not in the ground truth data; and N is the number of ground truth targets in the study area. The higher the value of the accuracy index, the better the result.

3. Results

In Figure 6, the visual results of our semi-automated ANFIS-wavelet approach to detecting large mammals are compared with the results gained with the thresholding method.

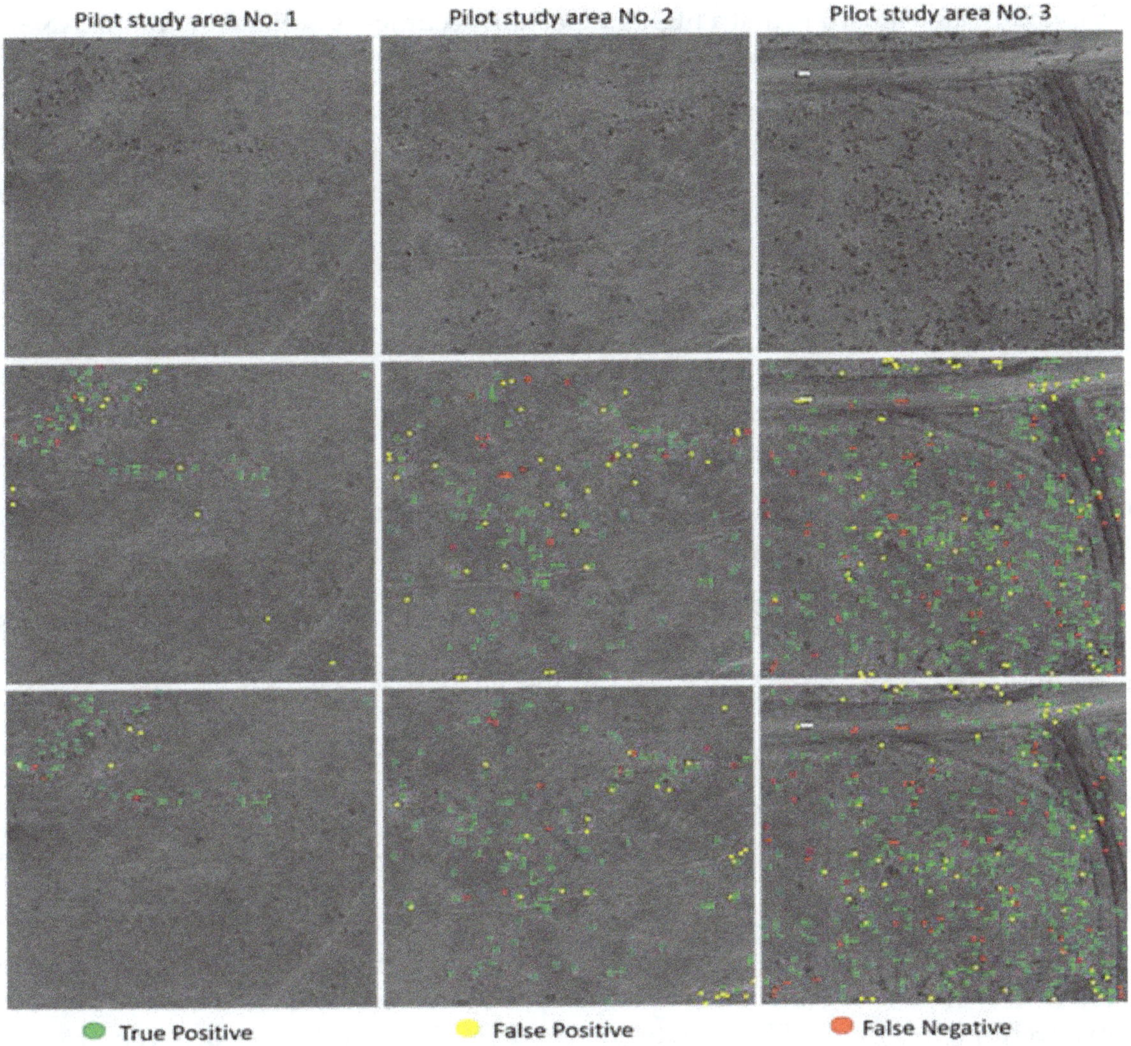

Figure 6. Results regarding large mammal detection in the three different pilot study areas. The columns show images of the three pilot areas: No. 1 is of low complexity, No. 2 of moderate complexity and No. 3 of high complexity. The first row contains the original panchromatic satellite images; the second row illustrates results based on the thresholding method; and the third row illustrates the results obtained using the method proposed in this study (i.e., ANFIS-wavelet). The green, red and yellow dots indicate true positive, false negative and false positive results, respectively.

The accuracy index regarding the proposed method for the low complexity study area (No. 1) was as high as 0.86 (Table 1). For the higher-complexity sites, the results also yielded acceptable accuracy indices: 0.79 and 0.72, respectively, for the moderately (No. 2) and highly (No. 3) complex sites. As shown in Table 2, the thresholding method produced accuracy indices of 0.64, 0.56 and 0.54, respectively, for the low, moderate and high complexity areas, with an average accuracy index of 0.58. The average accuracy index of our proposed method, depicted in Table 1, is 0.79, which is 0.21 higher than that of the thresholding method. Also, the calculated omission and commission errors of our approach (0.09 and 0.12, respectively) are lower than those of the thresholding method (0.15 and 0.24, respectively). It should also be noted that, if the study area is more complex, this does not necessarily mean that the detection is less accurate. As shown in Figure 6, specific ground features can introduce inaccuracies, such as the errors appearing in this study close to roads and edges of forests. In absolute terms of detected targets, the thresholding technique and our semi-automated ANFIS-wavelet approach showed different accuracies for each pilot study area. The statistical results regarding this study area illustrate that a higher detection accuracy is obtained with the ANFIS-wavelet method than with the threshold-based method.

Table 1. Accuracy assessment of the ANFIS-wavelet method for the three pilot study areas: No. 1, No. 2 and No. 3, with low, moderate and high complexity, respectively.

	Pilot Area No. 1	Pilot Area No. 2	Pilot Area No. 3	Average
Ground truth	50	128	416	198
True positive	47	118	370	178
False positive	4	17	64	28
False negative	3	10	56	23
Omission error	0.06	0.08	0.13	0.09
Commission error	0.08	0.13	0.15	0.12
Accuracy index	0.86	0.79	0.72	0.79

Table 2. Accuracy assessment of the threshold-based method for the three pilot study areas with low, moderate and high complexity, respectively.

	Pilot Area No. 1	Pilot Area No. 2	Pilot Area No. 3	Average
Ground truth	50	128	416	198
True positive	45	105	354	168
False positive	13	33	126	57
False negative	5	23	72	33
Omission error	0.10	0.18	0.17	0.15
Commission error	0.22	0.24	0.26	0.24
Accuracy index	0.64	0.56	0.54	0.58

4. Discussion

The results from this study demonstrate that it is feasible to use VHR panchromatic satellite imagery to detect and count large mammals in extensive open areas. In comparison with the traditional thresholding technique, our ANFIS-wavelet method produced a higher accuracy index and less commission/omission errors.

Although the thresholding method performs adequately when the targets share similar gray values and are dissimilar to their background, it is less accurate in more complex areas. There are two main reasons for the higher commission error found when using the thresholding method. Firstly, when the gray values of suspected objects (animals) are similar to those of the surroundings, they may be ignored by the threshold-based segmentation. In the ANFIS-wavelet method, the representation of the target is considered at different spatial scales. Suspected animals that do contrast with their immediate background, once different spatial scales are considered, will contribute to a higher weighted value in the preclassification results. Secondly, when animal objects and terrain have similar gray values, they cannot be altered simply by using thresholds: more information is required before further processing can be undertaken [32]. We statistically selected four geometric features to distinguish non-target objects from large mammals in the feature space. This approach proved more accurate than merely using a simple threshold value.

The commission error derived from our method was found to be three percentage points greater than the omission error, resulting in more non-target objects being incorrectly classified as large mammals than large mammals being incorrectly omitted. Further analysis revealed that commission errors always appeared near roads and vegetation. Bushes were confused with large mammals because of similarities in geometric features. Rough road surfaces or vehicles may result in discontinuous blobs and may thus also be recognized as large mammals by our method. Two reasons for omission include targets that are not clearly distinguishable from the background and targets that are too close to each other.

The geometric features chosen to distinguish an animal from its background were area, major axis length, minor axis length and bounding area. These features differ between target animals and non-targets such as shrubs or boulders. Even though some features were highly correlated, they can also help us in detecting animals. For example, defining both major and minor axis length can help to eliminate objects that do not have a correct length–width ratio.

The ANFIS-wavelet method has proved to be a feasible method for detecting animals in open savanna landscapes. This method is based on wavelet preclassification followed by ANFIS reclassification. The wavelet-based classification is able to highlight objects and maintain their geometric features. This is critical because the targets are dim and small, and as much useful information as possible needs to be retained. By using multiscale analysis, targets can be precisely located in poorer quality (i.e., low SNR) imagery without information loss. The ANFIS, which combines the advantages of machine learning and a fuzzy system, makes it possible to learn from data and concomitantly use existing expert knowledge, resulting in a method that is both efficient and stable.

5. Conclusions

We developed a novel semi-supervised object-based method that combines a wavelet algorithm and a fuzzy neural network for detecting and counting large mammals (e.g., wildebeests and zebras) from a single, very high resolution GeoEye-1 panchromatic image in open savanna. To discern large mammals from their surroundings and discriminate between animals and non-targets, we used the wavelet technique to highlight potential objects. To make full use of geometric attributes, we carefully trained the classifier, using the adaptive-network-based fuzzy inference system. We then compared our method with the traditional threshold-based method. The results showed that our proposed method (with an accuracy index of 0.79) significantly outperformed the traditional threshold-based method (with an accuracy index of 0.58) in detecting large mammals in open savanna. The greater availability of VHR images, and the advances in image segmentation techniques, mean that animal detection by means of remote sensing technology is a pragmatic alternative to direct animal counting. Further developments in image processing should eventually make it feasible to detect and monitor medium-sized and small animals remotely from space as well.

Acknowledgments: Yifei Xue was supported by the China Scholarship Council (CSC) and co-funded by the ITC Research Fund from the Faculty of Geo-Information Science and Earth Observation (ITC), University of Twente, the Netherlands. We acknowledge the satellite imagery support received from the DigitalGlobe Foundation. We also thank Festus Ihwagi and Tawanda Gara for their assistance with the visual interpretation of the GeoEye-1 satellite images.

Author Contributions: Yifei Xue, Tiejun Wang and Andrew K. Skidmore conceived and designed the experiment. Yifei Xue analyzed the data and wrote the paper. All authors contributed to the editing of manuscript.

Appendix A

ANFIS is a hybrid method which combines both least-square and backpropagation algorithms. The training datasets were used to construct the initial model, and the validation datasets were used for tuning. The training algorithm stops when either the training error goal value is satisfied, or the number of training epochs is reached. The training error goal was always being used as default value 0 when solving an unknown problem [91]. After a certain epoch number, the model will overfit the training data. To avoid overfitting, an optimal epoch number is required, but it is also difficult to determine. We evaluated the training error and the checking error (also known as validation error) with increasing the epoch number (Figure A1). Root-mean-square error (RMSE) is one of the most used indexes for performance indication [95]. The RMSE of the training data decreases along with the epoch number, but the tendency is slowed down after around 120 epochs and does not seem to have an obvious descent after 200 epochs. The RMSE of the checking data decreases along with the epoch number until around 75 epochs, and increases rapidly before around 120 epochs. According to this quantitative analysis, we found that it is proper to set the epoch number to around 75.

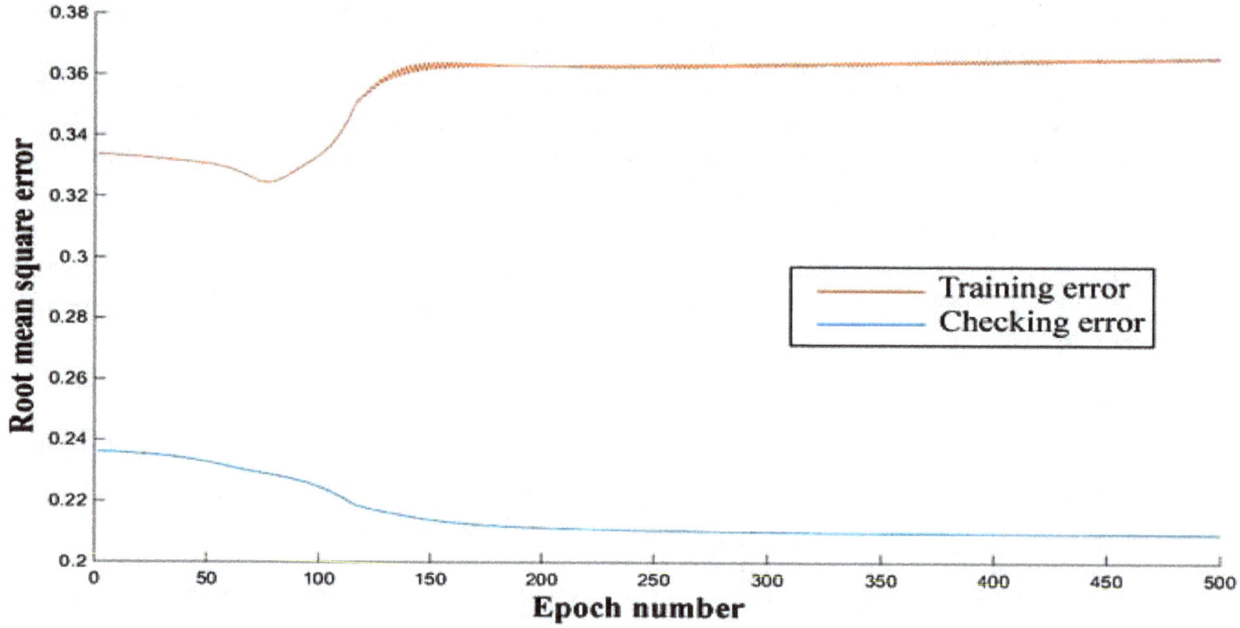

Figure A1. Indentification of optimum epoch number based on the root-mean-square error of both training error and checking error.

References

1. Skidmore, A.K.; Pettorelli, N.; Coops, N.C.; Geller, G.N.; Hansen, M.; Lucas, R.; Mücher, C.A.; O'Connor, B.; Paganini, M.; Pereira, H.M.; et al. Environmental science: Agree on biodiversity metrics to track from space. *Nature* **2015**, *523*, 403–405. [CrossRef] [PubMed]
2. Cuttelod, A.; Garcia, N.; Malak, D.A.; Temple, H.; Katariya, V. The Mediterranean: a biodiversity hotspot under threat. In *Wildlife in a Changing World: An Analysis of the 2008 IUCN Red List of Threatened Species*; Vié, J.-C., Hilton-Taylor, C., Stuart, S.N., Eds.; IUCN: Gland, Switzerland, 2009.
3. Carrington, D. Earth has lost half of its wildlife in the past 40 years, says WWF. Available online: https://www.theguardian.com/environment/2014/sep/29/earth-lost-50-wildlife-in-40-years-wwf (accessed on 30 August 2016).
4. Ramono, W.; Rubianto, A.; Herdiana, Y. Spatial distributions of Sumatran rhino calf at Way Kambas National Park based on its footprint and forest fire in one decade (2006 to 2015). In Proceedings of the Scientific Program of the 15th International Elephant & Rhino Conservation and Research Symposium, Singapore, 14–18 November 2016; p. 63.
5. Witmer, G.W. Wildlife population monitoring: Some practical considerations. *Wildl. Res.* **2005**, *32*, 259–263. [CrossRef]
6. Jones, G.P. *The Feasibility of Using Small Unmanned Aerial Vehicles for Wildlife Research*; University of Florida: Gainesville, FL, USA, 2003.
7. Gasaway, W.C.; DuBios, S.D.; Reed, D.J.; Harbo, S.J. *Estimating Moose Population Parameters from Aerial Surveys*; University of Alaska: Fairbanks, AK, USA, 1986.
8. Couturier, S.; Courtois, R.; Crépeau, H.; Rivest, L.-P.; Luttich, S.N. Calving photocensus of the Rivière George Caribou Herd and comparison with an independent census. *Rangifer* **1996**, *16*, 283–296. [CrossRef]
9. Pettorelli, N.; Côté, S.D.S.; Gingras, A.; Potvin, F.; Huot, J. Aerial surveys vs hunting statistics to monitor deer density: The example of Anticosti Island, Quebec, Canada. *Wildl. Biol.* **2007**, *3*, 321–327. [CrossRef]
10. Barnes, R.F.W. The problem of precision and trend detection posed by small elephant populations in West Africa. *Afr. J. Ecol.* **2002**, *40*, 179–185. [CrossRef]

11. Ransom, J.I.; Kaczensky, P.; Lubow, B.C.; Ganbaatar, O.; Altansukh, N. A collaborative approach for estimating terrestrial wildlife abundance. *Biol. Conserv.* **2012**, *153*, 219–226. [CrossRef]
12. Löffler, E.; Margules, C. Wombats detected from space. *Remote Sens. Environ.* **1980**, *9*, 47–56. [CrossRef]
13. Maglione, P. Very high resolution optical satellites: An overview of the most commonly used. *Am. J. Appl. Sci.* **2016**, *13*, 91–99. [CrossRef]
14. Fretwell, P.T.; LaRue, M.A.; Morin, P.; Kooyman, G.L.; Wienecke, B.; Ratcliffe, N.; Fox, A.J.; Fleming, A.H.; Porter, C.; Trathan, P.N. An emperor penguin population estimate: The first global, synoptic survey of a species from space. *PLoS ONE* **2012**, *7*. [CrossRef]
15. Stapleton, S.; LaRue, M.; Lecomte, N.; Atkinson, S.; Garshelis, D.; Porter, C.; Atwood, T. Polar bears from space: Assessing satellite imagery as a tool to track arctic wildlife. *PLoS ONE* **2014**, *9*. [CrossRef] [PubMed]
16. Yang, Z.; Wang, T.; Skidmore, A.K.; de Leeuw, J.; Said, M.Y.; Freer, J. Spotting East African Mammals in Open Savannah from Space. *PLoS ONE* **2014**, *9*, 1–16. [CrossRef] [PubMed]
17. Fretwell, P.T.; Staniland, I.J.; Forcada, J. Whales from space: Counting southern right whales by satellite. *PLoS ONE* **2014**, *9*, 1–9. [CrossRef] [PubMed]
18. Liu, J.G.; Mason, P.J. *Essential Image Processing and GIS for Remote Sensing*; John Wiley & Sons Ltd.: London, UK, 2009; ISBN 9780470510322.
19. Zhang, K.; Wang, M.; Yang, S.; Member, S.; Xing, Y.; Qu, R. Fusion of panchromatic and multispectral images via coupled sparse non-negative matrix factorization. *IEEE J. Sel. Top. Appl. Earth Obs. Remote Sens.* **2016**, *9*, 5740–5747. [CrossRef]
20. Peng, X.; Member, S.; Zhang, L.; Yi, Z.; Member, S. Constructing the L2-Graph for Subspace Learning and Subspace Clustering. *IEEE Trans. Cybern.* **2016**, *47*, 1053–1066. [CrossRef] [PubMed]
21. Otto, C.; Wang, D.; Jain, A. Clustering Millions of Faces by Identity. *IEEE Trans. Pattern Anal. Mach. Intell.* **2016**. [CrossRef] [PubMed]
22. Li, Z.; Itti, L. Saliency and gist features for target detection in satellite images. *IEEE Trans. Image Process.* **2011**, *20*, 2017–2029. [PubMed]
23. Wang, Q.; Yuan, Y.; Yan, P.; Li, X. Saliency detection by multiple-instance learning. *IEEE Trans. Cybern.* **2013**, *43*, 660–672. [CrossRef] [PubMed]
24. Wang, Z.; Du, L.; Wang, F.; Su, H.; Zhou, Y. Multi-Scale Target Detection in SAR Image Based on Visual Attention Model. In Proceedings of the 2015 IEEE 5th Asia-Pacific Conference on Synthetic Aperture Radar (APSAR), Singapore, 1–4 September 2015; pp. 704–709.
25. Wang, Q.; Lin, J.; Yuan, Y. Salient Band Selection for Hyperspectral Image Classification via Manifold Ranking. *IEEE Trans. Neural Netw. Learn. Syst.* **2016**, *27*, 1279–1289. [CrossRef] [PubMed]
26. Elhamifar, E.; Rene, V. Sparse Subspace Clustering: Algorithm, Theory, and Applications Ehsan. *IEEE Trans. Pattern Anal. Mach. Intell.* **2013**, *35*, 1–19. [CrossRef] [PubMed]
27. Yang, C.; Zhang, L.; Lu, H.; Ruan, X.; Yang, M.H. Saliency detection via graph-based manifold ranking. *Proc. IEEE Comput. Soc. Conf. Comput. Vis. Pattern Recognit.* **2013**, 3166–3173.
28. Gilmer, D.S.; Brass, J.A.; Strong, L.L.; Card, D.H. Goose counts from aerial photographs using an optical digitizer. *Wildl. Soc. Bull.* **1988**, *16*, 204–206.
29. Bajzak, D.; Piatt, J.F. Computer-aided procedure for counting waterfowl on aerial photographs. *Wildl. Soc. Bull.* **1990**, *18*, 125–129.
30. Glasbey, C.A.; Horgan, G.W.; Darbyshire, J.F. Image analysis and three-dimensional modelling of pores in soil aggregates. *J. Soil Sci.* **1991**, *42*, 479–486. [CrossRef]
31. Cunningham, D.J.; Anderson, W.H.; Anthony, R.M. An image-processing program for automated counting. *Wildl. Soc. Bull.* **1996**, *24*, 345–346.
32. Laliberte, A.S.; Ripple, W.J. Automated wildlife counts from remotely sensed imagery. *Wildl. Soc. Bull.* **2003**, *31*, 362–371.
33. Groom, G.; Krag Petersen, I.; Anderson, M.D.; Fox, A.D. Using object-based analysis of image data to count birds: Mapping of Lesser Flamingos at Kamfers Dam, Northern Cape, South Africa. *Int. J. Remote Sens.* **2011**, *32*, 4611–4639. [CrossRef]
34. Bai, X.; Zhang, S.; Du, B.; Liu, Z.; Jin, T.; Xue, B.; Zhou, F. Survey on dim small target detection in clutter background: Wavelet, inter-frame and filter based algorithms. *Procedia Eng.* **2011**, *15*, 479–483. [CrossRef]
35. Soni, T.; Zeidler, J.R.; Ku, W.H. Performance evaluation of 2-D adaptive prediction filters for detection of small objects in image data. *IEEE Trans. Image Process.* **1993**, *2*, 327–340. [CrossRef] [PubMed]

36. Shirvaikar, M.V. A neural network filter to detect small targets in high clutter backgrounds. *IEEE Trans. Neural Netw.* **1995**, *6*, 252–257. [CrossRef] [PubMed]
37. Casasent, D.; Ye, A. Detection filters and algorithm fusion for ATR. *IEEE Trans. Image Process.* **1997**, *6*, 114–125. [CrossRef] [PubMed]
38. Trathan, P.N.; Ratcliffe, N.; Masden, E.A. Ecological drivers of change at South Georgia: The krill surplus, or climate variability. *Ecography* **2012**, *35*, 983–993. [CrossRef]
39. Boccignone, G.; Chianese, A.; Picariello, A. Small target detection using wavelets. *Proc. Fourteenth Int. Conf. Pattern Recognit.* **1998**, *2*, 1776–1778.
40. Davidson, G.; Griffiths, H.D. Wavelet detection scheme for small targets in sea clutter. *Electron. Lett.* **2002**, *38*, 1128–1130. [CrossRef]
41. Kim, S. High-speed incoming infrared target detection by fusion of spatial and temporal detectors. *Sensors* **2015**, *15*, 7267–7293. [CrossRef] [PubMed]
42. Zhao, J.; Liu, F.; Mo, B. An algorithm of dim and small target detection based on wavelet transform and image fusion. In Proceedings of the 2012 Fifth International Symposium on Computational Intelligence and Design, Hangzhou, China, 28–29 October 2012; pp. 43–45.
43. Duk, V.; Ng, B.; Rosenberg, L. The potential of 2D wavelet transforms for target detection in sea-clutter. In Proceedings of the 2015 IEEE Radar Conference (RadarCon), Arlington, VA, USA, 10–15 May 2015; pp. 901–906.
44. Groom, G.; Stjernholm, M.; Nielsen, R.D.; Fleetwood, A.; Petersen, I.K. Remote sensing image data and automated analysis to describe marine bird distributions and abundances. *Ecol. Inform.* **2013**, *14*, 2–8. [CrossRef]
45. McNeill, S.; Barton, K.; Lyver, P.; Pairman, D. Semi-automated penguin counting from digital aerial photographs. In Proceedings of the 2011 IEEE International Geoscience and Remote Sensing Symposium (IGARSS), Vancouver, BC, Canada, 24–29 July 2011; pp. 4312–4315.
46. Descamps, S.; Béchet, A.; Descombes, X.; Arnaud, A.; Zerubia, J. An automatic counter for aerial images of aggregations of large birds. *Bird Study* **2011**, *58*, 302–308. [CrossRef]
47. Wang, T.J.; Skidmore, A.K.; Toxopeus, A.G. Improved understorey bamboo cover mapping using a novel hybrid neural network and expert system. *Int. J. Remote Sens.* **2009**, *30*, 965–981. [CrossRef]
48. Dagnino, A.; Allen, J.I.; Moore, M.N.; Broeg, K.; Canesi, L.; Viarengo, A. Development of an expert system for the integration of biomarker responses in mussels into an animal health index. *Biomarkers* **2007**, *12*, 155–172. [CrossRef] [PubMed]
49. Fernández-Delgado, M.; Cernadas, E.; Barro, S.; Amorim, D. Do we need hundreds of classifiers to solve real world classification problems? *J. Mach. Learn. Res.* **2014**, *15*, 3133–3181.
50. Schmidhuber, J. Deep Learning in neural networks: An overview. *Neural Netw.* **2015**, *61*, 85–117. [CrossRef] [PubMed]
51. LeCun, Y.; Bottou, L.; Bengio, Y.; Haffner, P. Gradient-based learning applied to document recognition. *Proc. IEEE* **1998**, *86*, 2278–2323. [CrossRef]
52. Johnson, J.; Karpathy, A.; Fei-Fei, L. DenseCap: Fully Convolutional Localization Networks for Dense Captioning. In Proceedings of the IEEE Conference Computer Vision and Pattern Recognition, Washington, WA, USA, 27–30 June 2016; pp. 4565–4574.
53. Krizhevsky, A.; Sutskever, I.; Hinton, G.E. ImageNet Classification with Deep Convolutional Neural Networks. *Adv. Neural Inf. Process. Syst.* **2012**, *25*, 1–9. [CrossRef]
54. Papandreou, G.; Kokkinos, I.; Savalle, P.A. Modeling local and global deformations in Deep Learning: Epitomic convolution, Multiple Instance Learning, and sliding window detection. *Proc. IEEE Comput. Soc. Conf. Comput. Vis. Pattern Recognit.* **2015**, 390–399. [CrossRef]
55. Wei, Y.; Xia, W.; Lin, M.; Huang, J.; Ni, B.; Dong, J.; Zhao, Y.; Yan, S. HCP: A Flexible CNN Framework for Multi-Label Image Classification. *IEEE Trans. Pattern Anal. Mach. Intell.* **2016**, *38*, 1901–1907. [CrossRef] [PubMed]
56. Wu, H.; Zhang, H.; Zhang, J.; Xu, F. Typical target detection in satellite images based on convolutional neural networks. In Proceedings of the IEEE International Conference on System, Man and Cybernetics, Hong Kong, China, 9–12 October 2015; pp. 2956–2961.
57. Ren, S.; He, K.; Girshick, R.; Sun, J. Faster R-CNN: Towards Real-Time Object Detection with Region Proposal Networks. *IEEE Trans. Pattern Anal. Mach. Intell.* **2017**, *39*, 1137–1149. [CrossRef] [PubMed]

58. LeCun, Y.; Yoshua, B.; Geoffrey, H. Deep learning. *Nature* **2015**, *521*, 436–444. [CrossRef] [PubMed]
59. Bengio, Y.; LeCun, Y. Scaling Learning Algorithms towards AI. In *Large-Scale Kernel Machines*; MIT Press: Cambridge, MA, USA, 2007; pp. 1–41, ISBN 1002620262.
60. Kala, R.; Shulkla, A.; Tiwari, R. Fuzzy Neuro Systems for Machine Learning for Large Data Sets. In Proceedings of the 2009 IEEE International Advance Computing Conference, Patiala, India, 6–7 March 2009; pp. 6–7.
61. Yager, R.R.; Zadeh, L.A. (Eds.) *An Introduction to Fuzzy Logic Applications in Intelligent Systems*; Springer Science & Business Media: Berlin, Germany, 2012.
62. Ma, H.; Ma, X.; Liu, W.; Huang, Z.; Gao, D.; Jia, C. Control flow obfuscation using neural network to fight concolic testing. *Lect. Notes Inst. Comput. Sci. Soc. Telecommun. Eng. LNICST* **2015**, *152*, 287–304.
63. Buckley, J.J.; Hayashi, Y. Fuzzy neural networks: A survey. *Fuzzy Sets Syst.* **1994**, *66*, 1–13. [CrossRef]
64. Hosseini, M.S.; Zekri, M. Review of Medical Image Classification using the Adaptive Neuro-Fuzzy Inference System. *J. Med. Signals Sens.* **2012**, *2*, 49–60. [PubMed]
65. McCulloch, W.S.; Pitts, W.H. A logical calculus of the idea immanent in nervous activity. *Bull. Math. Biophys.* **1943**, *5*, 115–133. [CrossRef]
66. Takagi, H.; Suzuki, N.; Koda, T.; Kojima, Y. Neural networks designed on approximate reasoning architecture and their applications. *IEEE Trans. Neural Netw.* **1992**, *3*, 752–760. [CrossRef] [PubMed]
67. Jang, J.-S.R. ANFIS: Adaptive-Network-Based Fuzzy Inference System. *IEEE Trans. Syst. Man Cybern.* **1993**, *23*, 665–685. [CrossRef]
68. Kurian, C.P.; George, V.I.; Bhat, J.; Aithal, R.S. Anfis model for the time series prediction of interior daylight illuminance. *ICGST Int. J. Artif. Intell. Mach. Learn.* **2006**, *6*, 35–40.
69. Yun, Z.; Quan, Z.; Caixin, S.; Shaolan, L.; Yuming, L.; Yang, S. RBF neural network and ANFIS-based short-term load forecasting approach in real-time price environment. *IEEE Trans. Power Syst.* **2008**, *23*, 853–858.
70. Boyacioglu, M.A.; Avci, D. An adaptive network-based fuzzy inference system (ANFIS) for the prediction of stock market return: The case of the Istanbul stock exchange. *Expert Syst. Appl.* **2010**, *37*, 7908–7912. [CrossRef]
71. Hiremath, S. Transmission rate prediction for cognitive radio using adaptive neural fuzzy inference system. Proceedings of 2010 5th International Conference on Industrial and Information Systems, ICIIS 2010, Mangalore, India, 29 July–1 August 2010; pp. 92–97.
72. Ford, A.T.; Fryxell, J.M.; Sinclair, A.R.E. Conservation challenges facing African savanna ecosystems. In *Antelope Conservation: From Diagnosis to Action*; Bro-Jørgensen, J., Mallon, D.P., Eds.; John Wiley & Sons Ltd.: London, UK, 2016; pp. 11–31. ISBN 9781118409572.
73. Hopcraft, J.G.C.; Sinclair, A.R.E.; Holdo, R.M.; Mwangomo, E.; Mduma, S.; Thirgood, S.; Borner, M.; Fryxell, J.M.; Olff, H. Why are wildebeest the most abundant herbivore in the Serengeti ecosystem? In *Serengeti IV: Sustaining Biodiversity in a Coupled Human-Natural System*; University of Chicago Press: Chicago, IL, USA, 2015; pp. 35–72.
74. Boone, R.B.; Thirgood, S.J.; Hopcraft, J.G.C. Serengeti Wildebeest Migratory Patterns Modeled from Rainfall and New Vegetation Growth. *Ecology* **2006**, *87*, 1987–1994. [CrossRef]
75. Pringle, R.M.; Doak, D.F.; Brody, A.K.; Jocque, R.; Palmer, T.M. Spatial pattern enhances ecosystem functioning in an african savanna. *PLoS Biol.* **2010**, *8*. [CrossRef] [PubMed]
76. Han, D. Comparison of commonly used image interpolation methods. In Proceedings of the 2nd International Conference on Computer Science and Electronics Engineering (ICCSEE 2013), Hangzhou, China, 22–23 March 2013; pp. 1556–1559.
77. Li, H.; Manjunath, B.S.; Mitra, S.K. Multisensor image fusion using the wavelet transform. *Graph. Model. Image Process.* **1995**, *57*, 235–245. [CrossRef]
78. Zhang, Y.; Dong, Z.; Wang, S.; Ji, G.; Yang, J. Preclinical diagnosis of magnetic resonance (MR) brain images via discrete wavelet packet transform with tsallis entropy and generalized eigenvalue proximate support vector machine (GEPSVM). *Entropy* **2015**, *17*, 1795–1813. [CrossRef]
79. Daubechies, I. The wavelet transform, time-frequency localization and signal analysis. *Inf. Theory IEEE Trans.* **1990**, *36*, 961–1005. [CrossRef]
80. Young, R.K. *Wavelet Theory and Its Applications*; Springer Science & Business Media: Berlin, Germany, 2012.

81. Addison, P.S. *The Illustrated Wavelet Transform Handbook: Introductory Theory and Applications in Science, Engineering, Medicine and Finance*; IOP Publishing: Bristol, UK, 2002.
82. Ye, X.; Wang, T.; Skidmore, A.K.; Fortin, D.; Bastille-Rousseau, G.; Parrott, L. A wavelet-based approach to evaluate the roles of structural and functional landscape heterogeneity in animal space use at multiple scales. *Ecography* **2015**, *38*, 740–750. [CrossRef]
83. Lee, B.Y.; Tarng, Y.S. Application of the discrete wavelet transform to the monitoring of tool failure in end milling using the spindle motor current. *Int. J. Adv. Manuf. Technol.* **1999**, *15*, 238–243. [CrossRef]
84. Ephraim, Y.; Malah, D. Speech enhancement using a minimum mean-square error short-time spectral amplitude estimator. *IEEE Trans. Acoust. Speech Signal Process.* **1984**, *32*, 1109–1122. [CrossRef]
85. Otsu, N. A threshold selection method from gray-level histograms. *IEEE Trans. Syst. Man Cybern.* **1979**, *9*, 62–66. [CrossRef]
86. Kohavi, R. A study of cross-validation and bootstrap for accuracy estimation and model selection. In Proceedings of the International Joint Conference on Artificial Intelligence, New York, NY, USA, 9–15 July 1995; pp. 1137–1143.
87. Bengio, Y.; Grandvalet, Y. No unbiased estimator of the variance of k-fold cross-validation. *J. Mach. Learn. Res.* **2004**, *5*, 1089–1105.
88. Dunn, J.C. A fuzzy relative of the ISODATA process and its use in detecting compact well-separated clusters. *Cybern. Syst.* **1973**, *3*, 32–57. [CrossRef]
89. Güler, C.; Thyne, G.D. Delineation of hydrochemical facies distribution in a regional groundwater system by means of fuzzy c-means clustering. *Water Resour. Res.* **2004**, *40*, 1–11. [CrossRef]
90. Shahi, A. An effective fuzzy C-Mean and Type-2 fuzzy logic for weather forecasting. *J. Theor. Appl. Inf. Technol.* **2009**, *5*, 556–567.
91. Ozkan, C. Surface interpolation by adaptive neuro-fuzzy inference system based local ordinary kriging. *Lect. Notes Comput. Sci.* **2006**, *3851*, 196–205.
92. Congalton, R.G. A review of assessing the accuracy of classification of remotely sensed data. *Remote Sens. Environ.* **1991**, *37*, 35–46. [CrossRef]
93. Yin, D.; Wang, L. How to assess the accuracy of the individual tree-based forest inventory derived from remotely sensed data: A review. *Int. J. Remote Sens.* **2016**, *37*, 4521–4553. [CrossRef]
94. Pouliot, D.A.; King, D.J.; Bell, F.W.; Pitt, D.G. Automated tree crown detection and delineation in high-resolution digital camera imagery of coniferous forest regeneration. *Remote Sens. Environ.* **2002**, *82*, 322–334. [CrossRef]
95. Dragomir, O.E.; Dragomir, F.; Stefan, V.; Minca, E. Adaptive Neuro-Fuzzy Inference Systems as a Strategy for Predicting and Controling the Energy Produced from Renewable Sources. *Energies* **2015**, *8*, 13047–13061. [CrossRef]

7

Sea Ice Concentration Estimation during Freeze-Up from SAR Imagery using a Convolutional Neural Network

Lei Wang, K. Andrea Scott * and David A. Clausi

Department of Systems Design Engineering, University of Waterloo, Waterloo, ON N2L 3G1, Canada; alphaleiw@gmail.com (L.W.); dclausi@uwaterloo.ca (D.A.C.)
* Correspondence: ka3scott@uwaterloo.ca.

Academic Editors: Qi Wang, Nicolas H. Younan, Carlos López-Martínez, Xiaofeng Li and Prasad S. Thenkabail

Abstract: In this study, a convolutional neural network (CNN) is used to estimate sea ice concentration using synthetic aperture radar (SAR) scenes acquired during freeze-up in the Gulf of St. Lawrence on the east coast of Canada. The ice concentration estimates from the CNN are compared to those from a neural network (multi-layer perceptron or MLP) that uses hand-crafted features as input and a single layer of hidden nodes. The CNN is found to be less sensitive to pixel level details than the MLP and produces ice concentration that is less noisy and in closer agreement with that from image analysis charts. This is due to the multi-layer (deep) structure of the CNN, which enables abstract image features to be learned. The CNN ice concentration is also compared with ice concentration estimated from passive microwave brightness temperature data using the ARTIST sea ice (ASI) algorithm. The bias and RMS of the difference between the ice concentration from the CNN and that from image analysis charts is reduced as compared to that from either the MLP or ASI algorithm. Additional results demonstrate the impact of varying the input patch size, varying the number of CNN layers, and including the incidence angle as an additional input.

Keywords: ice concentration; SAR imagery; convolutional neural network

---

1. Introduction

In the operational sea ice community, visual analyses of SAR imagery by expert ice analysts are a key contribution to ice charts, which are used to assist navigation and operations in ice-covered waters [1]. However, the generation of these analyses is time consuming. Upcoming and new satellite missions, such as the Canadian RADARSAT Constellation Mission (RCM), and the European Sentinel mission, will lead to significantly increased volumes of SAR imagery [2], increasing the need for automated methods to analyze the imagery.

There are several previous studies extracting information from SAR imagery using automated methods. Many of these studies use 'engineered' or hand-crafted features, which are features designed and selected to carry out a specific task. Examples include, the HH autocorrelation, normalized polarization difference and cross-polarization ratio all of which have been used in ice concentration estimation [3,4], grey level co-occurrence matrix features, Gabor filters and Markov random fields, which have been used to classify imagery into ice type and ice/water [5–8], and curvelet features used to locate the ice edge [9]. One of the challenges with using a set of engineered features to automatically extract information from SAR imagery is the difficulty of developing a set of robust features that can be applied to different geographic regions and seasons and for different imaging geometries. To capture various ice conditions, features may need to be designed for different locations or times of the year. For example, a large database of HH and HV backscatter values that represent typical signatures of

100% ice cover has been generated to retrieve ice observations for use in data assimilation. In the database, the backscatter values are estimated for each month as a function of incidence angle and windspeed on a region dependent-basis [10]. Such an extensive database may be necessary to assess the robustness of engineered features for large-scale applications, such as estimating ice concentration for assimilation in an operational prediction system [11]. Data assimilation requires high quality observations due to the nature of the assimilation cycle, in which erroneous observations will lead to an erroneous analysis, the influence of which will persist when the analysis is used to initialize the next assimilation cycle. For example, the open water regions that are estimated by Karvonen [4] as having an ice concentration of 10% or 15% would generate an incorrect analysis in a sea ice data assimilation system. A similar situation would arise upon assimilating a consolidated ice cover estimated with passive microwave data, in the event that the real ice cover has cracks and leads. Such openings in the ice are crucial for heat transfer from the ocean to the atmosphere. When the ice cover is used as a boundary condition for numerical weather prediction, an accurate estimate of the sea ice concentration is critical [12].

When an analyst estimates ice concentration from a SAR image, they combine their knowledge of ice conditions in the region with visual cues in the image. This may involve looking at the SAR image features over a range of scales. For example, at large scale, tonal changes across a region can be used to identify the region as either ice or open water, while at small scale visible ridges in the ice cover may indicate a region of high ice concentration, or small-scale ice floes may indicate a marginal ice zone. Thus, if it is desired to emulate the analyst's task, the goal can be viewed as emulating the human visual system's ability to assimilate information at various scales with prior knowledge. Convolutional neural networks (CNNs) are a known method to learn features from images, taking into account information at various scales. The training takes place by minimizing a difference between output of the CNN and training data, which represents prior knowledge. Remarkable similarities between CNNs and the human visual system have been demonstrated in numerous studies [13].

The present study uses a CNN trained with image analysis charts to estimate ice concentration from SAR imagery acquired over the Gulf of St. Lawrence during freeze-up in the winter of 2014. A previous study [14] has evaluated a similar architecture for the problem of ice concentration estimation in the Beaufort Sea for the 2010-2011 melt period. The present study builds on that work, addressing the following questions: (i) Can a CNN estimate sea ice concentration accurately during freeze-up, when the ice is very thin and may be difficult to distinguish from open water [15,16]? (ii) How is the performance of the CNN affected when some of the parameters (e.g., number of layers and input patch size) are modified? (iii) Can a CNN manage to interpret ice concentration for environments, such as the Gulf of St. Lawrence, where the ice characteristics are dynamic?

2. Background

Learning image features from SAR imagery to estimate ice concentration, as compared to first calculating engineered features from the image, builds on previous work in feature learning, which is a promising method to analyze complex and large volumes of data [17–21]. Deep learning is a type of feature learning method that can automatically extract complex data representations at high levels of abstraction [18,22,23]. For image recognition tasks, deep convolutional neural networks (CNN) are widely used due to their ability to model local image structures at multiple scales efficiently [24–27].

There has been limited research in using CNNs to learn features from satellite images. Related studies include using CNNs for road classification from aerial images [28] and the detection of vehicles [29] and buildings [30] from high resolution satellite images. Training of CNN models requires a large quantity of high quality training samples. For many remote sensing problems, gathering high quality ground truth is expensive and sometimes not feasible, due to the vast study area and diversity of surface conditions. This is in particular the case for ice concentration mapping. Due to harsh environmental conditions and in the interest of safety, obtaining adequate in situ samples coincident or near-coincident in time with a SAR scene is not usually feasible. Normally such in-situ studies are limited to small geographic regions and a limited time period.

Using other sources of satellite data or output from ice-ocean models may not be very suitable choices for training data. For example, many algorithms that compute ice concentration from passive microwave data are known to be biased over thin ice and in regions with low ice concentration levels [16,31,32]. Training a CNN with this data will lead to a CNN model that generates similar biases. Ice concentration estimated by image analysts is considered the best available ice concentration information [33]. Hence, the extensive image analysis database at the Canadian Ice Service (CIS) represents a promising archive that can be used to provide data to investigate the use of a CNN to estimate ice concentration from SAR imagery.

3. Data and Study Area

The study area is located in the Gulf of Saint Lawrence, which is situated on the east coast of Canada (Figure 1). The period of study extends from 17 January 2014 to 10 February 2014. This time of year corresponds to freeze-up in the Gulf of Saint Lawrence, with both ice concentration and thickness increasing from January into February. For the duration of the study, the ice cover is composed of new ice (less than 10 cm in thickness) and grey and grey-white ice (10–30 cm in thickness), with thicker first-year ice near Prince Edward Island. Definitions for the various ice types are provided by the World Meteorological Organization (WMO) [34].

A total of 25 RADARSAT-2 dual-pol (HH and HV) ScanSAR Wide [35] images are used for the present study. The full list of the SAR images used is provided in Table 1. The nominal pixel spacing of the acquired SAR images is 50 m by 50 m, and the incidence angle ranges from 20° to 49°. The image size is roughly 10 k × 10 k covering a spatial extent of about 500 km × 500 km. The outlines of all the SAR images in the dataset are shown in Figure 1.

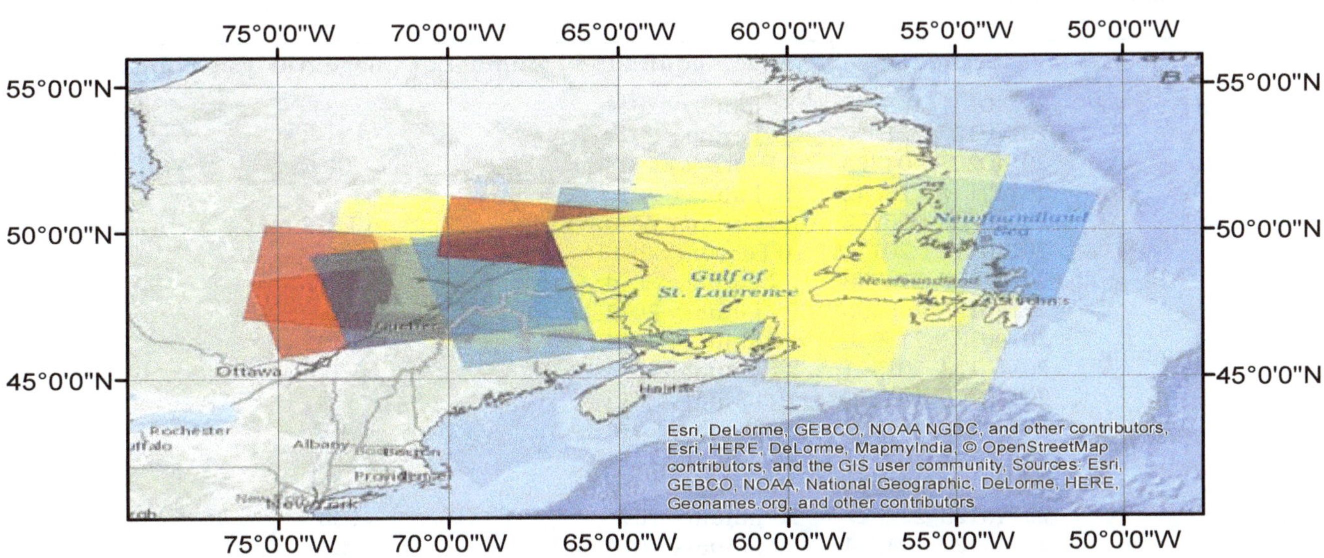

Figure 1. Study area and the dataset for the Gulf of Saint Lawrence. There are 25 scenes of dual-pol SAR images acquired between 16 January 2014 and 10 February 2014 in this area. The coverage for each scene is marked in a translucent polygon with different colors. Yellow scenes are used for training, red are used for validation and blue for testing.

Each SAR image has an accompanying image analysis chart, which is used to provide the training data to the ice concentration estimation methods. Compared to other types of ice charts (daily ice chart and regional ice chart), image analysis charts provide a more detailed interpretation of SAR images and are valid at the SAR image acquisition time [34]. Image analyses are prepared manually by a trained analyst who identifies regions (polygons) in which the ice conditions appear to be uniform, in terms of the total ice concentration and the relative mix of ice types. The ice types are defined according to their stage of development following World Meteorological Organization standards [34]. The ice concentration label given to a polygon is assigned in increments of 10%, hence the the precision of the image analyses cannot be higher than 10%. In addition, since each polygon of the image analysis is labeled with a single ice concentration value for the entire polygon, the actual ice concentration at the grid-point locations may be different from that indicated by the polygon label, depending on the spatial distribution of ice within the polygon.

As is the case with other sources of ice concentration data, it is difficult to quantify the accuracy of the image analysis charts. In comparing image analyses with other sources of data there are several factors that should be taken into account. First of all, the preparation of image analyses is subjective, and interpretation of image data by different analysts can lead to biases [36]. There are also errors due to converting continuous image data to discrete ice thickness categories, for example small scale details such as cracks in the ice or streaks of new ice are typically lost. Finally, the ice charts may have a slight tendency to over predict the ice concentration in the interest of marine safety.

The image analysis training data obtained from CIS in this study are grid-point data from the image analysis charts. The sampling interval is about 8 km in the north-south direction and 5 km in the east-west direction. The number of image analysis grid points for each SAR image varies from a few hundred to several thousand (Table 1) which depends on the area of sea surface in that scene. Note that while most of the validation data were acquired in February, these validation images overlay a large part of the study area. Visual inspection of the images reveals that they contain a variety of ice types, representative of those seen in the training and test data.

Table 1. Details of the Gulf of Saint Lawrence dataset. Each image analysis point covers an area of approximately 5 km × 8 km.

Set	Scene ID	Date Acquired	Number of Image Analysis Points
Training	20140131_103053	31 January 2014	8231
	20140127_221027	27 January 2014	1319
	20140203_104323	3 February 2014	3019
	20140116_223042	16 January 2014	530
	20140208_095758	8 February 2014	13,872
	20140210_220111	10 February 2014	8358
	20140207_214938	7 February 2014	612
	20140125_100500	25 January 2014	5200
	20140131_215240	31 January 2014	11,111
	20140124_103501	24 January 2014	6900
	20140120_105149	20 January 2014	829
	20140118_101002	18 January 2014	7492
	20140128_101751	28 January 2014	12,791
	20140130_222234	30 January 2014	1407
	20140123_222627	23 January 2014	950
	20140127_104734	27 January 2014	3427
	20140124_215646	24 January 2014	10,964
	20140121_214420	21 January 2014	15,897
Validation	20140122_095247	22 January 2014	5014
	20140206_221744	6 February 2014	3395
	20140209_223030	9 February 2014	545
	20140207_102631	7 February 2014	9228
Testing	20140210_103911	10 February 2014	2918
	20140130_110029	30 January 2014	425
	20140126_223850	26 January 2014	165
	20140117_103914	17 January 2014	2922

Corresponding daily AMSR2 ice concentration maps for each SAR scene are downloaded from the website of PHAROS group at the University of Bremen. These AMSR2 ice concentration maps are reprojected to their corresponding SAR image pixel grids with cubic interpolation, and are referred to as ASI ice concentration in the remainder of this paper, where ASI refers to the ARTIST sea ice concentration algorithm [31]. The ice cover during the study period is generally thin, with significant regions of thickness less than 30 cm. Based on previous studies [16] it is expected that the ice concentration calculated from passive microwave data will be underestimated in these regions of thin ice. However, the ASI ice concentration is based on the 89GHz channels of the AMSR2 sensor, and is known to have less of an underestimation than other products [16]. No modifications were made to the ASI algorithm, such as a recalibration of the algorithm tie-points, in order to compare our ice concentration against that from an available product. Note that the ASI algorithm contains a weather filter that on average removes all ice up to 15% concentration, and that the ASI data are daily averages whereas the CNN results and image analyses are snapshots valid at the image acquisition time.

4. Methodology

4.1. Preprocessing of SAR Images

All the SAR images are sub-sampled by 8×8 block averaging to reduce data volume while also reducing image speckle noise. Learning at this reduced scale requires a smaller spatial context window and therefore smaller neural networks. This is desired because of the limited number of training samples available (0.152 million image analysis sample points) for our study compared to model size ($\approx$3.9 million parameters). The sub-sampled images have 400 m pixel spacing with pixel values between 0 and 255. Input normalization is a common practice to improve the performance of CNNs [26,37]. In this study, the pixel values of the dual-polarized SAR images are normalized by first calculating the mean and standard deviation of pixel values over the entire dataset for each channel, then subtracting from each pixel value this mean, and dividing by the standard deviation.

If training sample patches are selected near land, when the patches are processed by the CNN, the land pixels may lead to signatures in the adjacent water regions that could be interpreted as ice. This may lead to overestimation (contamination) of ice concentration estimates near land. The size of land contaminated regions depends on the size of training sample patches. In our case, an image patch size 45 by 45 pixels is used, which corresponds to 18 km $\times$ 18 km ground distance. Therefore, land contamination can potentially affect regions within 18 km distance to the coast. Direct masking out land pixels to 0 is not used because the masked pixels may be confused with dark new ice or calm open water. Instead, a land mask is applied to the SAR images and land pixels are replaced by their corresponding mirrored ice or water pixels to reduce land contamination. By doing this, the estimated ice concentration only depends on local ice or water pixels. The actual ice concentration may be changed by the land mirroring process, depending on the shape of the coastline. However, in our testing, the mirroring was found to significantly reduce the effect of land on the estimated ice concentration. Therefore, no further investigation of alternative methods to mask land pixels is performed at this time.

The incidence angle for each SAR image pixel is calculated from the image meta data using linear interpolation and stored as incidence angle images. These incidence angle images are also normalized to have similar value ranges as the normalized SAR images. For the experiments that use incidence angle, each image patch is a three dimensional matrix of size $3 \times 45 \times 45$, while for the experiments that do not use incidence angle, each image patch is a two dimensional matrix of size $2 \times 45 \times 45$.

Each extracted patch and the ice concentration located at the patch center from the image analysis is one sample used to train the CNN. Polygon boundaries were not considered in selecting samples from the image analyses due to the limited number of samples available. Patches chosen that contain a polygon boundary are assigned the label corresponding to the polygon of the central pixel of the patch, but could be better described with a label the specifies the ice concentration as the mixture of the two polygons. These issues should be considered in a future study.

4.2. Overview and Structure of the CNN

CNN is a trainable architecture composed of multiple stages [38–40]. Each stage is composed of three consecutive operations (layers): convolutional filtering, non-linear transformation and sub-sampling (pooling). A CNN normally contains multiple stages that learn the image features, followed by a stack of fully connected (FC) layers [40]. The structure of the CNN used in this study is illustrated in Table 2. The CNN contains three convolutional layers followed by two fully connected layers. An excellent overview of CNNs can be found in [13].

Table 2. Structure and configuration of the CNN model used in the present study. Each row for a given layer corresponds to: the layer dimension (top row), the layer configuration (middle row) and the dimension the output (bottom row). For example for the layer *Conv1* there are 64 filters of dimension $3 \times 5 \times 5$ that are applied to an input patch of size $3 \times 45 \times 45$ with a stride of 1 and using a pad 2, to produce an output of dimension $64 \times 45 \times 45$.

Layer	
Data	$3 \times 45 \times 45$
Conv1	$64 \times 3 \times 5 \times 5$ stride 1, pad 2, ReLU $64 \times 45 \times 45$
Pool1	2×2 stride 2, pad 1, Max $64 \times 23 \times 23$
Conv2	$128 \times 64 \times 5 \times 5$ stride1, pad 2, ReLU $128 \times 23 \times 23$
Pool2	$128 \times 23 \times 23$ stride 2, pad 1, Max $128 \times 12 \times 12$
Conv3	$128 \times 128 \times 5 \times 5$ stride 1 , pad 2 , ReLU $128 \times 12 \times 12$
FC4	$1024 \times 128 \times 5 \times 5$ ReLU 1024×1
Dropout	$1024 \times 1 \times 1$ Drop rate: 0.5 1024×1
FC5	1×1024 Linear 1

In the convolutional layers, the layer input matrix $\boldsymbol{x}$ (width S_x pixels, height S_y pixels and number of channels S_z), which is a patch extracted from the SAR image, is convolved with K convolution filters of size (C_x, C_y, S_z), denoted by $\mathbf{C}^k, k = 1, \ldots, K$. Each filter is applied to the image patch with a step size (stride) P (convolution is carried out for locations that are P pixels apart). A total of K feature maps, denoted as $\boldsymbol{h}^k$ of dimension M_x and M_y will be generated as the output of this convolutional layer as described in Equation (1),

$$\boldsymbol{h}^k = (\boldsymbol{C}^k * \boldsymbol{x}) + b, \text{in which, } k = 1, \ldots, K \tag{1a}$$

$$M_x = \frac{S_x - C_x}{P} + 1 \tag{1b}$$

$$M_y = \frac{S_y - C_y}{P} + 1, \tag{1c}$$

where the operation of convolution is denoted by $*$ and the size of the feature maps ($M_x \times M_y$) is given for the case with zero padding. For a discussion of padding see [41]. Each convolutional layer is mainly characterized by the size and number of filters. The values of the filter weights and the bias term, b, are learned from the training data [42].

A convolutional layer is followed by a nonlinear transformation layer, which applies a nonlinear function to each element in the feature maps. This nonlinear function is also referred as the activation function, and is a well known feature used in neural networks to ensure the output is not simply a linear transformation of the input [43]. The rectified linear unit, ReLU is used as the activation function in the present study. ReLU activation has been demonstrated to lead to faster learning and better features than traditionally used sigmoid activation function, because ReLU activation does not saturate, as compared to sigmoid activation [26,44].

The nonlinear transformation layer is followed by the sub-sampling layer, also known as the pooling layer. Max pooling is used in the present study due to its simplicity and effectiveness [25,26,40,45]. It outputs the maximum value over each pooling window. For example, when pooling window size and step size are both set to 2, a max-pooling layer outputs the maximum value of every two by two non-overlapping window of its input.

The convolutional layers are followed by fully connected layers that serve as classification modules using the features extracted by the previous multiple stages. These layers have structure that is similar to that of a basic neural network [43]. Every neuron in a fully connected layer is connected to all the neurons of its input layer. The first fully connected layer takes a stack of feature maps, $\boldsymbol{h}^k$ as input. The feature maps are flattened to a vector and transformed to the output space by a weight matrix $\boldsymbol{W}$ and bias b. This is followed by the application of an activation function, f, to generate the output,

$$\boldsymbol{h} = f((\boldsymbol{W} * \boldsymbol{x}) + b). \tag{2}$$

4.3. Training and Testing

Our network is trained to output the ice concentration from SAR image patches. Instead of using softmax loss [26], which is commonly used in classification CNNs, the L_2 loss is used (3) for this regression problem to penalize the discrepancy between the CNN output and the ice concentration provided by the image analysis charts. The loss function is,

$$\mathbb{L}(F(\boldsymbol{x};\boldsymbol{\theta}), z) = \frac{1}{M}\sum_{m=1}^{M}(F(\boldsymbol{x};\boldsymbol{\theta})_m - z_m)^2, \tag{3}$$

where $F(\boldsymbol{x};\boldsymbol{\theta})$ is the network output given input $\boldsymbol{x}$ and parameterization $\boldsymbol{\theta}$, z_m is the ice concentration for the mth sample from image analyses, and M is the number of samples used in each training sample batch. For batch sizes larger than 1, the overall loss of this mini-batch is the average loss of all samples in that mini-batch.

Backpropagation and mini-batch stochastic gradient descent (SGD) [46] are used as the training algorithm. This method uses the derivatives of loss function (3) with respect to the network parameters

$$\frac{\partial \mathbb{L}}{\partial \boldsymbol{\theta}} = \frac{2}{M}\sum(F(\boldsymbol{x};\boldsymbol{\theta})_m - z_m)\frac{\partial F(\boldsymbol{x};\boldsymbol{\theta})_m}{\partial \boldsymbol{\theta}}, m = 1, ..., M. \tag{4}$$

The derivatives are backpropagated through each pixel in the predictions. The network parameters are updated according to the derivative of the loss to the parameters over each mini-batch, which is described by (5).

$$\boldsymbol{V}_{t+1} = \alpha \cdot \boldsymbol{V}_t - r \cdot \epsilon \cdot \boldsymbol{\theta}_t - \epsilon \frac{\partial \mathbb{L}}{\partial \boldsymbol{\theta}} |_{\boldsymbol{\theta}_t} \tag{5a}$$

$$\boldsymbol{\theta}_{t+1} = \boldsymbol{\theta}_t + \boldsymbol{V}_{t+1}. \tag{5b}$$

The weights $\boldsymbol{\theta}$ are updated by V_{i+t} at iteration $t+1$ with learning rate $\epsilon = 10^{-3}$ and weight decay of $r = 2 \times 10^{-5}$ with momentum, α, of 0.9. The setting of the training parameters for SGD is similar to the published setting by Krizhevsky et al. [26]. Adjustments are made by tuning the training parameters sequentially. ϵ is first tuned due to its significant effect on the training results. Then r and α are tuned. Similar to Krizhevsky et al. [26], the parameters of the CNN are initialized by uniform random sampling between -0.05 and 0.05. Stochastic gradient descent is used to iteratively update the model weights using the gradient of loss with respect to the model parameters calculated using a subset of the training samples (mini-batch). The gradients of the loss with respect to the network parameters ($\partial \mathbb{L} / \partial \boldsymbol{\theta}$) are calculated and averaged over the mini-batch. An epoch training scheme [46] is adopted. For each epoch, all the training samples are iterated once by the training algorithm. The learning rate is reduced by a factor of 10 for every 20 thousand mini-batches (about 17 epochs). To accelerate the training process, the training is set to stop when the score of the loss function is changing less than 0.001 for 20 consecutive epochs, in case the training converges early (which is typical [47]).

Overfitting is a common problem with CNNs. It is common practice to use a validation dataset to validate the CNN model during training time [26]. The derived CNN model is evaluated after each training epoch by calculating the loss function on the validation dataset using the current model. The CNN model with the smallest validation error will be selected as the trained CNN. Note that validation is used for model selection and it is therefore part of the training scheme. In this case, the 25 scenes are randomly divided to 17 training images, 4 testing images and 4 validation images, as described in Table 1.

To further reduce overfitting, training sample augmentation and dropout are used. Training sample augmentation artificially enlarges the training dataset by label-preserving transformations, such as rotation and flipping [26,48]. In our experiment, training samples are augmented on-the-fly by random rotating and flipping. These transformed SAR image patches are used for forward-propagation, which corresponds to increasing the training set by a factor of several hundred times. Dropout is a different and complementary technique used to reduce overfitting. A dropout layer randomly sets the outputs of neurons (also referred as units) in a layer to zero with predefined probability [49]. Those dropped neurons are not contributing to the forward pass and therefore are not updated in the backpropagation. The use of dropout can reduce the co-adaptations between neurons because a neuron cannot rely on the presence of other neurons [26,49]. The network is therefore forced to learn more representative features. A dropout layer with drop rate 0.5, i.e., half of the neurons are randomly chosen and their outputs are set to zero, is used in the present study.

Once the CNN model is trained, ice concentration for each pixel location is estimated by applying the trained model on the target SAR images. Since the CNN can only predict a single location in one forward-propagation, the CNN model is used on input images with stride 1, i.e., the input window moves one pixel every time.

4.4. Implementation

Caffe [50], a popular C++ open-source deep learning package, is used in this study. It provides a ready-to-use implementation of the CNN. SAR image preprocessing and patching are implemented in Python. A data layer is implemented using C++ under Caffe to read the image patches and their corresponding image analyses ice concentration values. In-situ training sample augmentation is also implemented in the data layer.

5. An MLP for Ice Concentration Estimation

For the purpose of evaluation, a fully connected neural network, known as MLP (multilayer perceptron) has also been developed to estimate sea ice concentration from the set of SAR images. The structure of this MLP is similar to that of a fully connected layer, and is described in [43]. The MLP used here is a variation of that used in the ice concentration estimation algorithm developed by Karvonen [4]. Karvonen's method [4] uses a preliminary ice concentration estimated from the autocorrelation of HH pol SAR images by a segmentation based approach [3] and four other SAR image features (HV, HV/HH, (HH-HV)/HH, and incidence angle) as input to an MLP with one hidden layer of 10 units. The MLP developed in [4] was trained using data from Finnish Ice Service (FIS) ice charts.

In our implementation, the ice concentration is estimated on a pixel-by-pixel basis using an MLP with one hidden layer of size 40. Ten GLCM features are used in addition to the four features used by Karvonen (HV, HV/HH, (HH-HV)/HH, and incidence angle). These ten GLCM features are identified as the most important ten SAR image features from a pool of 172 SAR image features used to distinguish ice and water [7] and should also benefit the ice concentration estimation task. The features input to the MLP are listed in Table 3. In Leigh et al. [7], image features are extracted from 4 by 4 block averaged SAR images. For consistency with the 8 by 8 block averaged SAR images used here, the image features are first calculated from 4 by 4 block averaged SAR images as done in [7], and are then averaged for every 2 by 2 block.

Table 3. Image features used for method MLP40.

#	Pol	Feature
1	HV	GLCM mean 25 by 25 step 5
2	HH	GLCM correlation 51 by 51 step 5
3	HH	GLCM mean 25 by 25 step 1
4	HH	GLCM dissimilarity 51 by 51 step 20
5	HH	GLCM second moment 101 by 101 step 5
6	HH	Intensity
7	HV	Average 25 by 25 window
8	HH	Average 5 by 5 window
9	HH	GLCM dissimilarity 51 by 51 step 5
10	HH	GLCM mean 101 by 101 step 20
11	HV	Intensity
12	HH, HV	HV/HH
13	HH, HV	(HH-HV)/HH
14	HH	Intensity autocorrelation
15		Incidence angle

Due to the larger number of input image features in our MLP as compared to [4], the number of hidden neurons needs to be increased. The resulting MLP has higher ratio of hidden neurons to input features (40/15) as compared to Karvonen's implementation (10/6). Note that Karvonen [4] made a correction to the images to account for the variation of backscatter with incidence angle, while in our implementation such a correction was not applied due to the fact that such a correction depends on whether the underlying surface is ice or water [10], and also varies with ice type, none of which can be assumed known in advance. The same training scheme used by Karvonen is used to train the MLP [4].

6. Results

6.1. Evaluation

The ice concentration estimated from the SAR images using the CNN described in Section 4, as well as ice concentration from ASI and MLP40 are evaluated against image analyses in the SAR image space. In other words, each image analysis sample point is compared to the ice concentration of its nearest pixel in the associated SAR scene, which means the image analysis samples are used at a finer spatial resolution than what the analyst intended. The mean error (E_{sgn}), mean absolute error (E_{L1}), error standard deviation (E_{std}) and root mean squared error (E_{rmse}) are calculated for evaluation purposes using (6)

$$E_{sgn} = mean(IC - ImA) \tag{6a}$$
$$E_{L1} = mean(|IC - ImA|) \tag{6b}$$
$$E_{std} = std(IC - ImA) \tag{6c}$$
$$E_{rmse} = \sqrt{(mean[(IC - ImA)^2])}. \tag{6d}$$

The term IC denotes the ice concentration estimated using the CNN and ImA denotes the ice concentration from the image analysis charts.

While the ice concentration derived from the image analysis is a discrete number (0–10) scaled between 0 and 1 (0, 0.1, ..., 1.0), the ice concentration from the CNN is determined as a real number between 0 and 1. This difference may introduce errors into the evaluation statistics. To investigate this, the ice concentration estimates are also quantized by rounding to 11 levels between 0 and 1 and re-evaluated against the image analyses. The evaluation results are similar with slight improvement after quantization, and are therefore not shown.

The evaluation results for training, testing and validation datasets are given in Table 4. The E_{rmse} is lower for the ice concentration estimated by the CNN than that from either MLP or ASI. The statistical significance of the E_{rmse} for each of the test datasets is assessed using a z-test, with the E_{rmse} assumed to follow a chi-squared distribution [51]. For Table 4, the null hypothesis is that the E_{rmse} of the CNN and MLP have the same distribution. The calculated p-value is $<<0.001$, indicating that the difference between the two is statistically significant for significance level of 0.01. Similar tests were done for the other experiments (discussed in Sections 6.3.2 and 6.3.3) and in all cases the p-value is $<<0.001$, with the exception of the experiment comparing two convolutional layers with three convolutional layers, in which case the p-value is 0.0019. For each experiment it was the E_{rmse} of the test dataset that was evaluated.

Table 4. Average error statistics across different methods for Gulf of Saint Lawrence dataset.

Method	Set	E_{sgn}	E_{L1}	E_{std}	E_{rmse}
ASI	Training	−0.2423	0.2605	0.3207	0.4020
	Validation	−0.3416	0.3768	0.3693	0.5031
	Testing	−0.2717	0.2877	0.3097	0.4121
MLP40	Training	0.0002	0.1460	0.2050	0.2049
	Validation	−0.0410	0.2381	0.2986	0.3015
	Testing	−0.0819	0.1727	0.2325	0.2466
CNN	Training	−0.0039	0.0845	0.1506	**0.1507**
	Validation	−0.0123	0.1253	0.2056	0.2059
	Testing	−0.0274	0.1295	0.2197	0.2214

In Table 4 it can be seen that ASI underestimates ice concentration by around 24% when compared with image analyses (Table 4). Since the CNN is trained using image analysis charts, while ASI ice concentration is not, it is expected to have lower error than ASI when the error is calculated with respect to image analysis charts. Previous studies reported that the ASI ice concentration normally has errors less than 10% for intermediate and high ice concentrations [31]. The large underestimation of ice concentration observed in this study is mainly caused by the large regions of thin ice, and the magnitude of the error is consistent with that reported in other studies [16]. The underestimation of ice concentration is improved by the CNN compared to MLP40. Note that the error standard deviation (E_{std}) for testing is at the same level as training and validation for the CNN, which indicates a low level of over-fitting for the trained CNN model. The validation errors are larger than testing errors for MLP40. This might be caused by the insufficient testing samples used, which could lead to different distributions of image surface types for validation and testing images.

Figure 2 shows the mean value of the estimated ice concentration $\pm$ one standard deviation of the ice concentration estimate errors for different ice concentration bins from the image analysis charts. Results are shown separately for training, validation and testing datasets. There is a clear trend between image analyses and ice concentration estimates generated from SAR images for all three sets in general. ASI shows underestimation for almost all ice concentration levels, with larger underestimation for higher ice concentration values. MLP40 overestimates ice concentration for water regions by about 15% for all three datasets, and underestimates ice concentration by 20% to 40% for training, testing and validation in the highest ice concentration bin. The CNN has relatively less overestimation for water regions and less underestimation for ice regions compared to MLP40. For water, CNN overestimates ice concentration on average by approximately 5% for training and 10% for testing and validation. For ice (where ice concentration is equal to 1), CNN underestimates ice concentration by less than 10% on average for all three sets. The estimation of pure water or ice generally has smaller error standard deviation than the estimate for intermediate ice concentration levels. This might be caused by the abundant water samples and ice samples in the training dataset (Figure 3), or the better quality (less errors) of ice/water samples than samples of intermediate ice concentration levels. It is reasonable to assume that the ice concentration estimates could be improved by using more training samples of intermediate ice concentration levels.

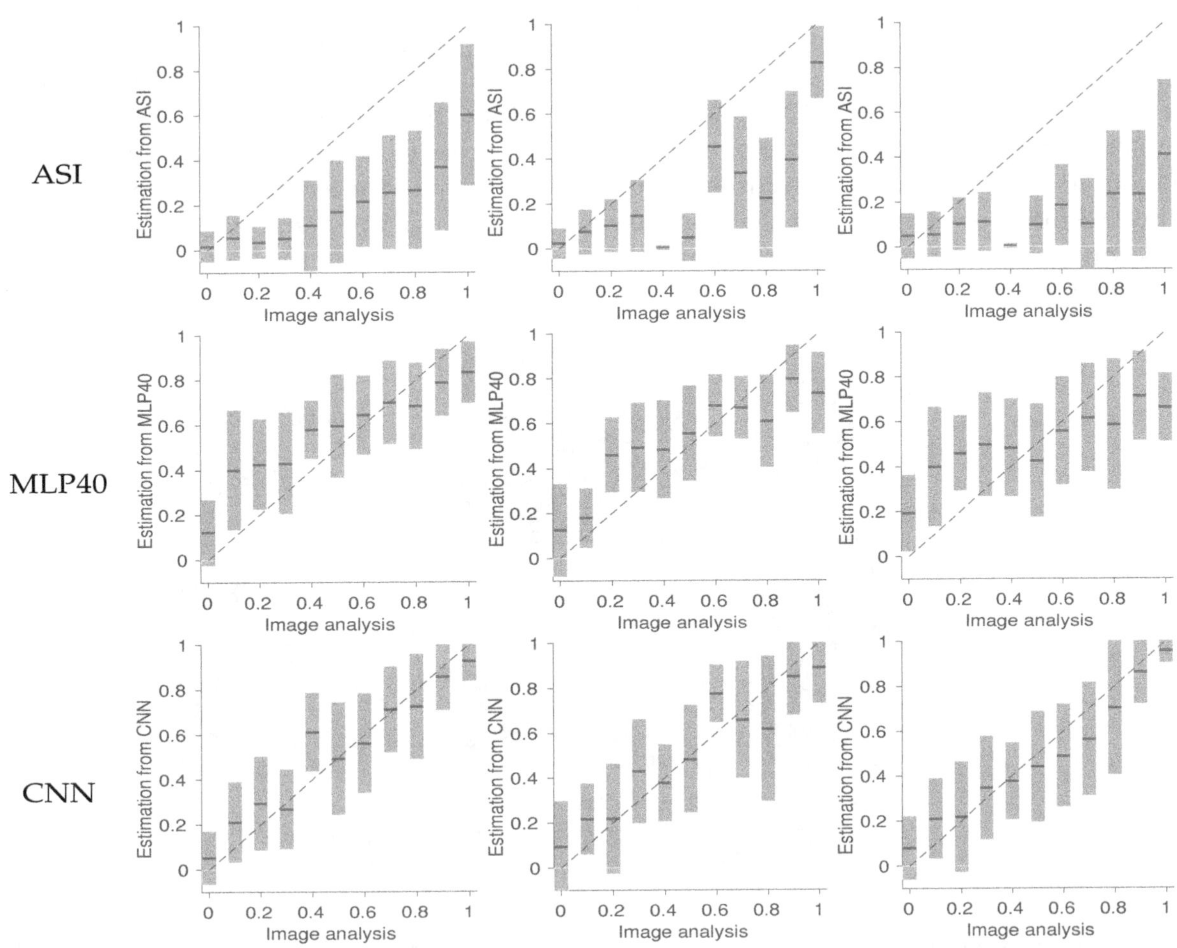

Figure 2. Errors at different ice concentration levels for ASI (1st row), MLP40 (2nd row), and CNN (3rd row) for training (1st column), validation (2nd column) and testing (3rd column) sets. The red lines represent the mean ice concentration, and half length of a bar represents the error standard deviation.

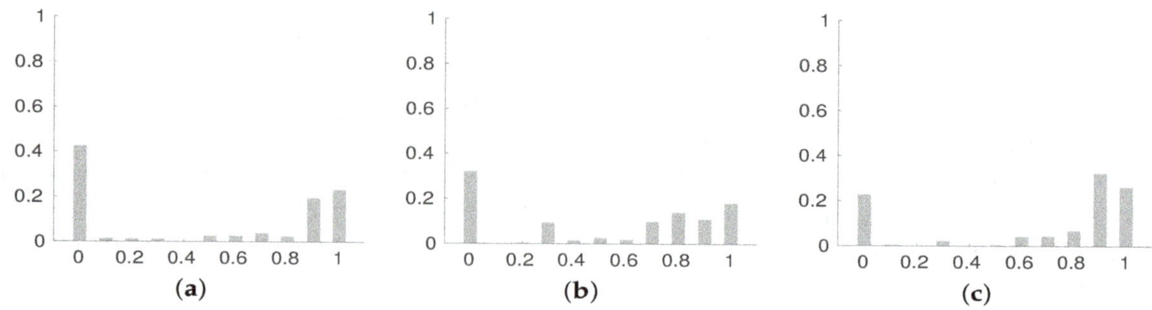

Figure 3. Histogram of the percentage of samples from each 10% interval of the image analyses for training, validation and testing dataset of the Gulf of Saint Lawrence. The training samples are strongly biased since the majority of the training samples are either water or ice. (**a**) Training; (**b**) Validation; (**c**) Testing.

6.2. Comparison between MLP and CNN

All SAR based algorithms produce ice concentration estimates with more details and sharper ice-water boundaries than the ASI data (see Figures 4 and 5), which may be due to the higher resolution of SAR images, and the fact that regions of thin ice are reasonably well captured in the training data used for the SAR based methods. Figure 4e,f shows that MLP40 is more sensitive to backscatter changes in SAR images than the CNN. Therefore, MLP40 produces more details in the ice concentration estimates, as well as an ice cover that appears noisy (e.g., spurious ice can be seen over open water regions). This can sometimes introduce errors, noted in the lower left portion of Figure 6d. The ice concentration estimates by the CNN contain fewer visible errors in assignment of ice concentration than the result of MLP40, but more details than the image analysis charts, especially in low ice concentration regions and marginal ice zones (Figure 6). These differences may be caused by the difficulty to manually identify accurate boundaries of low ice concentration regions by ice analysts or the limited number of polygons they can use for each image analysis, or simply the fact that the ice charts contain an estimate of ice concentration in 10% intervals over a region (polygon) identified as homogeneous.

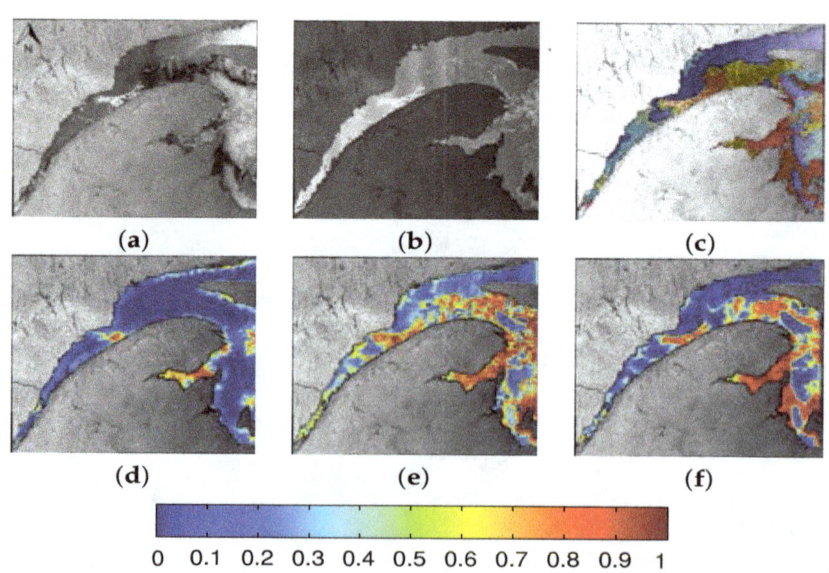

Figure 4. Ice concentration estimated by CNN compared to that from other methods. The HH and HV images are shown in panels (**a**) and (**b**) repectively. Panel (**c**) is the image analysis, (**d–f**) are the ice concentration from ASI, MLP40 and CNN, repectively. Scene shown is 20140117_103914, which is used for testing. Scene centered at 47.99°N, 66.85°W with extent of 500 km by 500 km.

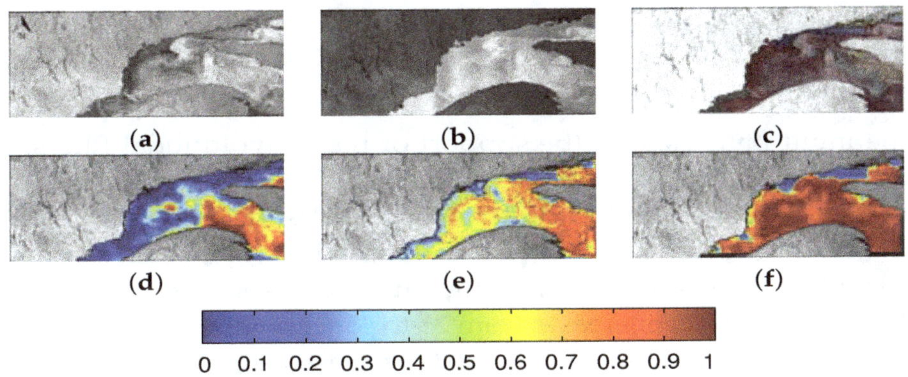

Figure 5. Ice concentration estimated by CNN compared to that from other methods. The HH and HV images are shown in panels (**a**) and (**b**) respectively. Panel (**c**) is the image analysis. Panels (**d–f**) are ice concentration from ASI, MLP40 and CNN respectively. Scene shown is 20140210_103911, which is used for testing. Scene centered at 49.90°N, 66.42°W with extent of 500 km by 230 km.

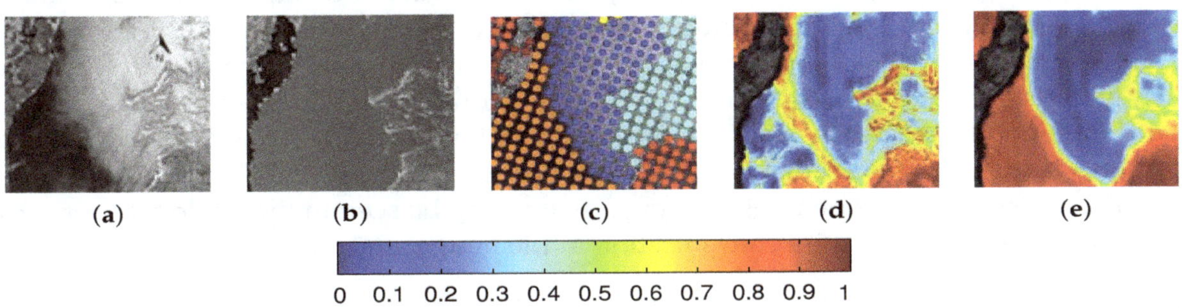

Figure 6. An example shows the details for a region with new ice and water. The ASI result is mainly water for this region. It can be seen MLP40 (**d**) produces noisy ice concentration estimates with new ice in the bottom left identified as water with some ice of low ice concentration. The CNN (**e**) is able to correctly identify new ice and water with higher accuracy. Subscene of dimension 60 km × 60 km from 20140117_103914 centered at 47.60°N, 64.13°W. The HH image, HV image and image analysis are shown in panels (**a–c**) respectively.

Strong banding in the HV channel of the RADARSAT-2 imagery may cause overestimation of ice concentration for water regions. Such an example is given in Figure 7, where MLP and CNN overestimate ice concentration for water regions with strong banding in the HV pol. The level of overestimation is reduced slightly when a larger patch size (55 vs. 45) is used for the CNN.

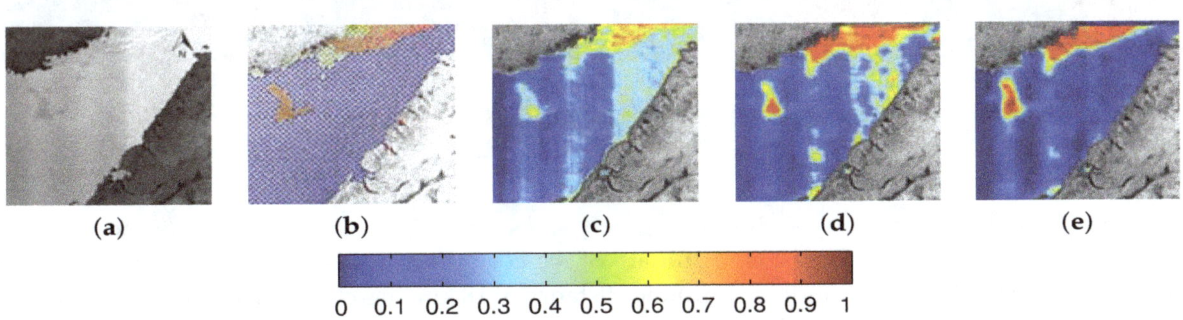

Figure 7. Example of water misidentified as ice for both MLP40 and CNN due to the banding effect in HV pol. Water in the right part of HV pol (**a**) is obviously brighter than water in the left. Water regions are estimated incorrectly for MLP40 (**c**), and CNN with patch size 45 (**d**). Results from the CNN are improved when a patch size of 55 is used, as shown in panel (**e**), although the features are also less sharp. Subscene centered at 49.72°N, 59.11°W of dimension 200 km × 200 km from 20140121_214420. Image analysis is shown in panel (**b**).

6.3. *Evaluation of CNN Architecture and Parameters*

6.3.1. Patch Size

The size of the input patches, and the support of the convolutional filters, are related to the intrinsic scale and complexity of the problem. The impact of patch size was evaluated by examining the output of the CNN for patch sizes of 25, 35, 45 and 55, corresponding to 10 km (25 × 400 m), 14 km (35 × 400 m), 18 km (45 × 400 m) and 22 km (55 × 400 m). With larger patch size, the model is a better fit to the training data and the E_{rmse} of the training data decreases. The E_{rmse} for test and validation data decreases when the patch size increases from 25 to 45. However, when the patch size increased from 45 to 55, the E_{rmse} for the test and validation data increased slightly, which could be an indication of slight overfitting for the dataset used. Therefore, a patch size of 45 was used in this study. Note that for a different dataset, the patch size selected may be different.

The impact of patch size on the estimated ice concentration can be seen in the regions contaminated with either banding or wind roughened open water. Examples are shown in Figure 8. The smaller patch sizes (Figure 8e,f) lead to spurious ice in water regions due to wind and banding. These results suggest that the separation of water and ice requires spatial context information over a larger region. This is also seen in studies using GLCM statistics to separate ice from water, in which case the separation of the two generally improves when larger patches are considered [7]. In contrast, ice is generally well identified for all tested patch sizes. Using small patch sizes tends to slightly underestimate ice concentration, leading to ice cover that is less homogeneous, as compared to larger patch sizes. For the patch size of 25, in some cases openings (i.e., open water) can be seen in the ice cover (not shown) for polygons corresponding to 100% ice concentration in the image analysis chart.

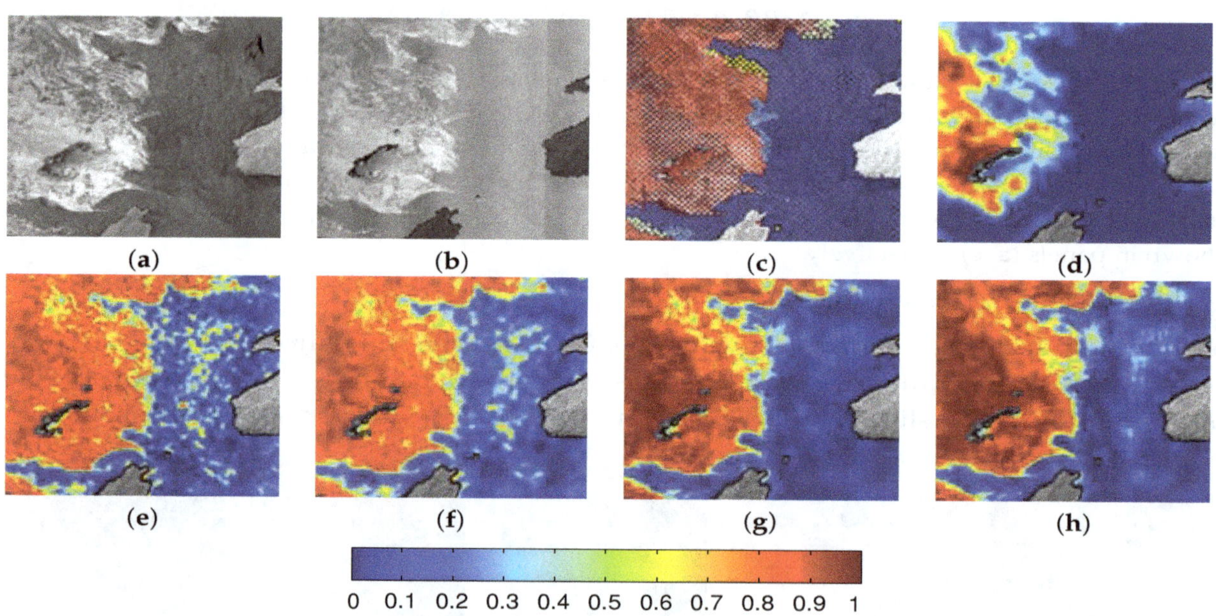

Figure 8. Visual comparison of different patch sizes, (**e**) 25 × 25 pixels, (**f**) 35 × 35 pixels, (**g**) 45 × 45 pixels, (**h**) 55 × 55 pixels. Estimate of ice concentration is improved when patch size increases. Patch size 45, corresponding to ground distance of 18 km, has cleaner water estimates than the others. Subscene of dimension 270 km × 270 km from 20140124_215646 centered at 47.86°N, 60.94°W. Panels (**a**–**d**) are the HH image, HV image, image analysis chart and ASI ice concentration respectively.

6.3.2. Use of Incidence Angle Data

The results shown in previous sections used input image patches consisting of HH pol, HV pol and incidence angle. To investigate the impact of including incidence angle on the estimated ice concentration, CNNs are trained, validated and tested, with HH pol and HV pol only. The network structure used is the same as that for the CNN without incidence angle (Table 2). The ice concentration from the CNN is evaluated against image analysis charts, results are given in Table 5. The errors are higher in all cases when incidence angle is included. This is in part likely due to the fact that including the incidence angle information leads to greater dependency of the CNN on the HH channel. This may also be due to the fact that with a third channel of input, the model is larger (there are more weights that need to be trained), and is therefore has more potential to overfit the training data.

Table 5. The average error statistics for networks trained with or without incidence angle data using CNN on the Gulf of Saint Lawrence data.

	Set	E_{sgn}	E_{L1}	E_{std}	E_{rmse}
with incidence angle	Training	−0.0039	0.0845	0.1506	0.1507
	Validation	−0.0123	0.1253	0.2056	0.2059
	Testing	−0.0274	0.1295	0.2197	0.2214
without incidence angle	Training	0.0052	0.0817	0.1434	**0.1435**
	Validation	0.0035	0.1183	0.1837	**0.1836**
	Testing	−0.0119	0.1220	0.2031	**0.2035**

Due to the reduced dependency of the CNN on the HV pol, and more significant extraction of information from the HH pol with the use of incidence angle, the banding effect from HV is reduced, but the ice concentration estimates appear to be more sensitive to wind roughening. New ice is more likely to be correctly identified when incidence angle is used (Figure 9), in particular for cases when there are features visible in the HH image that appear to indicate a region of new ice.

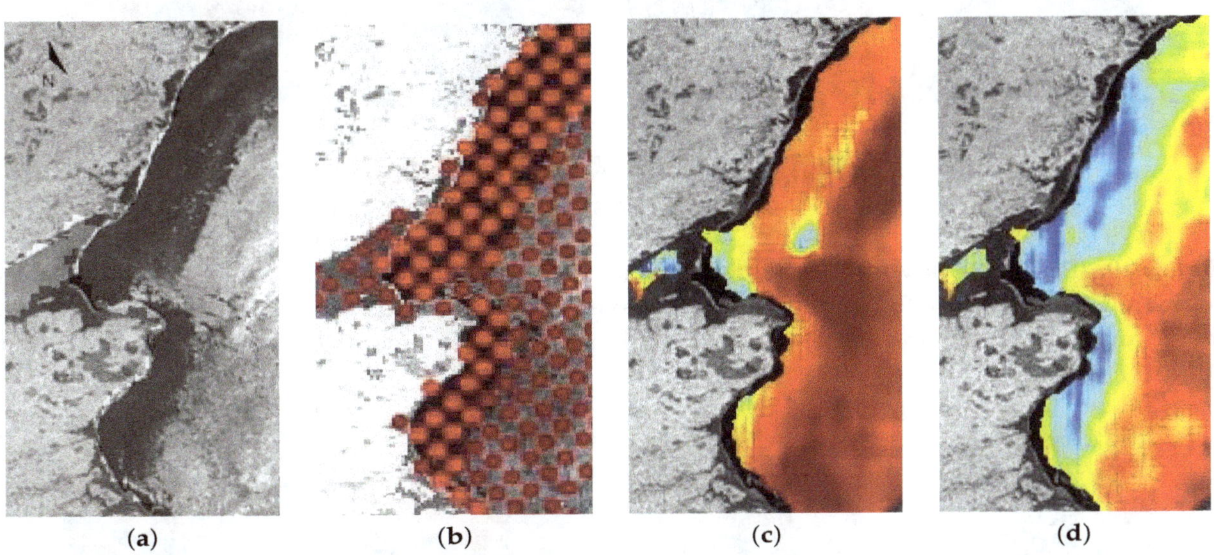

Figure 9. New ice can be seen in the HH image as the dark regions along the coast (**a**). This ice is correctly identified when incidence angle data are used (**c**), as compared to when the incidence angle data is not used (**d**). Subscene of dimension 120 km × 52 km from 20140206_221744 centered at 47.12°N, 64.72°W. Image analysis is shown in panel (**b**).

6.3.3. Network Depth

Network depth is the number of convolutional layers in the CNN, where each layer contains a filtering, non-linear activation and pooling operation. The network depth is an important parameter that determines the level of abstraction used for classification or regression. Here, CNN models with two and three convolutional layers are trained and evaluated. In both cases, there are two fully connected layers after the convolutional layers. The error statistics against image analyses are illustrated in Table 6. Although the use of two or three convolutional layers in the networks generates similar error statistics, visually, the network with three convolutional layers produces smoother and more reasonable ice concentration estimates, as shown in Figure 10. This makes sense as deeper networks extract more abstract features so that the results are less sensitive to raw pixel values. The ice-covered regions in Figure 10a that are incorrectly identified by the network with two convolutional layers (Figure 10c) are correctly identified by the network with three convolutional layers (Figure 10d). Regions that can be visually identified as open water in Figure 11a look cleaner when three layers are used, as shown in Figure 11d. While similar results (meaning sharper features with increasing layers) are obtained when more convolutional layers are used, as adding more layers leads to increased computational complexity, the three-convolutional-layer structure is deemed adequate.

Table 6. Average error statistics for networks with two convolutional layers and three convolutional layers on the Gulf of Saint Lawrence dataset.

	Two Convolutional Layers				Three Convolutional Layers			
Set	E_{sgn}	E_{L1}	E_{std}	E_{rmse}	E_{sgn}	E_{L1}	E_{std}	E_{rmse}
Training	−0.0055	0.0874	0.1266	**0.1269**	−0.0039	0.0845	0.1506	0.1507
Validation	−0.0028	0.1229	0.1933	**0.1934**	−0.0123	0.1253	0.2056	0.2059
Testing	0.0054	0.1556	0.2300	0.2302	−0.0274	0.1295	0.2197	**0.2214**

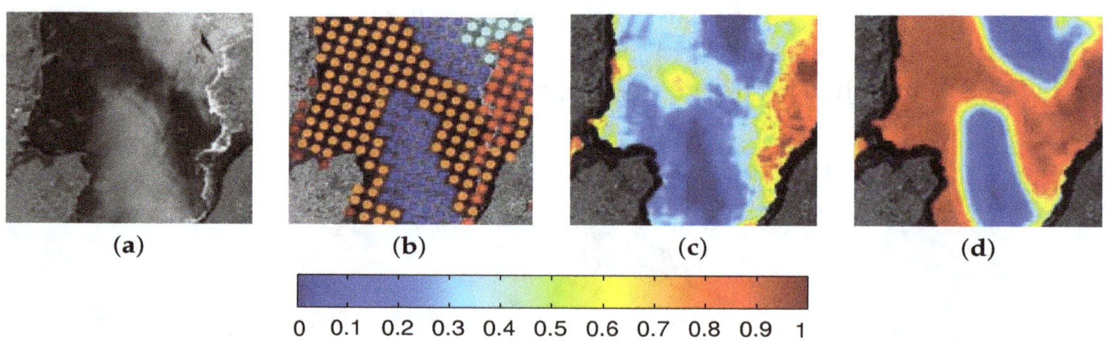

Figure 10. The network with three convolutional layers (**d**), improves the estimation for new ice compared to network with two convolutional layers (**c**). Panel (**a**) is the HH image, and (**b**) is the image analysis chart. Subscene of dimension 8 km × 8 km centered at 47.06°N and 64.46°W from 20140117_103914.

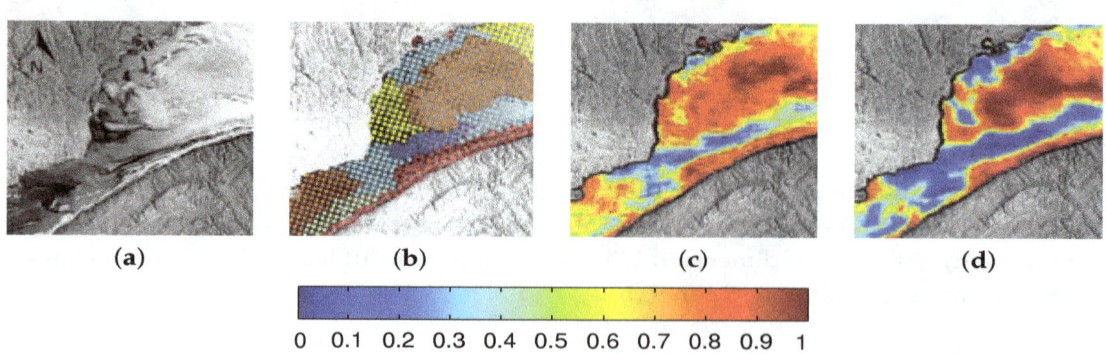

Figure 11. Comparison of results produced by networks with two-convolutional-layer (**c**) and three-convolutional-layer structures (**d**) for a sample location centered at 49.57°N, 66.59°W with size 200 km × 200 km in scene 20140127_104734 in Gulf of Saint Lawrence. Estimate by the two-convolutional-layer network is noisier. The three-convolutional-layer network produces smoother and more reasonable results. Panel (**a**) is the HH image and (**b**) is the image analysis chart for the subregion.

7. Discussion

In this study, a CNN has been applied to estimate sea ice concentration from dual-polarized SAR images in the Gulf of St. Lawrence. State-of-the-art ice concentration estimates with finer details than the image analysis chart are generated. Experiments using HV pol or HH pol only have also been carried out (results not shown here). Using dual-pol SAR imagery leads to improved ice concentration estimates as compared to using HH pol or HV pol only. When using HH pol only, the results are strongly affected by the incidence angle, which causes overestimation of ice concentration for water regions at low incidence angles. Using only HV pol shows banding in the estimated ice concentration. Similar results have been demonstrated in previous studies [4,52].

Sea ice concentration from image analysis charts was selected as the training data for this study. These charts contain regions (polygons) labelled by a trained analyst as having homogeneous ice conditions. When the image analysis charts were sampled, a single pixel from the image analysis (representing an area of 8 km $\times$ 5 km) was associated with a patch from the SAR image (representing an area of 18 km $\times$ 18 km). This means the SAR image patches could overlap polygon boundaries. While it may have been more appropriate to sample the image analysis charts to avoid this overlap, the accuracy of the polygon boundaries is not known. The image analyses are also subjective manual analyses, and are known to contain errors [36], as is the case with any ice concentration analysis. Even if it can be assumed the polygon boundaries are accurate, the use of spatially discrete polygons to represent the ice concentration over an image of continuous grey levels, introduces sampling errors in the ice concentration estimates. Preliminary work on the impact of errors in the training data, and alternative methods to train a CNN to estimate sea ice concentration, can be found in [53]. Learning a sparse representation of the data could improve the ice concentration estimates when training sample quality and quantity are not sufficient [54].

Testing demonstrates that the CNN is robust to the changes in image tone with incidence angle, even without explicitly including incidence angle data as an input. When the incidence angle data was included, an increased dependence on the HH pol image was observed in the ice concentration. Windspeed information could also be included as an additional input, which could help reduce the spurious ice that appears in some cases over open water when it appears to be wind-roughened. This would require accurate windspeed information at a sufficiently high spatial resolution, which is not presently available.

A linear activation function has been chosen for the last fully connected layer, which means that ice concentration values can be estimated that are greater than 1 or less than zero. For comparison between the different methods these ice concentration values were truncated to remain in the range of [0,1]. A sigmoid activation would be a more intuitive choice, as it naturally bounds the output of the CNN to 0 and 1. However, in our experiment, sigmoid activation was found to produce saturated ice concentration predictions close to 0 or 1, and large errors for intermediate ice concentration levels.

8. Conclusions

The CNN has been found to generate ice concentration estimates with improved details and accuracy as compared to ASI passive microwave ice concentration products when IA charts are used as the verification data. Our CNN ice concentration is also improved as compared to that from a method that uses an MLP to regress ice concentration from a set of engineered SAR image features. Because of the shallow network structure, MLP40 is more sensitive to the SAR image backscatter values than the CNN, which causes noisy ice concentration estimates. The small model used by MLP40 does not have the large learning capacity as the CNN. Some complex cases, such as dark new ice, are not recognized correctly. This causes systematic errors in the results, which cannot be corrected by segmentation based post-processing. Therefore, the deeper and larger CNNs used here can generate more accurate ice concentration estimates than MLP40. Note that while a multilayer version of MLP40 could be developed, maintaining full connectivity between the weights in these layers would require many weights to be learned, making such networks prone to overfitting [13]. Compared to standard fully connected neural networks with similar number of units, CNNs are able to model local spatial information more efficiently with fewer trainable parameters, which also makes them easier to train [26,55]. The success of CNNs as multi-layer networks is due to weight sharing and the local connectivity between adjacent layers [38], and methods developed to reduce overfitting [26], such as training sample augmentation and dropout, which have been implemented in the present study.

We note that there are alternative approaches to using a CNN for this problem that may be more efficient than that presented here. For example, methods that predict dense labelling as compared to a label [30,56] at a single pixel location (as has been done in the present study). Preliminary work using such an architecture for the GSL data has been presented in [53], and will be investigated further in a future study.

Acknowledgments: This study was funded through the Chinese Scholarship Council, the National Science and Engineering Research Council of Canada, ArcticNet Networks of Centres of Excellence, and the Grants and Contributions program through Environment Canada. The authors would like to thank Lynn Pogson and Alain Caya for providing the SAR imagery and image analysis charts.

Author Contributions: Lei Wang carried out the data processing and analysis, and contributed to the research design and manuscript writing. Andrea Scott led manuscript writing, contributed to the research design and co-supervised this study. David Clausi contributed to manuscript writing and research design, and co-supervised this study.

References

1. Carrieres, T.; Greenan, B.; Prinsenberg, S.; Peterson, I. Comparison of Canadian daily ice charts with surface observations off Newfoundland, winter 1992. *Atmos. Ocean* **1996**, *34*, 207–226.
2. Arkett, M.; Braithwaite, L.; Pestieau, P.; Carrieres, T.; Pogson, L.; Fabi, C.; Geldsetzer, T. Preparation by the Canadian Ice Service for the operational use of the RADARSAT Constellation Mission in their ice and oil spill monitoring programs. *Can. J. Remote Sens.* **2015**, *41*, 380–389.
3. Karvonen, J. Baltic sea ice concentration estimation based on C-band HH-polarized SAR data. *Sel. Top. Appl. Earth Obs. Remote Sens.* **2012**, *5*, 1874–1884.
4. Karvonen, J. Baltic sea ice concentration estimation based on C-band dual-polarized SAR data. *IEEE Trans. Geosci. Remote Sens.* **2014**, *52*, 5558–5566.
5. Clausi, D.A. Comparison and fusion of co-occurrence, Gabor, and MRF texture features for classification of SAR sea ice imagery. *Atmos. Ocean* **2001**, *39*, 183–194.
6. Deng, H.; Clausi, D.A. Unsupervised segmentation of synthetic aperture radar sea ice imagery using a novel Markov random field model. *IEEE Trans. Geosci. Remote Sens.* **2005**, *43*, 528–538.
7. Leigh, S.; Wang, Z.; Clausi, D.A. Automated ice-water classification using dual polarization SAR satellite imagery. *IEEE Trans. Geosci. Remote Sens.* **2014**, *52*, 5529–5539.
8. Zakhvatkina, N.Y.; Alexandrov, V.Y.; Johannessen, O.M.; Sandven, S.; Frolov, I.Y. Classification of Sea Ice Types in ENVISAT Synthetic Aperture Radar Images. *IEEE Trans. Geosci. Remote Sens.* **2013**, *51*, 2587–2600.
9. Liu, J.; Scott, K.; Gawish, A.; Fieguth, P. Automatic detection of the ice edge in SAR imagery using curvelet transform and active contour. *Remote Sens.* **2016**, *8*, doi:10.3390/rs8060480.
10. Pogson, L.; Geldsetzer, T.; Buehner, M.; Carrieres, T.; Ross, M.; Scott, K. A collection of empirically-derived characteristic values from SAR across a year of sea ice environments for use in data assimilation. *Mon. Weather Rev.* **2016**, in press.
11. Buehner, M.; Caya, A.; Pogson, L.; Carrieres, T.; Pestieau, P. A new Environment Canada regional ice analysis system. *Atmos. Ocean* **2013**, *51*, 18–34.
12. Drusch, M. Sea ice concentration analyses for the Baltic Sea and their impact on numerical weather prediction. *J. Appl. Meteorol. Climatol.* **2006**, *45*, 982–994.
13. LeCun, Y.; Bengio, J.; Hinton, G. Deep Learning. *Nature* **2015**, *521*, 436–444.

14. Wang, L.; Scott, K.A.; Xu, L.; Clausi, D.A. Sea Ice Concentration Estimation During Melt From Dual-Pol SAR Scenes Using Deep Convolutional Neural Networks: A Case Study. *IEEE Trans. Geosci. Remote Sens.* **2016**, *54*, 4524–4533.
15. Geldsetzer, T.; Yackel, J. Sea ice type and open water discrimination using dual co-polarized C-band SAR. *Can. J. Remote Sens.* **2009**, *35*, 73–84.
16. Ivanova, N.; Tonboe, R.; Pedersen, L.T. *SICCI Product Validation and Algorithm Selection Report (PVASR)—Sea Ice Concentration*; Technical Report; European Space Agency: Paris, France, 2013.
17. Dumbill, E. *Strata 2012: Making Data Work*; O'Reilly: Santa Clara, CA, USA, 2012.
18. Najafabadi, M.M.; Villanustre, F.; Khoshgoftaar, T.M.; Seliya, N.; Wald, R.; Muharemagic, E. Deep learning applications and challenges in big data analytics. *J. Big Data* **2015**, *2*, doi:10.1186/s40537-014-0007-7.
19. National Research Council. *Frontiers in Massive Data Analysis*; The National Academies Press: Washington, DC, USA, 2013.
20. Domingos, P. A Few Useful Things to Know About Machine Learning. *Commun. ACM* **2012**, *55*, 78–87.
21. Bengio, Y.; Courville, A.; Vincent, P. Representation learning: A review and new perspectives. *IEEE Trans. Pattern Anal. Mach. Intell.* **2013**, *35*, 1798–1828.
22. Hinton, G.E. Learning multiple layers of representation. *Trends Cogn. Sci.* **2007**, *11*, 428–434.
23. Bengio, Y. Learning deep architectures for AI. *Found. Trends Mach. Learn.* **2009**, *2*, 1–127.
24. Lee, H.; Grosse, R.; Ranganath, R.; Ng, A.Y. Convolutional deep belief networks for scalable unsupervised learning of hierarchical representations. In Proceedings of the 26th Annual International Conference on Machine Learning, Montreal, QC, Canada, 14–18 June 2009; pp. 609–616.
25. Ciresan, D.C.; Meier, U. Flexible, high performance convolutional neural networks for image classification. In Proceedings of the Twenty-Second International Joint Conference On Artificial Intelligence, Barcelona, Spain, 16–22 July 2011; pp. 1237–1242.
26. Krizhevsky, A.; Sutskever, I.; Hinton, G.E. ImageNet classification with deep convolutional neural networks. In Proceedings of the 25th International Conference on Neural Information Processing Systems, Lake Tahoe, NV, USA, 3–6 December 2012; pp. 1097–1105.
27. Karpathy, A.; Toderici, G.; Shetty, S.; Leung, T.; Sukthankar, R.; Fei-Fei, L. Large-scale video classification with convolutional neural networks. In Proceedings of the 2014 IEEE Conference on Computer Vision and Pattern Recognition, Columbus, OH, USA, 23–28 June 2014; pp. 1725–1732.
28. Mnih, V.; Hinton, G.E. Learning to detect roads in high-resolution aerial images. In *Computer Vision-ECCV 2010*; Springer: Berlin/Heidelberg, Germany, 2010; pp. 210–223.
29. Chen, X.; Xiang, S.; Liu, C.L.; Pan, C.H. Vehicle Detection in Satellite Images by Hybrid Deep Convolutional Neural Networks. *IEEE Geosci. Remote Sens. Lett.* **2014**, *11*, 1797–1801.
30. Maggiori, E.; Tarabalka, Y.; Charpiat, G.; Alliez, P. Fully convolutional neural networks for remote sensing image classification. In Proceedings of the 2016 IEEE Geoscience and Remote Sensing Symposium, Beijing, China, 10–15 July 2016.
31. Spreen, G.; Kaleschke, L.; Heygster, G. Sea ice remote sensing using AMSR-E 89-GHz channels. *J. Geophys. Res.* **2008**, *113*, doi:10.1029/2005JC003384.
32. Agnew, T.; Howell, S. The use of operational ice charts for evaluating passive microwave ice concentration data. *Atmos. Ocean* **2003**, *41*, 317–331.
33. Karvonen, J.; Vainio, J.; Marnela, M.; Eriksson, P.; Niskanen, T. A comparison between high-resolution EO-based and ice analyst-assigned sea ice concentrations. *J. Sel. Top. Appl. Earth Obs. Remote Sens.* **2015**, *8*, 1799–1807.
34. Fequest, D. *MANICE: Manual of Standard Procedures for Observing and Reporting Ice Conditions*; Environment Canada: Ottawa, ON, Canada, 2002.
35. Slade, B. *RADARSAT-2 Product Description*; MacDonald, Dettwiler and Associates Ltd.: Richmond, BC, Canada, 2009.
36. Moen, A.; Doulgeris, P.; Anfinsen, S.; Renner, A.; Hughes, N.; Gerland, S.; Eltoft, T. Comparison of feature based segmentation of full polarimetric SAR satellite sea ice images with manually drawn ice charts. *Cryosphere* **2013**, *7*, 1693–1705.
37. De Andrade, A. *Best Practices for Convolutional Neural Networks Applied to Object Recognition in Images*; Technical Report; Department of Computer Science, University of Toronto: Toronto, ON, USA, 2014.

38. LeCun, Y.; Bottou, L.; Bengio, Y.; Haffner, P. Gradient-based learning applied to document recognition. *Proc. IEEE* **1998**, *86*, 2278–2324.
39. LeCun, Y.; Huang, F.J.; Bottou, L. Learning methods for generic object recognition with invariance to pose and lighting. In Proceedings of the 2004 IEEE Computer Society Conference on Computer Vision and Pattern Recognition, Washington, DC, USA, 27 June–2 July 2004.
40. LeCun, Y.; Kavukcuoglu, K.; Farabet, C. Convolutional networks and applications in vision. In Proceedings of 2010 IEEE International Symposium on Circuits and Systems (ISCAS), Paris, France, 30 May–2 June 2010; pp. 253–256.
41. Chen, L.C.; Papandreou, G.; Kokkinos, I.; Murphy, K.; Yuille, A.L. Semantic image segmentation with deep convolutional nets and fully connected CRFs. *arXiv* **2016**, arXiv:1412.7062.
42. Hastie, T.; Tibshirani, R.; Friedman, J. *The Elements of Statistical Learning*; Springer: Berlin/Heidelberg, Germany, 2009; Volume 2.
43. Bishop, C.M. *Pattern Recognition and Machine Learning*; Springer: Berlin/Heidelberg, Germany, 2006.
44. Nair, V.; Hinton, G.E. Rectified linear units improve restricted Boltzmann machines. In Proceedings of the 27th International Conference on Machine Learning (ICML-10), Haifa, Israel, 21–24 June 2010; pp. 807–814.
45. Scherer, D.; Müller, A.; Behnke, S. Evaluation of pooling operations in convolutional architectures for object recognition. In *Artificial Neural Networks–ICANN 2010*; Springer: Berlin/Heidelberg, Germany, 2010; pp. 92–101.
46. LeCun, Y.; Bottou, L.; Orr, G.; Müller, K. Efficient backprop. In *Neural Networks: Tricks of the Trade*; Springer: Berlin/Heidelberg, Germany, 2012.
47. Prechelt, L. Early stopping-but when? In *Neural Networks: Tricks of the Trade*; Springer: Berlin/Heidelberg, Germany, 2012; pp. 53–67.
48. Chatfield, K.; Simonyan, K.; Vedaldi, A.; Zisserman, A. Return of the Devil in the Details: Delving Deep into Convolutional Nets. *arXiv* **2014**, arXiv:1405.3531.
49. Hinton, G.E.; Srivastava, N.; Krizhevsky, A.; Sutskever, I.; Salakhutdinov, R.R. Improving neural networks by preventing co-adaptation of feature detectors. *arXiv* **2012**, arXiv:1207.0580.
50. Jia, Y.; Shelhamer, E.; Donahue, J.; Karayev, S.; Long, J.; Girshick, R.; Guadarrama, S.; Darrell, T. Caffe: Convolutional architecture for fast feature embedding. In Proceedings of the ACM International Conference on Multimedia, Orlando, FL, USA, 3–7 November 2014; pp. 675–678.
51. Watkins, J.C. Probability Theory. In *Course Note for Probability Theory*; University of Arizona: Tucson, AZ, USA, 2006.
52. Wang, L.; Scott, K.; Clausi, D. Automatic feature learning of SAR images for sea ice concentration estimation using feed-forward neural networks. In Proceedings of the 2014 IEEE Geoscience and Remote Sensing Symposium, Quebec City, QC, Canada, 13–18 July 2014; pp. 3969–3971.
53. Wang, L. Learning to Estimate Sea Ice Concentration from SAR Imagery. Ph.D. Thesis, University of Waterloo, Waterloo, ON, Canada, 2016.
54. Wang, Q.; Lin, J.; Yuan, Y. Salient band selection for hyperspectral image classification via manifold ranking. *IEEE J. Neural Netw. Learn. Syst.* **2016**, *27*, 1279–1289.
55. LeCun, Y. Generalization and network design strategies. In *Connections in Perspective*; Elsevier: Amsterdam, The Netherlands, 1989; pp. 143–155.
56. Maggiori, E.; Tarabalka, Y.; Chariot, G.; Alliez, P. High-resolution semantic labeling with convolutional neural networks. *arXiv* **2016**, arXiv:1611.01962v1.

8

Classification for High Resolution Remote Sensing Imagery using a Fully Convolutional Network

Gang Fu [1,2,*], Changjun Liu [2], Rong Zhou [3], Tao Sun [2] and Qijian Zhang [4]

1 Department Of Engineering Physics, Tsinghua University, Beijing 100084, China
2 China Institute of Water Resources and Hydropower Research (IWHR), Beijing 100038, China; lcj2005@iwhr.com (C.L.); sunt@iwhr.com (T.S.)
3 Beijing Soil and Water Conservation Center, Beijing 100036, China; zhour@bjwater.gov.cn
4 Water Resources Information Center of Henan Province, Zhengzhou 450003, China; zqj@hnsl.gov.cn
* Correspondence: gangfu2008@hotmail.com.

Academic Editors: Qi Wang, Nicolas H. Younan, Carlos López-Martínez, Lenio Soares Galvao and Prasad S. Thenkabail

Abstract: As a variant of Convolutional Neural Networks (CNNs) in Deep Learning, the Fully Convolutional Network (FCN) model achieved state-of-the-art performance for natural image semantic segmentation. In this paper, an accurate classification approach for high resolution remote sensing imagery based on the improved FCN model is proposed. Firstly, we improve the density of output class maps by introducing Atrous convolution, and secondly, we design a multi-scale network architecture by adding a skip-layer structure to make it capable for multi-resolution image classification. Finally, we further refine the output class map using Conditional Random Fields (CRFs) post-processing. Our classification model is trained on 70 GF-2 true color images, and tested on the other 4 GF-2 images and 3 IKONOS true color images. We also employ object-oriented classification, patch-based CNN classification, and the FCN-8s approach on the same images for comparison. The experiments show that compared with the existing approaches, our approach has an obvious improvement in accuracy. The average precision, recall, and Kappa coefficient of our approach are 0.81, 0.78, and 0.83, respectively. The experiments also prove that our approach has strong applicability for multi-resolution image classification.

Keywords: deep learning; convolutional neural network (CNN); fully convolutional network (FCN); classification; remote sensing; high resolution

1. Introduction

Classification is a fundamental task for remote sensing imagery analysis. Applying intelligent methods, such as pattern recognition and statistical learning, is an effective way to obtain class information of ground objects. It is always the main focus of research and commercial development. Early classification was mainly for low spatial resolution (10–30 m) images and pixel-leveled images, including unsupervised classification (also known as clustering, such as K-means [1]) and supervised classification (such as Neural Networks [2,3] and Support Vector Machines [4,5]). These methods often use only spectral information of the images, and have formed general modules in commercial software, and have been successfully applied in land resources, environment, agriculture, and other fields. In recent years, some new approaches have appeared that are much superior to the traditional approaches. For example, Yuan Yuan et al. [6] and Qi Wang et al. [7] applied the latest achievements in the machine learning field, such as Manifold Ranking and Sparse Representation, to hyperspectral image classification.

High resolution (2 m spatial resolution and higher) remote sensing images contain more ground details. Many applications tend to obtain attributes of a ground object (such as a single building) rather than pixels. However, the pixel-level classification methods are sensitive to noise, and lack semantic meaning of the objects, and are difficult for obtaining object-level information. Therefore, object-oriented classification [8] is proposed, and it has made great achievements in high resolution image classification. At present, eCognition [9], ENVI [10], and other commercial software have developed object-oriented classification modules. Most of the object-oriented approaches perform a "segmentation-classification" mode. In the segmentation stage, Multi-Resolution (MR) [11], Full-Lambda Schedule (FLS) [12], Mean-Shift [13], Quadtree-Seg [14], and other image segmentation approaches are used to generate image segments, which we called image objects. In the classification stage, object features (color, texture, and geometric features) are calculated, which are taken as inputs of supervised or unsupervised classification, or a manually designed rule set for feature filtering, to achieve the final class discrimination.

Land-cover has various types, and is affected by noise, illumination, season, and many other factors, and brings great difficulties to classification using high resolution images. Even using the object-oriented approaches, accurate classification is still very difficult. From the pattern recognition perspective, selection/extraction of representative features is the bottleneck to improving accuracy. That is, the use of a specific set of features cannot be achieved on the classification for all kinds of ground objects. Therefore, learning features automatically from a remote sensing data set rather than using manually designed features, and then performing classification on the learned features, is an effective way to improve the accuracy of classification.

Deep learning theory was explicitly proposed by Hinton et al. [15] in 2006. It is a branch of machine learning based on a set of algorithms that attempt to model high level abstractions in data [16]. The basic motivation of deep learning is to establish a deep neural network to simulate the leaning and analysis mechanism of the human brain. Compared with the traditional machine learning theories, the most significant difference of deep learning is emphasizing automatic feature learning from a huge data set through the organization of multi-layer neurons. In recent years, various deep learning architectures such as Deep Belief Networks (DBN) [17], Convolutional Neural Networks (CNN) [18], and Recurrent Neural Networks (RNN) [19] have been applied to fields like computer vision [20,21], speech recognition, natural language processing, audio recognition, and bioinformatics, and they have been shown to produce state-of-the-art results in these domains.

In deep learning techniques, CNN has achieved remarkable results in image classification, recognition, and other vision tasks, and has the highest score on many visual databases such as ImageNet, Pattern Analysis, Statistical Modeling and Computational Learning Visual Object Classes (PASCAL VOC), and Microsoft Common Objects in Context (MS-COCO). For image classification, the basic structure of the standard CNN is stacks of "convolutional-pooling" layers as multi-scale feature extractors, and subsequent numbers of fully connected layers as classifiers. Many works on CNN-based remote sensing image analysis emerged in recent years. Nguyen et al. [22] presented an approach for satellite image classification using a five-layered network and achieved classification accuracy higher than 75%. Wang et al. [23] used a CNN structure with three layers and Finite State Machine (FSM) for road network extraction for long-term path planning. Marco Castelluccio et al. [24] explored the use of CNNs for the semantic classification of remote sensing scenes. Similarly, Hu et al. [25] also classified different scenes from high resolution remote sensing imagery using a pre-trained CNN model. Weixun Zhou et al. [26] employed CNN architecture as a deep feature extractor for high-resolution remote sensing image retrieval (HRRSIR). Volodymyr Mnih [27] proposed a CNN-based architecture to learn large scale contextual features for aerial image labeling. The model produces a dense classification patch, instead of outputting a single value image category. Martin Lagkvist et al. [28] presented a novel remote sensing imagery classification method based on CNNs for five classes (vegetation, ground, road, building, and water), outperforming the existing classification approaches. Besides the CNN family approaches, Yuan Yuan et al. [6] used a Stacked

AutoEncoder classifier for a classification experiment after using the Manifold Ranking based salient band selection.

The standard CNN is in an "image-label" manner and its output is the probability distribution over different classes. However, most of the remote sensing image classification expects a dense class map as the output, which has the same dimensions as the original image. A class map is a 2-D distribution of class labels with pixel correspondence, which is in a "pixel-label" mode. In the study of Martin Lagkvist et al. [28], a "per-pixel" classification is considered using overlapped patches and average post-processing. However, the use of the overlapped patches introduces too much redundant computations, and the averaging processing may easily lose useful edge information. Based on the standard CNN, Jonathan Long et al. [29] proposed the Fully Convolutional Network (FCN) model in 2015. By replacing fully connected (FC) layers in the standard CNN with convolutional layers, the FCN model maintains the 2-D structure of images, and firstly carries out CNN-based image semantic segmentation. In order to obtain a dense class map, Liang-Chieh Chen et al. [30] used the "atrous" convolution instead of the ordinary convolution, increasing the density of the predicted class labels, and then performed the Conditional Random Fields (CRFs) as post-processing to refine the region boundaries. The CRFs-based boundary refinement is also used in the works of Sakrapee et al. [31]. In order to integrate the CRFs procedure into the training stage, Shuai Zheng et al. [32] applied the idea of RNN to image segmentation, implementing an "end-to-end" training procedure. In the remote sensing society, several studies employ FCN-based approaches for dense class map generation. Jamie Sherrah [33] analyzed the down-sampling and up-sampling mechanism in CNNs, and adopted an FCN architecture for aerial image semantic labelling. The down-sampling mechanism of standard FCN is removed by involving deconvolution. D. Marmanis et al. [34] also used FCN and subsequent deconvolution architecture to perform a semantic segmentation for aerial images. Emmanuel Maggiori et al. [35–37] addressed the dense classification problem, and compared the patch-based CNN dense classification using CNN with FCN. With the advantages of FCN, the author proposed an end-to-end framework for large-scale remote sensing classification. A multi-scale mechanism was also considered by designing a specific neuron module that processes its input at multiple scales.

In this paper, we perform a FCN-based classification on high spatial resolution remote sensing imagery with 12 classes (bare land, grass, tree, water, building, cement ground, parking lot, playground, city road, trail, shadow, and others). These classes are typical ground objectives in city areas, and some of them (such as building, cement ground, road, and parking lot) are easily confused in traditional classification tasks. The class configurations were arranged to test the effectiveness of our approach in a complex environment. We fine-tuned the model parameters of the ImageNet-pretrained VGG-16 [37] network using GF-2 satellite images, to adapt it to our remote sensing imagery classification task. The VGG network has a more compact structure of convolutional and pooling layers, and achieved the highest classification accuracy for ImageNet ILSVRC-2014. To overcome the noise caused by pixel-level classification, we refine the region boundaries using fully connected CRFs, following the procedure of Liang-Chieh Chen et al. [30] and Sakrapee et al. [31]. The refined output is more readily applied to an object-oriented analysis.

We compare our approach with the object-oriented approach with MR segmentation [11] and SVM classification, patch-based CNN classification proposed in [27], and the FCN-8s approach proposed in [29], which achieved success for high resolution imagery classification or natural image segmentation. The result shows that our approach achieves higher accuracy in the classification. For those objectives which are difficult to be classified, our approach has lower confusion rates.

2. Methods

Similar to other supervised classification, our approach generally has two stages: the training stage and the classification stage, which is illustrated in Figure 1. In the training stage (the upper part of Figure 1), image-label pairs, with pixel-class correspondence, are input into the FCN network as training samples. The error between predicted class labels and ground truth (GT) labels is calculated and back-propagated through the network using the chain rule, and then the parameters of the FCN network are updated using the gradient descent method. The above iteration will be stopped when the error is less than a given threshold. In the classification stage (the lower part of Figure 1), the trained FCN network is performed on an input image to generate a rough class prediction. The rough class prediction, with the input image, is then input into the CRFs post-processing module to generate the final refined classification. The details of the training stage and classification stage are presented in Sections 2.2 and 2.3, respectively.

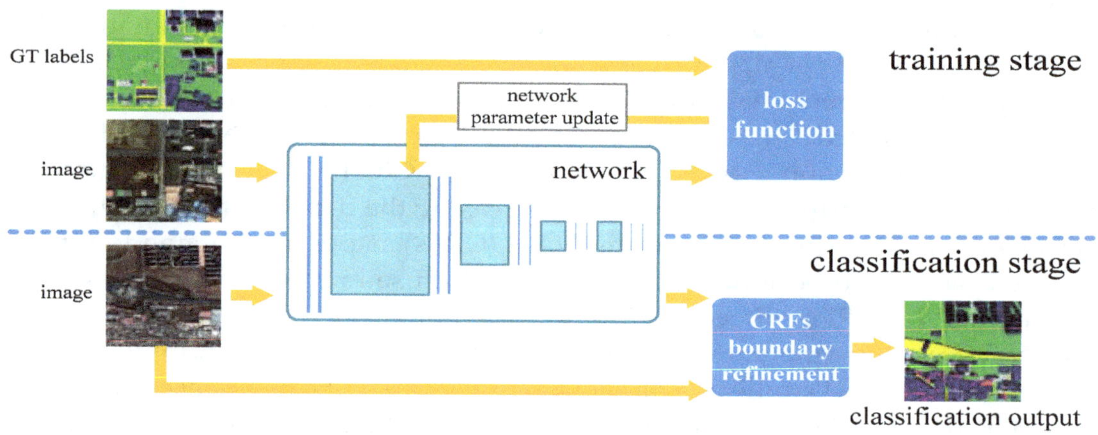

Figure 1. The general pipeline of our approach: The training stage and the classification stage are illustrated in the upper and lower parts, respectively.

2.1. *Network Architecture*

CNN currently is the state-of-the-art in visual recognition such as classification and detection. Simonyan et al. [38] developed the very deep CNN networks (VGG) by increasing the depth to 16–19 weight layers. To reduce the number of parameters in the networks, small 3×3 filters are used in all the convolutional layers. VGG models won the runner-up in ImageNet ILSVRC-2014. Although the subsequently emerged deeper models, such as ResNet [39] and Inception-V4 [40], achieved a higher score in many vision tasks, VGG networks have clear structures and compact memory requirements, which can be easily extended and applied, so we chose the 16-layered VGG network as our basic network architecture. Based on the VGG network, we constructed the FCN model by replacing the last three fully connected layers (two layers with 4096 neurons and one with 1000 neurons) with convolutional layers. Then following the idea of Liang-Chieh Chen et al. [30], we use "atrous" convolution (also known as "dilation" convolution in other studies) instead of the ordinary convolution to increase the feature density, and build the multi-scale classification model by adding the skip-layer network architecture.

2.1.1. Fully Convolutional Network

In classification tasks, the last structures in standard CNN are always several Fully Connected (FC) layers (see Figure 2a for illustration). These layers play the role of classifier like standard BP neural networks (For example, in Figure 2a, the 3 FC layers are similar to a 3-layered BP network with one hidden layer). From the first FC layer, the 2-D structure of the input image maintained by the convolutional-pooling layers is lost. The output of standard CNN is a 1-D distribution over classes (for a Softmax regression). It works in an "image-label" manner. In other words, given an image, it predicts one class label (a scalar) for it. The "image-label" mode has great advantages in single scene classification. The effectiveness has been presented in studies of Marco Castelluccio et al. [24] and Hu et al. [25].

However, in most remote sensing applications, a 2-D dense class map is required as an output. To maintain the 2-D structure, some approaches were presented based on the common CNN structures. The most typical one is the patch-based CNN approach [27,28]. The basic idea of patch-based CNN is: separate the large image into small patches, and apply the common CNN model on each patch to predict the class label(s) centered at the corresponding patch. Finally, the class labels will be arranged in a 2-D layout as the output. Jonathan Long et al. [29] proposed the FCN model, which is a convolutionalized version of CNN. FCN replaces all the FC layers with convolutional layers. Thus, the important 2-D structure of the image is maintained. Figure 2b is the illustration of the FCN model.

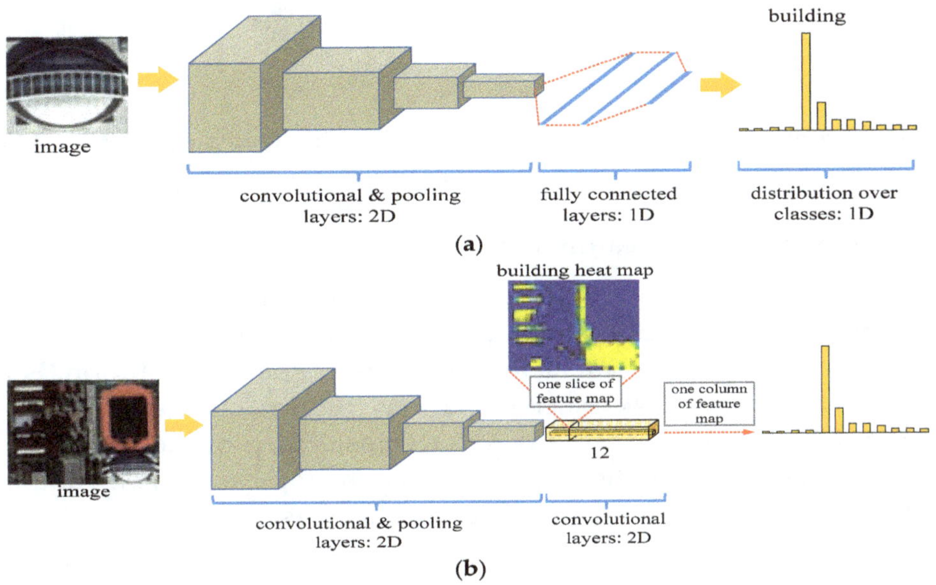

Figure 2. Network architectures for standard Convolutional Neural Network (CNN) and Fully Convolutional Network (FCN). (**a**) Architecture of standard CNN: stacks of convolutional-pooling layers and fully connected (FC) layers. Given an image, the distribution over classes is predicted. The class with the largest distribution value is considered as the class of a given image; (**b**) Architecture of FCN: FC layers are replaced by convolutional layers. FCN maintains the 2-D structure of the image.

Compared with patch-based CNN, the advantages of the FCN model are obvious for

- *Easy implementation*: The FCN architecture is designed brilliantly by replacing the FC layers by convolutional layers, which enables us to take arbitrary sized images as inputs. Additionally, by training entire images at a time instead of patch cropping, FCN does not have to rearrange the output labels together to obtain the label predictions and thus reduces the implementation complexity.
- *Higher accuracy*: Under the patch-based CNN learning framework, only the "intra-patch" context information is taken into account. Nevertheless, correlations among patches are ignored, which might lead to obvious gaps between patches. Unlike the patch-based CNN, FCN performs the classification in a single-loop manner, and considers the context information overall and seamlessly. Please refer to Section 4.2 for more details.
- *Less expensive computation*: In patch-based CNN, when using overlapped patches for dense class label generation, such as the study of Martin Lagkvist et al. [28], it introduces too much redundant computations (especially convolutions) on the overlapped regions. By performing a single loop operation, the FCN model makes remarkable progress and allows the large image classification to be implemented in a more effective way.

We adopt the FCN model for remote sensing imagery classification. The output number (channels) of the last convolutional layer (also called feature maps) is equal to the class number of our task (so in this paper, it is 12 for 12-class classifications). The feature maps can be seen as a stack of heat maps for all classes. A 2-D slice along the channel axis represents the heap map (score distribution) of the corresponding class (For example in Figure 2b and in Figure 3c, we extract the heap map for the building).

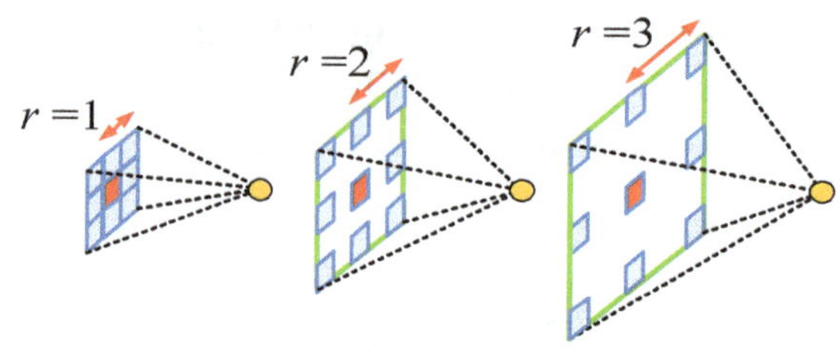

Figure 3. "Atrous" convolutions with $r = 1, 2,$ and 3. The first convolution ($r = 1$) is actually the ordinary convolution.

2.1.2. Atrous Convolution for Dense Feature Extraction

The repeated combination of pooling and striding at consecutive layers significantly reduces the spatial resolution of the resulting feature map. Typically in our VGG-16 model, 5 max-pooling layers with 1/2 down-sampling cause 1/32 total factor reduction in spatial resolution. For high resolution remote sensing image classification tasks, such operations lead to a serious loss of spatial information. Liang-Chieh Chen et al. [30], inspired by the Wavelet Transform, proposed the "atrous" convolution for generating dense feature maps. In the 1-D case, given the input signal $x[i]$, and the convolutional kernel w, the output of "atrous" convolution $\mathrm{y}[i]$ is calculated as:

$$\mathrm{y}[i] = \sum_{k=1}^{K} x[i + r \cdot k] w[k] \tag{1}$$

where r denotes the *rate* parameter corresponding to the stride. In the 2-D cases, "atrous" convolutions (use 3×3 kernel) with rate $r = 1$, 2, and 3 are demonstrated in Figure 3.

In order to further illustrate the effect of "atrous" convolution, we compare it with standard convolution using a simple example in Figure 4. Firstly, represented by the red route, we take an image patch (300×300) as an input, and perform 1/2 down-sampling and 10×10 standard convolution (horizontal Gaussian derivative kernel) on it, which is used to simulate a pooling-convolution combination in standard CNNs. The receptive field corresponding to the original image is 20×20, and only 1/4 of the image positions are involved in calculating the feature map. The obtained low resolution feature map is then enlarged by an up-sampling operation with a factor of 2. Secondly, as a comparison, we perform "atrous" convolution with rate $r = 2$ on the original image. The size of the receptive field is unchanged, but the density of the feature map is increased by two times, which means half of the image positions are considered for generating the feature map. Compared with the standard convolution, the "atrous" convolution generates a high resolution feature map, while keeping the size of receptive field. Besides, there is no extra parameter involved. The "atrous" convolution for dense feature map generation is illustrated by the blue route in Figure 4.

The "atrous" convolution is generally applicable and allows us to efficiently compute dense CNN feature maps at any target subsampling rate without introducing any approximations and extra parameters. Theoretically, the "atrous" convolution can be applied to each convolutional layer of the network to maintain the resolution, but this ends up being too costly, and the advantage for translation invariant brought by the down-sampling operation could also be weakened. So we modify the basic VGG-16 network to adapt it to our classification task. We take this modified network as our primary architecture (we add multi-scale functionality, which is described in Section 2.1.3).

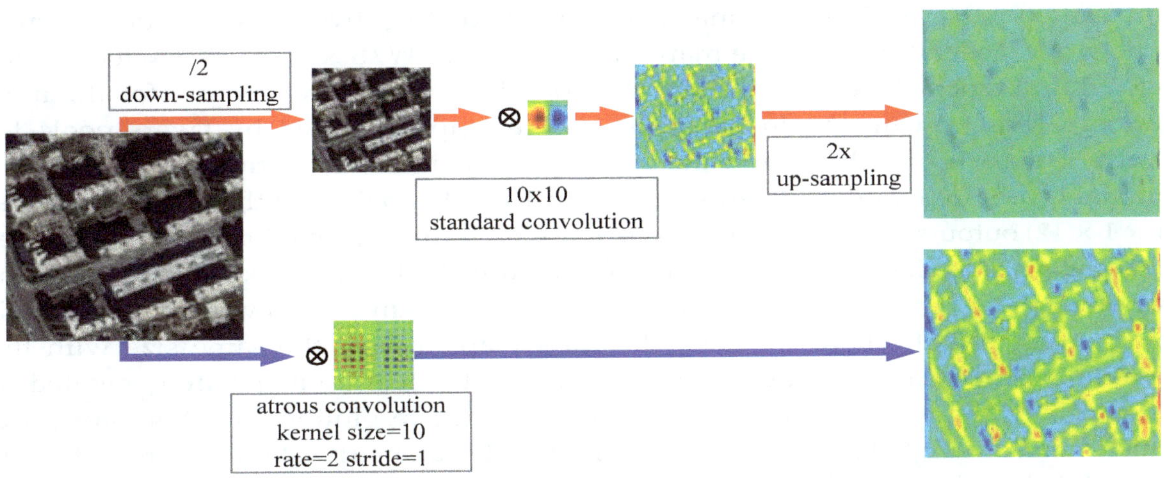

Figure 4. Illustration of atrous convolution for dense feature map generation. Red route: standard convolution performed on a low resolution feature map. Blue route: dense feature map generated using atrous convolution with rate $r = 2$ on a high resolution input feature map.

2.1.3. Network Architecture for Multi-Scale Classification

The variant of resolution will affect the classification accuracy. Single-scale classification has great limitation in its applicability. Therefore, many works considered multi-scale classification in their approaches [29–32]. A simple method for a multi-scale classification is training the model on datasets that contain objects of varying sizes. However, this approach needs the times of sample storage and training time (more iteration to traverse all the samples). A good idea for CNN-based multi-scale segmentation and detection is using the skip-layer network architecture [29,41]. In this architecture, links are added to incorporate the feature responses from different levels of the primary network stream, and these responses are then combined in a shared output layer [42]. Our multi-scale network architecture is illustrated in Figure 5.

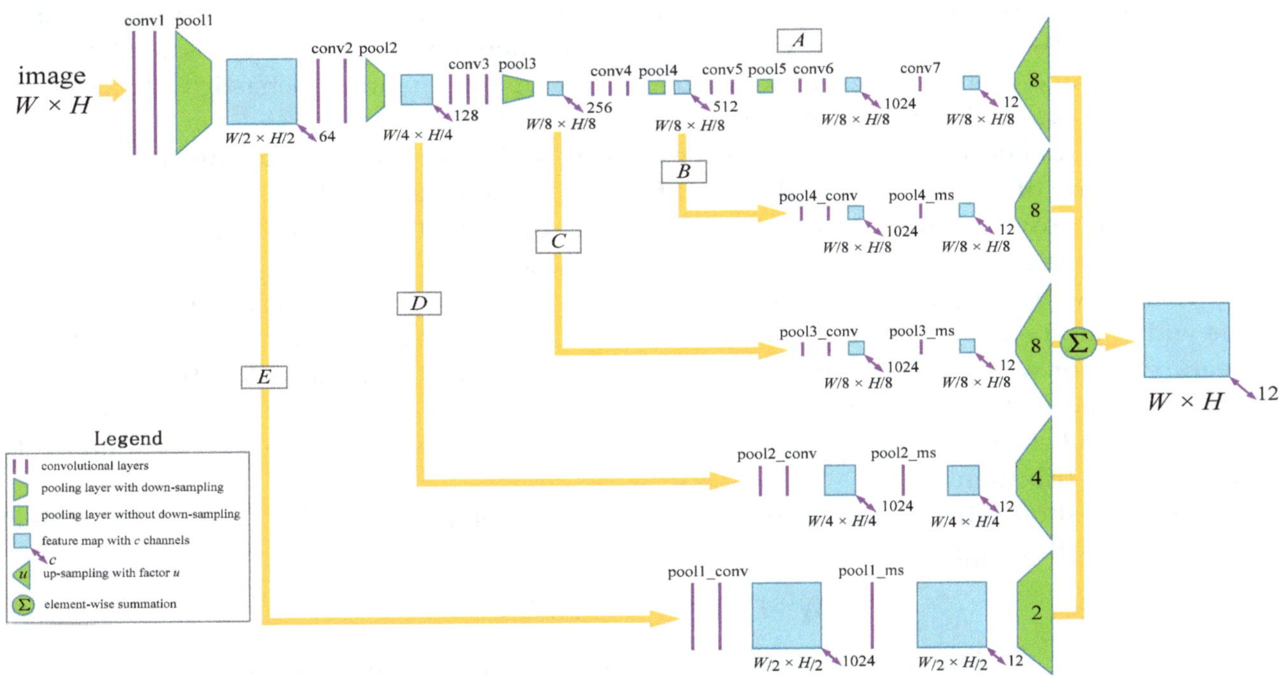

Figure 5. Multi-scale network architecture.

As presented in Figure 5, feature maps are generated along five streams. The stream A is our primary network, generating a feature map with dimension $W/8 \times H/8 \times 12$, which is described in Section 2.1.2. Branch streams B to E are the added skip-layer architecture for the multi-scale classification. These streams begin from the feature map generated by layers pool4 to pool1, respectively. For each branch stream, the subsequent architecture is the layer group with two convolutional layers, generating a feature map with 1024 channels, and then a convolutional layer (kernel $1 \times 1 \times 12$) outputs a 12-channeled feature map. Each stream, including the primary stream and the branch streams, introduce down-sampling effects caused by the max-pooling operation (the factor is $1/8$ for stream A to C, $1/4$ for stream D, and $1/2$ for stream E). However, in the applications of remote sensing classification, we need the class map to have the same size with the input image. So we perform the up-sampling operation after the feature maps are generated by these streams to recover the feature maps at the original image resolution. In this paper, we adopt Liang-Chieh Chen et al.'s [30] approach, and use simple bilinear interpolation to increase the resolution by a factor of 8, 4, and 2 at negligible computational cost. The up-sampled feature maps are then combined using summation in an element-wise manner. The output of this network architecture is a feature map with dimension $W \times H \times 12$. Our multi-scale network architecture captures three levels of resolution, represented by stream A to C, stream B, and stream E.

2.2. *Network Training*

Our training dataset is collected from two GF-2 high resolution remote sensing images (true color fusion images with 0.8 meter resolution) of northeastern Beijing, China.

The images were taken in 5 December 2014 and 2 September 2015, respectively. The reason why we chose images with different imaging times is to increase the anti-interference abilities of our model, such as the change of seasons, to enhance its applicability. In our training dataset, there are a total of 74 images (size 1024×1024). We manually labeled all images at the pixel level as ground truth (GT) label data. In other words, for each image, there exists a 1024×1024 label map, having a pixel-class (row-col indexed) correspondence with it. We used 70 images for training, and the remaining 4 images for testing. Three image-GT label pair examples are illustrated in Figure 6.

The general procedure of our training stage is: Image-GT label pairs are input into the multi-scale classification network as training samples. The *Softmax* function is performed on the output feature map generated by the network to predict the class distribution. Then the cross entropy loss is calculated and back-propagated, and finally the network parameters are updated using Stochastic Gradient Descent (SGD) with momentum. The general procedure is shown in Figure 7.

The softmax function is used to *probabilize* the output feature map of our multi-scale network. However, the mode of softmax here is different from that in the standard CNNs: it is performed on each location with row-column coordinate $(i,\ j), 0 \leq i < H \ and\ 0 \leq j < W$, and it outputs a dense distribution over the classes. Figure 8 illustrates this function.

Figure 8 shows that the output of our multi-scale network is a $H \times W \times 12$ feature map, which has the same width and height as the original image. A "drill hole" along the channel axis at location $(i,\ j)$ is the feature vector with 12 elements corresponding to the pixel at the same location. The softmax function is adopted on this feature vector to generate a 12-D probabilized vector, which is the discrete distribution over 12 classes at location $(i,\ j)$. The softmax function will traverse each location to obtain the dense class distribution.

The SGD method with momentum is used for parameter updates in our training, which is described by the following:

$$W^{(n+1)} = W^{(n)} - \Delta W^{(n+1)} \tag{2}$$

where $W^{(n)}$ and $W^{(n+1)}$ denote the old parameters and new parameters, respectively, and $\Delta W^{(n+1)}$ is the increment for the current iteration, which is a combination of old parameters, gradient, and historical increment:

$$\Delta W^{(n+1)} = \eta \cdot \left(d_w \cdot W^{(n)} + \frac{\partial J(W)}{\partial W^{(n)}} \right) + m \cdot \Delta W^{(n)} \tag{3}$$

where $J(W)$ is the loss function, η is the learning rate for step length control, and d_w and m denote the weight decay and momentum, respectively.

We employ the VGG-16 network which has been pre-trained on ImageNet for fast convergence. We use a "step" policy for learning rate adjustment ($gamma = 0.1$, $step\_size = 15{,}000$) so that closer to the error minimum, the smaller the step length is. The base learning rate is 0.0001. The basic parameters for calculating increments are: $m = 0.9$, and $d_w = 0.0005$. The max iteration in our training is 60,000. In the training procedure, we first randomly shuffle the samples, and then feed them into the network in batches. Each batch contains 10 images. We also crop and rotate samples randomly in each batch to increase the diversity and variability of the samples.

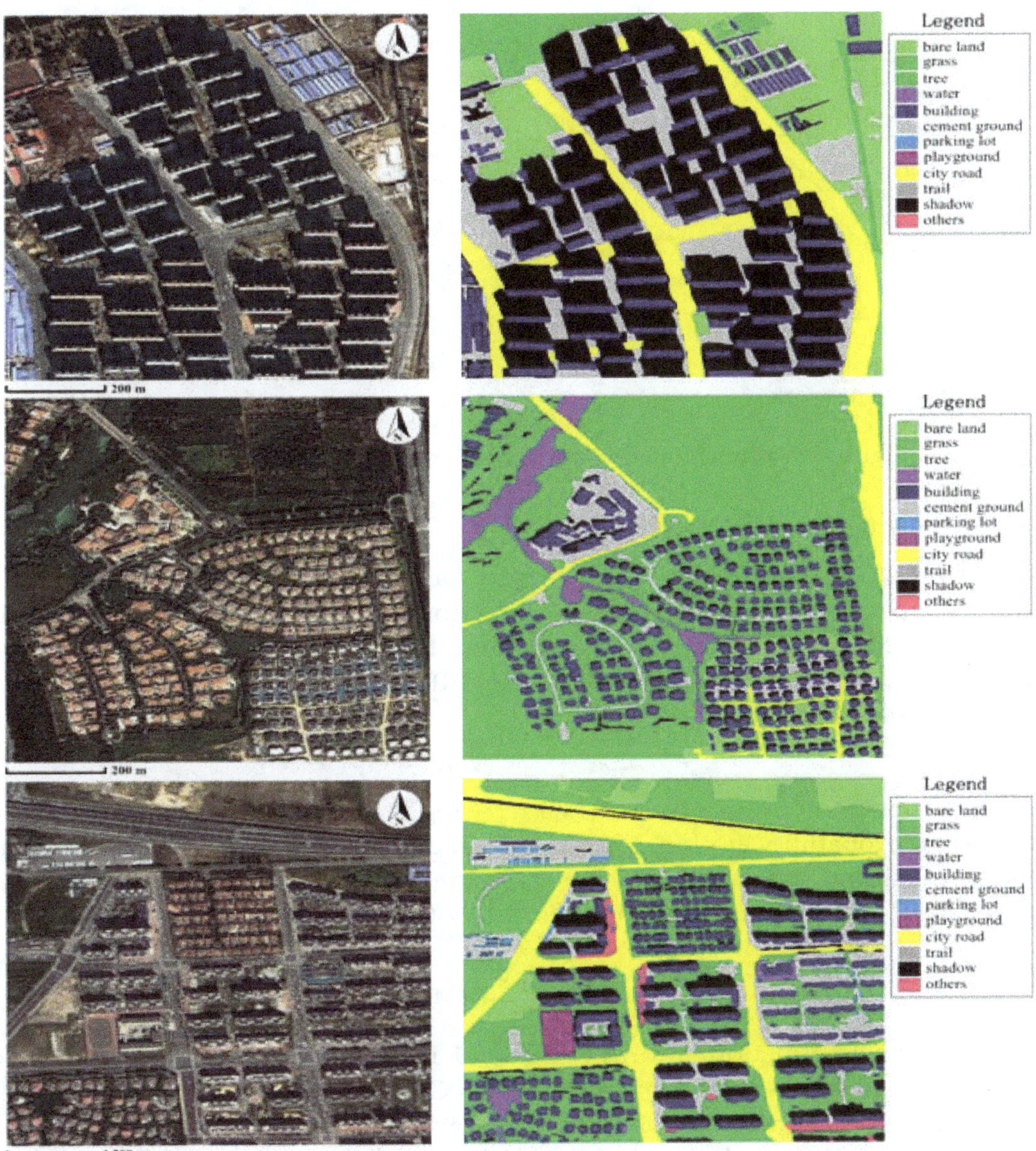

(a) (b)

Figure 6. Three sample examples for our classification training. (**a**) Original images; (**b**) Ground truth (GT) labels corresponding to the images in (**a**).

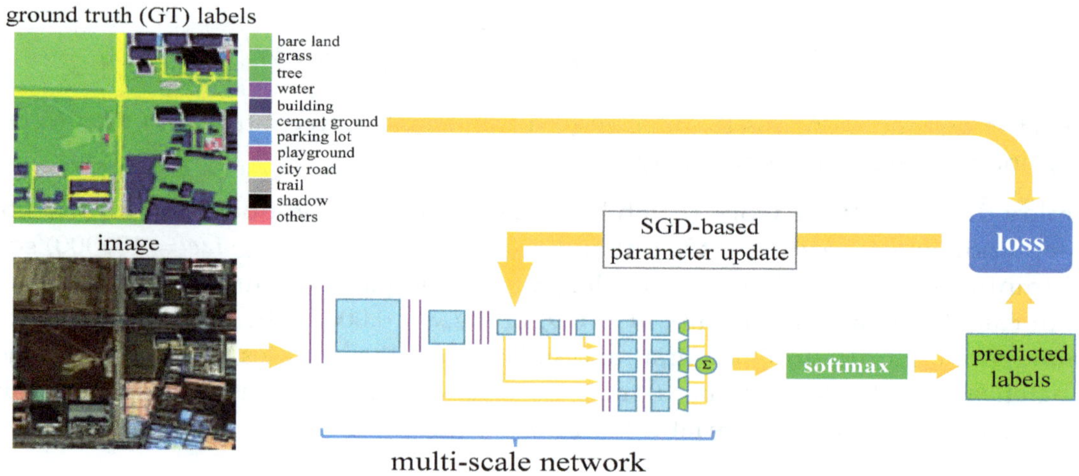

Figure 7. General procedure of network training.

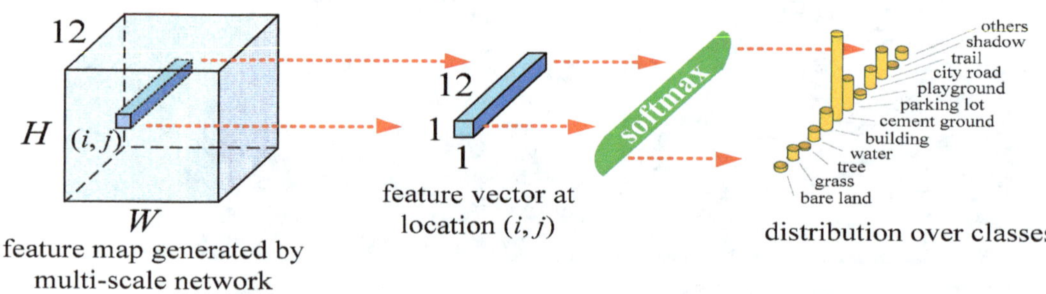

Figure 8. Softmax function performed on the output feature map.

2.3. Classification Using the Trained Network

The trained network is adopted on an image for classification. However, our multi-scale network involves up-sampling operations, leading to the blurring of classification boundaries. Several works [29–32] use CRFs as post-processing to refine the image segmentation results. So following their idea, we adopt the fully connected CRFs for our rough class prediction. The model employs the energy function:

$$E(\boldsymbol{x}) = \sum_i \theta_i(x_i) + \sum_{ij} \theta_{ij}(x_i, x_j) \tag{4}$$

where x is the label assignment for pixels. $\theta_i(x_i) = -\log P(x_i)$ is the *unary potential*, where $P(x_i)$ is the label assignment probability at pixel i as the output of our multi-scale network after the softmax function. $\theta_{ij}(x_i, x_j)$ is the *pairwise potential* represented by a fully connected graph, connecting all pairs of image pixels i and j. We use the following definition of the pairwise potential [43]

$$\theta_{ij}(x_i, x_j) = \mu(x_i, x_j) \sum_{m=1}^{K} w_m \cdot k^m \left(\boldsymbol{f}_i, \boldsymbol{f}_j\right) \tag{5}$$

where $\mu(x_i, x_j)$ is the sign function, and $\mu(x_i, x_j) = 1$ if $x_i \neq x_j$, and is zero otherwise. $\mu(x_i, x_j)$ removes the self-connected links from the graph. k^m is a Gaussian kernel function that takes feature as input (denoted by $\boldsymbol{f}_i$ and $\boldsymbol{f}_j$ extracted for pixel i and j). Each Gaussian kernel is weighted by w_m. In our study, the bilateral position and color terms is adopted as the kernel function

$$w_1 \cdot \exp\left(-\frac{\|p_i - p_j\|^2}{2\sigma_\alpha^2} - \frac{\|I_i - I_j\|^2}{2\sigma_\beta^2}\right) + w_2 \cdot \exp\left(-\frac{\|p_i - p_j\|^2}{2\sigma_\gamma^2}\right) \tag{6}$$

where p_i, p_j denote the locations, and p_i, p_j denote the color of pixel i, j. So the first kernel depends on both pixel positions and color, and the second kernel only depends on pixel positions. σ_α, σ_β, and σ_γ are the hyper parameters that control the scale of the Gaussian kernels. The classification pipeline is illustrated in Figure 9.

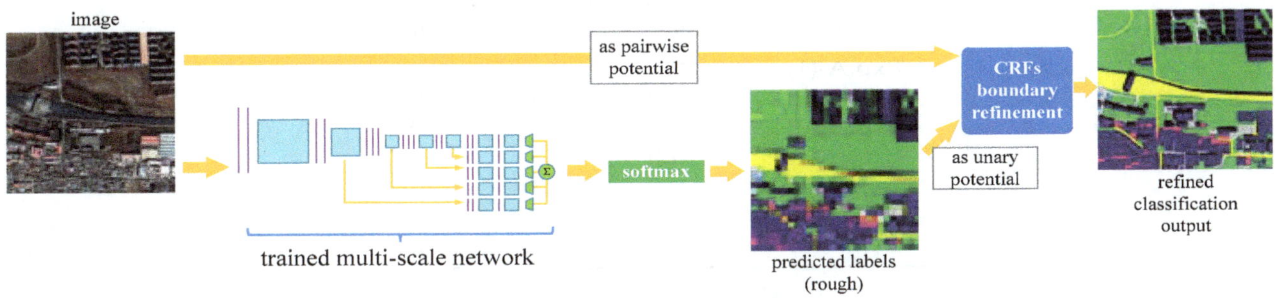

Figure 9. General procedure of image classification using the trained network.

In CRFs post-processing, the rough class distribution predicted by the multi-scale network is input as the unary potential, and the original image provides the pairwise potential with position and color information. The CRFs is solved using mean field approximation [43]. The class labels are adjusted and refined under the position-color constraints. The weight parameters we adopt in this paper are $w_1 = 4, w_2 = 3$, which are the default configuration of [30]. Following the idea of [43], we use $\sigma_\alpha = 54, \sigma_\beta = 5$, and $\sigma_\gamma = 4$ through a cross-validation on the training set. We employ 10 mean field iterations for solving CRFs.

3. Experiment and Comparison

In the following section, the experiment and comparison will be presented to evaluate our classification approach. Our algorithm is implemented using Microsoft Visual C++ 11, and is performed on the Windows 7 operating system installed NVIDIA GeForce GTX980M graphic device with 8G byte graphic memory.

3.1. Comparison Setup

We conduct two groups of experiment (denoted as Experiment A and B) on GF-2 and IKONOS true color images, respectively. We compare our approach with object-oriented classification using MR segmentation [11], SVM classification (MR-SVM), patch-based CNN classification proposed in [27], and the FCN-8s approach proposed in [29].

3.1.1. MR-SVM

For Multi-Resolution and Support Vector Machine (MR-SVM) object-oriented classification, the first step is MR segmentation [11] to generate image objects. The quality of image objects directly affects the classification results. We believe that the high quality image objects are neither over-covered nor over-segmented. Ideally, each image object contains only a single-class ground object. The MR segmentation is controlled by the scale, shape, and compactness parameters. In order to obtain high-quality image objects, we determine the parameters through the times of experiments by different settings, to achieve the ideal segmentation as much as possible. The parameters we used in MR segmentation are listed in Table 1.

Once the image objects are obtained, we construct the initial feature space using 60 common features involving spectral, geometric, and texture aspects:

- *Spectral features*: mean, standard deviation, brightness, and max difference for each band.
- *Geometric features*: area, length, width, length-width ratio, border length, compactness, elliptic fit, rectangular fit, density, shape index, main direction, and symmetry.

- *Texture features*: Features calculated from the Gray Level Co-occurrence Matrix (GLCM) and the Gray Level Difference Vector (GLDV) with all directions, etc.

Table 1. Scale, shape, and compactness parameters used in the Multi-Resolution (MR) segmentation.

Experiment	Scale	Shape	Compact
Exp.A-(1)	115	0.5	0.5
Exp.A-(2)	140	0.3	0.8
Exp.A-(3)	105	0.4	0.5
Exp.A-(4)	100	0.4	0.7
Exp.B-(1)	120	0.3	0.5
Exp.B-(2)	80	0.5	0.4
Exp.B-(3)	85	0.5	0.7

To select the most representative features for the following classification, we seek significant features for optimal class separation using the Separability and Thresholds (SEaTH) method [44]. According to the SEaTH method, we optimize the 60-D initial feature space, and obtain a 10-D sub feature space including: mean value and brightness for each band; density and length-width ratio of the image object; GLCM-mean value for each band; GLDV-mean for the first band. In the classification stage, we select almost 25% of the image objects from each image as training samples, and input their features to the SVM classifier implemented using the LibSVM library [45]. The kernel function we used in SVM is the Radial Basis Function (RBF), and the objective function type is the C-Support Vector Classification (C-SVC). To determine the optimal penalty factor C and kernel function parameter γ, we employ a simple grid search for all training samples on the $C - \gamma$ domain that minimize the classification error. The search range of C and γ are $[0.4, 1.6]$ and $[0.02, 0.14]$ according to the experience [45]. The step lengths are 0.2 and 0.01, respectively. According to the grid search, the optimal parameters we used for the SVM classifier are $C = 1.2$ and $\gamma = 0.08$.

3.1.2. Patch-Based CNN

In the patch-based classification experiment, the general procedure is illustrated in Figure 10.

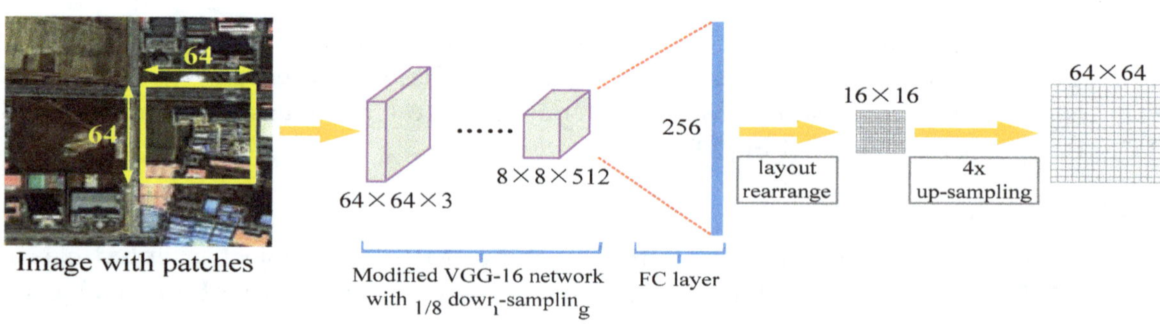

Figure 10. General procedure of our patch-based CNN classification experiment.

Different from the architecture used in [27], we employ the VGG-16 network as the main structure for its high performance in the previous vision tasks. In order to prevent excessive reduction of the resolution, we modified the stride and padding values of the last two pooling layers (the stride and padding values we used are all 1) so that the architecture has a 1/8 down-sampling effect. Following the idea of Volodymyr Mnih [27], the last 3 FC layers are modified to a single FC layer with output number 256 representing a 16×16 prediction area. So for 64×64 input patches, the overall architecture causes a 1/4 down-sampling. Finally, we perform an up-sampling post-processing with a factor of 2 to increase the resolution.

3.1.3. FCN-8s

For the FCN model, we directly employ the FCN-8s model proposed by Jonathan Long et al. [29]. The architecture of the model is also the VGG-16 network with skip-layer structure. The final prediction is fused from the output of three branches (from the primary network, the pool4 layer, and the pool3 layer, respectively) after the up-sampling operation. In the training phase, by modifying the number of outputs from 21 to 12, we fine-tuned the network based on the ImageNet pre-trained model. The training parameters for FCN-8s in the experiment are the same as ours. In the testing stage, except for the CRF-based post-processing, we use the same classification parameters as our approach. Please refer to [29] for detailed information.

3.2. Experiments and Comparison

In Experiment A, we adopt our trained model on four GF-2 true color images (0.8 m resolution) for the classification (In the following section, they will be abbreviated as Exp.A-(1) to Exp.A-(4)). All the image sizes are 1024 × 1024. These images are the testing images that are not involved in training. Figure 11 is the illustration of the results and the comparison. In Experiment B, we adopt the same trained model on three IKONOS true color images (1.0 m resolution) for the classification (Abbreviated as Exp.B-(1) and Exp.B-(3) in the following section) to test the applicability. All the image sizes are also 1024 × 1024. Figure 12 illustrates the classification results and comparison.

We employ precision, recall, and Kappa coefficient as the indicators to evaluate our approach. These indexes are calculated from the confusion matrix C, where the precision is calculated as $\frac{1}{12}\sum_i C_{ii} / \sum_j C_{ij}$ that denotes the average proportion of pixels being classified to one class that are correct, and the recall is computed as $\frac{1}{12}\sum_i C_{ii} / \sum_i C_{ij}$ that represents the average proportion of pixels that are correctly classified, and the Kappa coefficient measures the consistency of the predicted classes with the GT classes. The comparisons are listed in Table 2.

Table 2. Comparison between approaches using MR-SVM, patch-based CNN, FCN-8s, and our approach.

Approach	Index	Exp.A-(1)	Exp.A-(2)	Exp.A-(3)	Exp.A-(4)	Exp.B-(1)	Exp.B-(2)	Exp.B-(3)	Mean
MR-SVM	Precision	0.67	0.72	0.67	0.66	0.65	0.73	0.64	0.68
	Recall	0.52	0.59	0.52	0.63	0.39	0.51	0.74	0.56
	Kappa	0.55	0.66	0.62	0.65	0.54	0.64	0.64	0.61
Patch-based CNN	Precision	0.68	0.64	0.71	0.55	0.73	0.76	0.70	0.68
	Recall	0.61	0.61	0.70	0.73	0.47	0.58	0.74	0.63
	Kappa	0.64	0.69	0.62	0.70	0.63	0.71	0.75	0.68
FCN-8s	Precision	0.83	0.84	0.68	0.66	0.81	0.78	0.83	0.78
	Recall	0.71	0.79	0.80	0.80	0.66	0.66	0.79	0.74
	Kappa	0.73	0.80	0.81	0.80	0.76	0.81	0.82	0.79
Ours	Precision	**0.86**	**0.87**	**0.74**	**0.68**	**0.84**	**0.78**	**0.92**	**0.81**
	Recall	**0.83**	**0.78**	**0.81**	**0.82**	**0.70**	**0.68**	**0.84**	**0.78**
	Kappa	**0.79**	**0.85**	**0.84**	**0.83**	**0.78**	**0.84**	**0.89**	**0.83**

The above statistics show our approach obtains the best performance compared with the others. Approaches using carefully-designed MR-SVM and patch-based CNN achieve similar accuracy levels, and the FCN-8s approach performs much better than those two. Some ground objects such as building, city road, and cement ground, have similar spectral and geometrical features, which are hard to distinguish. For example, in Exp.A-(2), when using MR-SVM, the recall for "cement ground" is 0.41. That means that more than half of the pixels are wrongly classified. The proportions that are incorrectly classified as "building" and "road" are 0.26 and 0.19. It means that in that case, the object-oriented classification has almost no effect on distinguishing these classes.

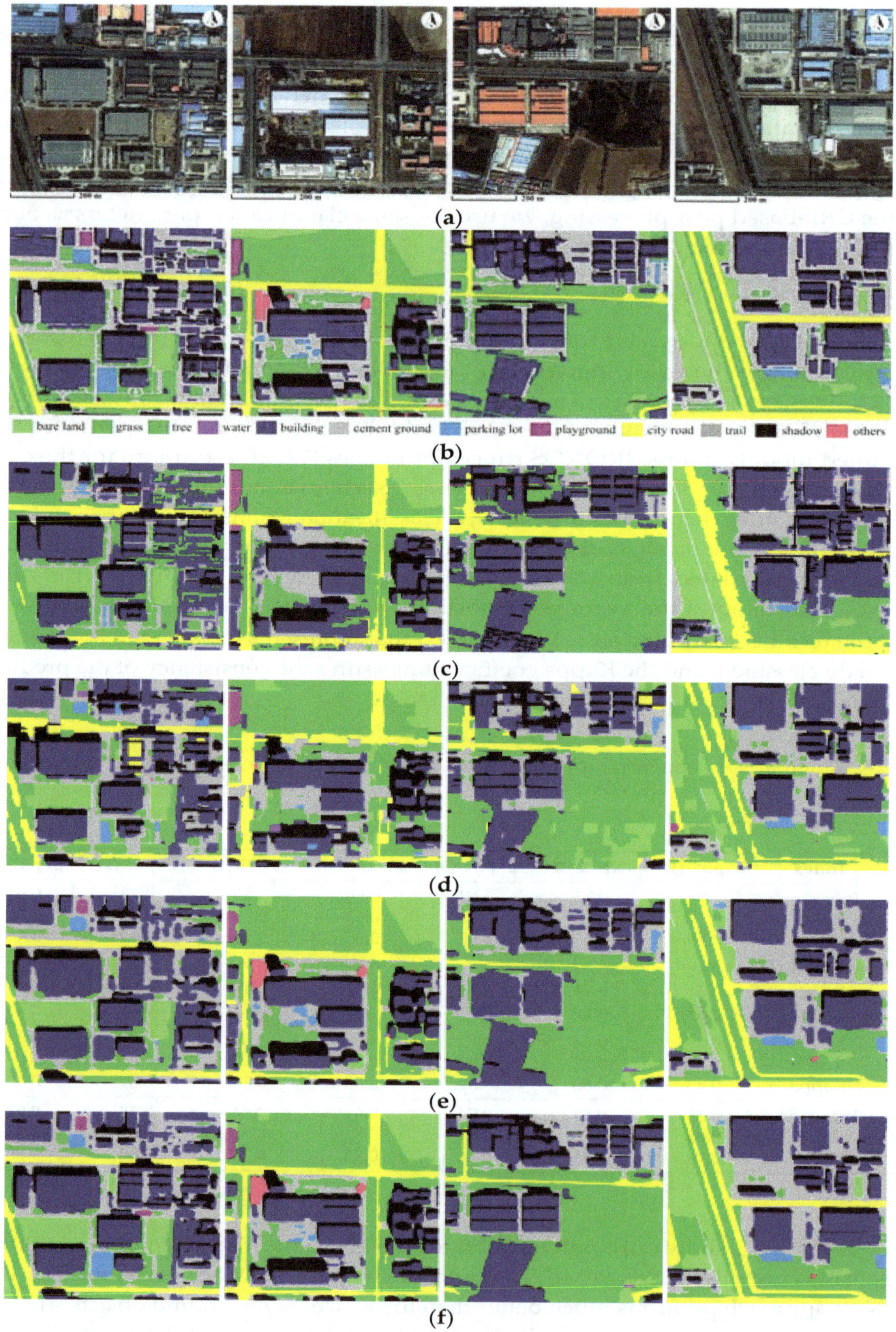

Figure 11. Classification results on GF-2 images (Experiment A). (**a**) Original images; (**b**) GT labels corresponding to the images in (**a**); (**c**–**e**) Results of the MR-SVM object-oriented classification, patch-based CNN classification, and FCN-8s classification corresponding to the images in (**a**), respectively; (**f**) Our classification results corresponding to the images in (**a**).

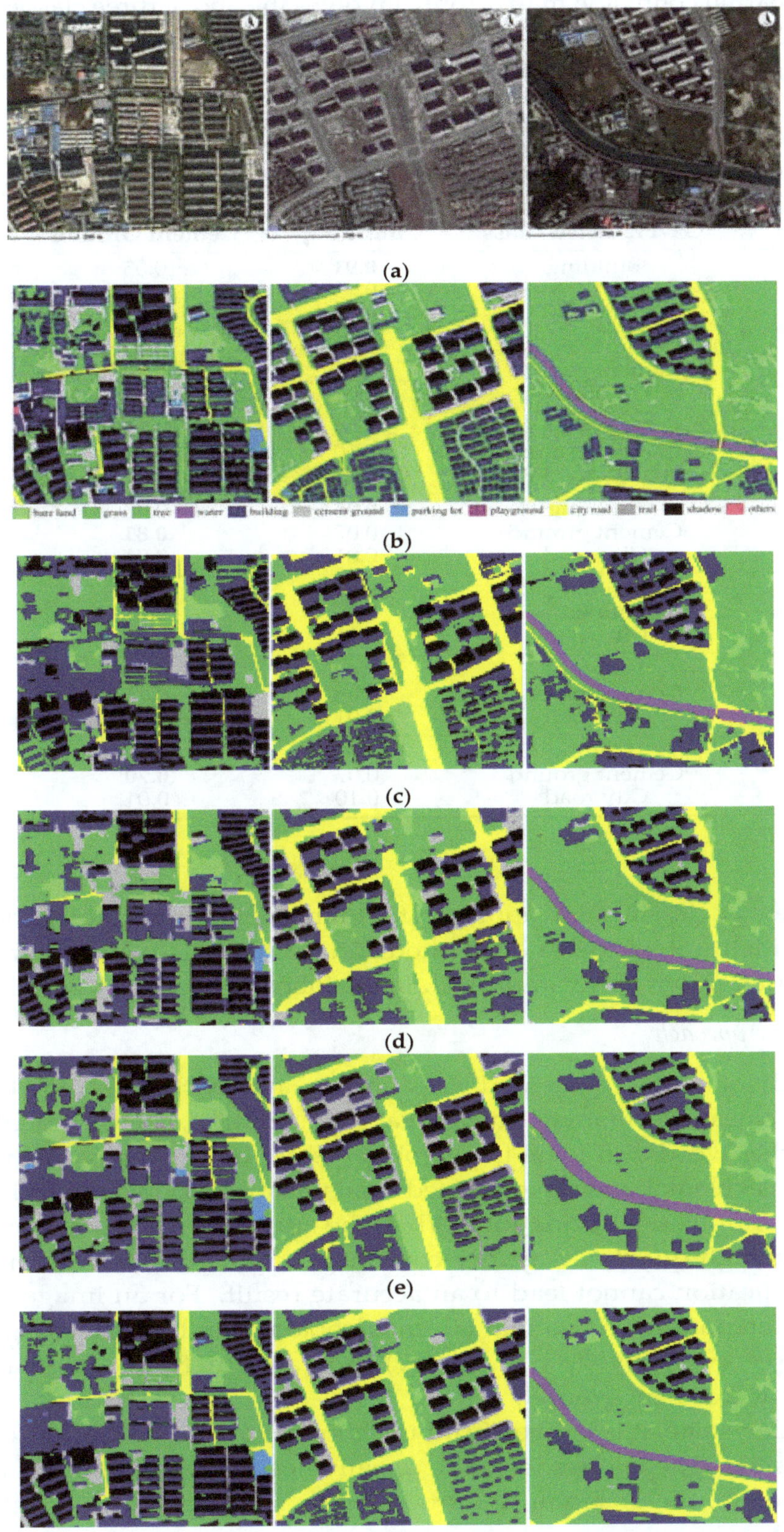

Figure 12. Classification result on IKONOS images (Experiment B). (**a**) Original images; (**b**) GT labels corresponding to the images in (**a**); (**c**–**e**) Results of the MR-SVM object-oriented classification, patch-based CNN classification, and FCN-8s classification corresponding to the images in (**a**), respectively; (**f**) Our classification results corresponding to the images in (**a**).

Table 3 lists the partial confusion matrix (only involves the above three classes) of our classification results. From the table, we can see that our approach achieves higher classification performance. In the above example, our recall for "cement ground" is 0.79. The proportions that are wrongly classified as "building" and "city road" are 0.05 and 0.06, respectively.

Table 3. Partial confusion matrix of our approach for "building", "cement ground", and "city road".

Experiment	GT/Predicted Class	Building	Cement Ground	City Road
Exp.A-(1)	Building	**0.91**	0.05	0.02
	Cement ground	0.13	**0.76**	0.02
	City road	0.02	0.01	**0.95**
Exp.A-(2)	Building	**0.92**	0.03	0.03
	Cement ground	0.05	**0.79**	0.06
	City Road	0.01	0.04	**0.89**
Exp.A-(3)	Building	**0.91**	0.02	0.05
	Cement ground	0.10	**0.82**	0.03
	City road	0.05	0.04	**0.82**
Exp.A-(4)	Building	**0.95**	0.03	0.00
	Cement ground	0.07	**0.81**	0.05
	City road	0.01	0.01	**0.93**
Exp.B-(1)	Building	**0.90**	0.02	0.01
	Cement ground	0.26	**0.65**	0.01
	City road	0.11	0.03	**0.84**
Exp.B-(2)	Building	**0.83**	0.01	0.00
	Cement ground	0.08	**0.75**	0.15
	City road	0.01	0.01	**0.96**
Exp.B-(3)	Building	**0.87**	0.06	0.01
	Cement ground	0.03	**0.70**	0.04
	City road	0.10	0.01	**0.87**

4. Discussion

This paper presents a classification approach for high resolution images using the improved FCN model. Compared with the object-oriented method and two typical deep learning-based approaches, the classification accuracy is obviously improved. In the following sections, we will discuss the reasons.

4.1. MR-SVM vs. Our Approach

Most of the traditional object-oriented classification approaches employ their classification in a "segmentation-classification" manner. In an ideal segmentation, each segment represents a single ground object. In other words, an ideal image object is neither over-covered nor over-segmented. However, most of the segmentation was conducted in an unsupervised way, which relies only on image information, but no prior class information. When the spectral and geometric features are similar, it is difficult to obtain high-quality image objects. Once the image objects are incorrect, subsequent object-oriented classification cannot lead to an accurate result. For an image, it is difficult to find universal segmentation parameters so that all image objects can be correctly generated. Figure 13 shows one image object (with a yellow boundary) generated by MR segmentation that incorrectly covers both building and cement ground.

In the classification stage, it is very difficult to choose expressive features for an image object as the input of the classifier. The feature selection usually needs many attempts and largely depends on experience. Therefore, the uncertainty introduced by the two stages, together affects the final classification accuracy.

In our FCN-based approach, the class information, which is the ultimate objective for classification, is taken as the supervisory signal that controls the whole process including both feature extraction and classification. Our approach combines the segmentation and classification stages, and achieves high quality classification in an end-to-end way. This is also the most obvious advantage of the deep learning theory.

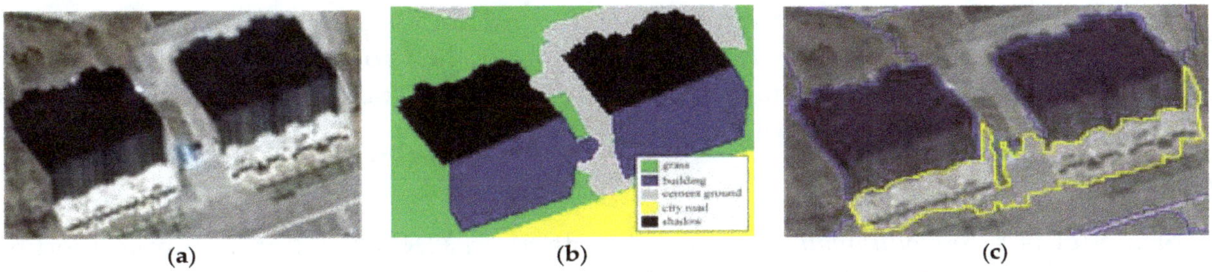

Figure 13. Incorrect image object generated by MR segmentation. (**a**) Original images; (**b**) GT labels corresponding to the images in (**a**); (**c**) Incorrect image object covers both the building and cement ground (with yellow boundary).

4.2. Patch-Based CNN vs. Our Approach

In the patch-based CNN approach, each image patch is input to the model independently, which means that only the "intra-patch" context information is considered. However, correlations between patches are not taken into account, which might lead to obvious gaps between patches. Especially for objects with strong continuity, such as road and building edges, the problem is more serious. Figure 14 shows the differences between patch-based CNN and our approach for building heat map generation.

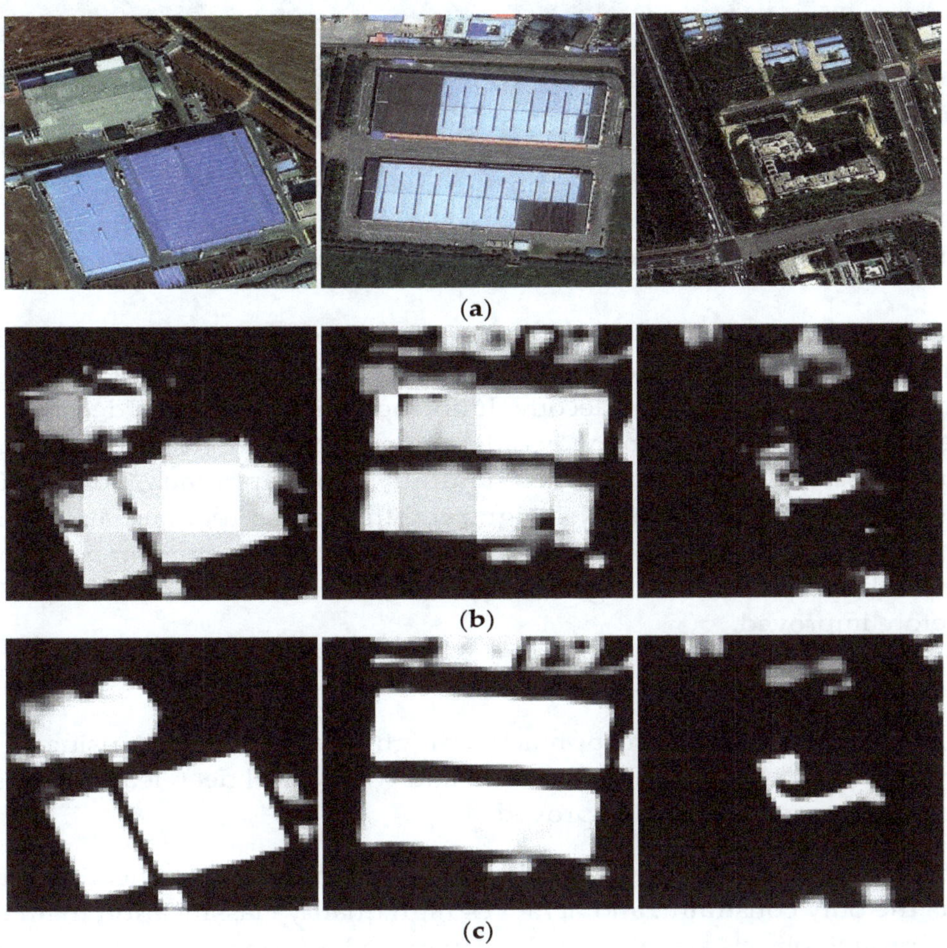

Figure 14. Heat map for the building generated by patch-based CNN and our approach. (**a**) Original images; (**b**) Heat map generated by patch-based CNN classification using 128 × 128 patches; (**c**) Heat map generated by the FCN model.

Compared with the patch-based approaches, our model takes the whole image as the input, and performs the classification in a single-loop manner, which considers the context information overall and seamlessly. Our model eliminates the discontinuities at the patch boundaries. This is also the most remarkable advantage of FCN.

4.3. FCN-8s vs. Our Approach

FCN model is a *convolutionalized* version of standard CNN through a simple modification. The most significant feature of the FCN model is: on the one hand, FCN inherits the high accuracy feature for image-label classification from standard CNN. On the other hand, it maintains the 2-D spatial information of the input image, thus achieving dense class prediction. However, pooling operations cause serious reduction of the resolution. The output is not fine enough, which will result in the loss of valuable detail information. As can be seen from Figure 15, our approach outperforms FCN-8s in terms of detail preserving. Therefore, the classification accuracy is greatly improved.

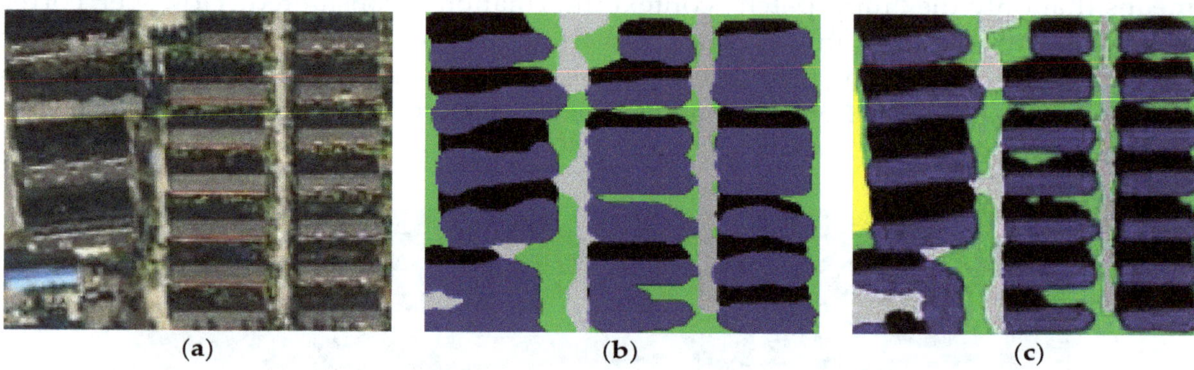

Figure 15. Detail comparison between FCN-8s and our approach. (**a**) Original images; (**b**) Classification result from FCN-8s; (**c**) Classification result from our approach.

As the most accurate model in the FCN family, FCN-8s combines the feature maps with different resolutions from different pooling stages, to obtain a more intensive class prediction. In FCN models, the lost resolution is compensated by the deconvolution operation. However, deconvolution is difficult for efficiently restoring the resolution by way of learning. Benefiting from the "atrous" convolution, the resolution of the feature map is maintained naturally in our approach. Besides, FCN models do not consider the relationship between pixels, ignoring the spatial regularization that is commonly employed in remote sensing image analysis. In our approach, the relationship between pixels is taken into account by CRF-based post-processing. The class map predicted by FCN is further refined, and the accuracy is therefore improved.

5. Conclusions

This paper presents a classification approach for high resolution images using an improved FCN model. Compared with the object-oriented method and two typical deep learning-based approaches, the classification accuracy is obviously improved.

Our FCN-based classification combines the segmentation and classification stages, taking the class accuracy as the only constraint, and achieves high quality classification in an end-to-end way. The GT classes of ground objects are taken as the supervised information that guides both the feature extraction and the region generation. The classification results of using "atrous" convolution and CRF-based post-processing allows us to obtain a high resolution class prediction. In addition, due to the use of a multi-scale model, the model trained from the GF-2 images also has high classification accuracy on the IKONOS images. It is proven that our approach has a strong applicability for images with different resolutions.

The main limitation of our approach is that it needs a large number of high quality GT-labels for the model training, which relies on professional interpretation experiences and lots of manual work. Therefore, the main aspect of our future work is training the model in a weak supervision way, to further enhance its applicability.

Acknowledgments: This work was jointly supported by the National Natural Science Foundation of China (Grant No. 41571414. Title: Eco-environment assessment of Yanhe watershed based on temporal-spatial entropy), Beijing Municipal Science and Technology Project (Grant No. Z161100001116102. Title: Research on Remote Sensing Technology in Soil and Water Conservation and Demonstrative Application in Beijing), and the Fundamental Research Project in China Institute of Water Resources and Hydropower Research (Grant No. JZ0145B2017. Title: Research on Spatial-Temporal Variable Source Runoff Model and Mechanism).

Author Contributions: Gang Fu and Changjun Liu proposed and designed the technique roadmap, and performed the programming works; Gang Fu and Tao Sun designed and performed the experiments; Rong Zhou, Tao Sun, and Qijian Zhang collected the training and testing image data, and analyzed the experimental results.

References

1. MacQueen, J.B. Some Methods for classification and Analysis of Multivariate Observations. Proceedings of 5th Berkeley Symposium on Mathematical Statistics and Probability; University of California Press: Berkeley, CA, USA, 1967.; pp. 281–297.
2. Miller, D.M.; Kaminsky, E.J.; Rana, S. Neural network classification of remote-sensing data. *Comput. Geosci.* **1995**, *21*, 377–386.
3. Mas, J.; Flores, J. The application of artificial neural networks to the analysis of remotely sensed data. *Int. J. Remote Sens.* **2008**, *29*, 617–663.
4. Camps-Valls, G.; Bruzzone, L. Kernel-based methods for hyperspectral image classification. *IEEE Trans. Geosci. Remote Sens.* **2005**, *43*, 1351–1362.
5. Mountrakis, G.; Im, J.; Ogole, C. Support vector machines in remote sensing: A review. *ISPRS J. Photogramm. Remote Sens.* **2011**, *66*, 247–259.
6. Yuan, Y.; Lin, J.; Wang, Q. Hyperspectral Image Classification via Multitask Joint Sparse Representation and Stepwise MRF Optimization. *IEEE Trans. Cybern.* **2016**, *46*, 2966–2977.
7. Wang, Q.; Lin, J.; Yuan, Y. Salient Band Selection for Hyperspectral Image Classification via Manifold Ranking. *IEEE Trans. Neural Netw. Learn. Syst.* **2016**, *27*, 1279–1289.
8. Walter, V. Object-based classification of remote sensing data for change detection. *ISPRS J. Photogramm. Remote Sens.* **2004**, *58*, 225–238.
9. Definients Image. *eCognition User's Guide 4*; Definients Image: Bernhard, Germany, 2004.
10. Feature Extraction Module Version 4.6. In *ENVI Feature Extraction Module User's Guide*; ITT Corporation: Boulder, CO, USA, 2008.
11. Baatz, M.; Schäpe, A. Multiresolution Segmentation: An Optimization Approach for High Quality Multi-scale Image Segmentation. In *Angewandte Geographische Information Sverarbeitung XII*; Herbert Wichmann Verlag: Heidelberg, Germany, 2000; pp. 12–23.
12. Robinson, D.J.; Redding, N.J.; Crisp, D.J. *Implementation of a Fast Algorithm for Segmenting SAR Imagery*; DSTO Electronics and Surveillance Research Laboratory: Edinburgh, Australia, 2002.
13. Cheng, Y. Mean shift, mode seeking, and clustering. *IEEE Trans. Pat.* **1995**, *17*, 790–799.
14. Fu, G.; Zhao, H.; Li, C.; Shi, L. Segmentation for High-Resolution Optical Remote Sensing Imagery Using Improved Quadtree and Region Adjacency Graph Technique. *Remote Sens.* **2013**, *5*, 3259–3279.
15. Hinton, G.; Osindero, S.; Welling, M.; Teh, Y.-W. Unsupervised Discovery of Nonlinear Structure Using Contrastive Backpropagation. *Science* **2006**, *30*, 725–732.
16. Deep Learning. Available online: https://en.wikipedia.org/wiki/Deep_learning (accessed on 3 May 2017).
17. Hinton, G.E.; Osindero, S.; Teh, Y.W. A Fast Learning Algorithm for Deep Belief Nets. *Neural Comput.* **2006**, *18*, 1527–1554.

18. Convolutional Neural Networks (LeNet)—DeepLearning 0.1 Documentation. DeepLearning 0.1. LISA Lab. Retrieved 31 August 2013. Available online: http://deeplearning.net/tutorial/lenet.html (accessed on 5 May 2017).
19. Graves, A.; Liwicki, M.; Fernandez, S.; Bertolami, R.; Bunke, H.; Schmidhuber, J. A Novel Connectionist System for Improved Unconstrained Handwriting Recognition. *IEEE Trans. Pattern Anal. Mach. Intell.* **2009**, *31*, 855–868.
20. Krizhevsky, A.; Sutskever, I.; Hinton, G.E. Imagenet classification with deep convolutional neural networks. In Proceedings of the Neural Information Processing Systems (NIPS) Conference, La Jolla, CA, USA, 3–8 December 2012.
21. Ciresan, D.; Meier, U.; Schmidhuber, J. Multi-column deep neural networks for image classification. In Proceedings of the Conference on Computer Vision and Pattern Recognition (CVPR), Providence, RI, USA, 16–21 June 2012; pp. 3642–3649.
22. Nguyen, T.; Han, J.; Park, D.C. Satellite image classification using convolutional learning. In Proceedings of the AIP Conference, Albuquerque, NM, USA, 7–10 October 2013; pp. 2237–2240.
23. Wang, J.; Song, J.; Chen, M.; Yang, Z. Road network extraction: A neural-dynamic framework based on deep learning and a finite state machine. *Int. J. Remote Sens.* **2015**, *36*, 3144–3169.
24. Castelluccio, M.; Poggi, G.; Sansone, C.; Verdoliva, L. Land Use Classification in Remote Sensing Images by Convolutional Neural Networks. *arXiv*, **2015**, arXiv:1508.00092.
25. Hu, F.; Xia, G.S.; Hu, J.; Zhang, L. Transferring deep convolutional neural networks for the scene classification of high-resolution remote sensing imagery. *Remote Sens.* **2015**, *7*, 14680–14707.
26. Zhou, W.; Newsam, S.; Li, C.; Shao, Z. Learning Low Dimensional Convolutional Neural Networks for High-Resolution Remote Sensing Image Retrieval. *arXiv*, **2016**, arXiv:1610.03023.
27. Mnih, V. Machine Learning for Aerial Image Labeling. Ph.D. Thesis, University of Toronto, Toronto, ON, Canada, 2013.
28. Lagkvist, M.; Kiselev, A.; Alirezaie, M.; Loutfi, A. Classification and Segmentation of Satellite Orthoimagery Using Convolutional Neural Networks. *Remote Sens.* **2016**, *8*, 329.
29. Long, J.; Shelhamer, E.; Darrell, T. Fully Convolutional Networks for Semantic Segmentation. In Proceedings of the Conference on Computer Vision and Pattern Recognition (CVPR), Boston, MA, USA, 7–12 June 2015.
30. Chen, L.C.; Kokkinos, I.; Murphy, K.; Yuille, A.L. Semantic Image Segmentation with Deep Convolutional Nets and Fully Connected CRFs. In Proceedings of the International Conference on Learning Representations (ICLR), San Diego, CA, USA, 7–9 May 2015.
31. Paisitkriangkrai, S.; Sherrah, J.; Janney, P.; Van den Hengel, A. Effective Semantic Pixel labelling with Convolutional Networks and Conditional Random Fields. In Proceedings of the Conference on Computer Vision and Pattern Recognition (CVPR), Boston, MA, USA, 7–12 June 2015.
32. Zheng, S.; Jayasumana, S.; Romera-Paredes, B.; Vineety, V.; Su, Z. Conditional Random Fields as Recurrent Neural Networks. In Proceedings of the Conference on Computer Vision and Pattern Recognition (CVPR), Las Vegas, NV, USA, 26 June–1 July 2016.
33. Sherrah, J. Fully Convolutional Networks for Dense Semantic Labelling of High-Resolution Aerial Imagery. *arXiv*, **2016**, arXiv:1606.02585.
34. Marmanis, D.; Wegner, J.D.; Galliani, S.; Schindler, K.; Datcu, M.; Stilla, U. Semantic segmentation of aerial images with an ensemble of CNSS. In Proceedings of the ISPRS Annals of the Photogrammetry, Remote Sensing and Spatial Information Sciences, Prague, Czech Republic, 12–19 July 2016.
35. Maggiori, E.; Tarabalka, Y.; Charpiat, G.; Alliez, P. Fully Convolutional Neural Networks for Remote Sensing Image Classification. In Proceedings of the 2016 IEEE International Geoscience and Remote Sensing Symposium (IGARSS), Beijing, China, 10–15 July 2016; pp. 5071–5074.
36. Maggiori, E.; Tarabalka, Y.; Charpiat, G.; Alliez, P. Convolutional Neural Networks for Large-Scale Remote Sensing Image Classification. *IEEE Trans. Geosci. Remote Sens.* **2017**, *2*, 645–657.
37. Maggiori, E.; Tarabalka, Y.; Charpiat, G.; Alliez, P. High-Resolution Semantic Labeling with Convolutional Neural Networks. *arXiv*, **2016**, arXiv:1611.01962.
38. Simonyan, K.; Zisserman, A. Very deep convolutional networks for large-scale image recognition. *arXiv*, **2014**, arXiv:1409.1556.
39. He, K.; Zhang, X.; Ren, S.; Sun, J. Deep residual learning for image recognition. *arXiv*, **2015**, arXiv:1512.03385.

40. Szegedy, C.; Ioffe, S.; Vanhoucke, V.; Alemi, A. Inception-v4, Inception-ResNet and the Impact of Residual Connections on Learning. *arXiv*, **2016**, arXiv:1602.07261.
41. Bertasius, G.; Shi, J.; Torresani, L. Deepedge: A multiscale bifurcated deep network for top-down contour detection. In Proceedings of the Conference on Computer Vision and Pattern Recognition (CVPR), Boston, MA, USA, 7–12 June 2015.
42. Xie, S.; Tu, Z. Holistically-Nested Edge Detection. *arXiv*, **2015**, arXiv:1504.06375.
43. Krahenbuhl, P.; Koltun, V. Efficient Inference in Fully Connected CRFs with Gaussian Edge Potentials. In Proceedings of the 24th International Conference on Neural Information Processing Systems (NIPS), Granada, Spain, 12–15 December 2011.
44. Nussbaum, S.; Niemeyer, I.; Canty, M.J. SEATH—A new tool for automated feature extraction in the context of object-based image analysis. In Proceedings of the 1st International Conference on Object-Based Image Analysis, Salzburg, Austria, 4–5 July 2006. XXXVI-4/C42.
45. Chang, C.C.; Lin, C.J. LIBSVM: A library for support vector machines. *ACM Trans. Intell. Syst. Technol.* **2011**, 2.

9

Convolutional Neural Networks Based Hyperspectral Image Classification Method with Adaptive Kernels

Chen Ding *, Ying Li, Yong Xia, Wei Wei, Lei Zhang and Yanning Zhang

Shaanxi Key Lab of Speech & Image Information Processing (SAIIP), School of Computer Science and Engineering, Northwestern Polytechnical University, Xi'an 710129, China; lybyp@nwpu.edu.cn (Y.L.); yxia@nwpu.edu.cn (Y.X.); weiweinwpu@nwpu.edu.cn (W.W.); zhanglei211@mail.nwpu.edu.cn (L.Z.); ynzhang@nwpu.edu.cn (Y.Z.)

* Correspondence: dingchen@mail.nwpu.edu.cn.

Academic Editors: Qi Wang, Nicolas H. Younan, Carlos López-Martínez and Prasad S. Thenkabail

Abstract: Hyperspectral image (HSI) classification aims at assigning each pixel a pre-defined class label, which underpins lots of vision related applications, such as remote sensing, mineral exploration and ground object identification, etc. Lots of classification methods thus have been proposed for better hyperspectral imagery interpretation. Witnessing the success of convolutional neural networks (CNNs) in the traditional images based classification tasks, plenty of efforts have been made to leverage CNNs to improve HSI classification. An advanced CNNs architecture uses the kernels generated from the clustering method, such as a K-means network uses K-means to generate the kernels. However, the above methods are often obtained heuristically (e.g., the number of kernels should be assigned manually), and how to data-adaptively determine the number of convolutional kernels (i.e., filters), and thus generate the kernels that better represent the data, are seldom studied in existing CNNs based HSI classification methods. In this study, we propose a new CNNs based HSI classification method where the convolutional kernels can be automatically learned from the data through clustering without knowing the cluster number. With those data-adaptive kernels, the proposed CNNs method achieves better classification results. Experimental results from the datasets demonstrate the effectiveness of the proposed method.

Keywords: hyperspectral image classification; automatic cluster number determination; adaptive convolutional kernels

1. Introduction

Different from traditional images (e.g., RGB image), hyperspectral image (HSI) contains a continuous spectrum at each pixel, which is beneficial for identifying different imaged land covers. With such abundant spectral information, hyperspectral image (HSI) classification that aims at assigning each pixel a pre-defined class label has facilitated various applications, such as mineral exploration, ground object identification, survey of agriculture and monitoring of geology, etc. Therefore, plenty of efforts have been made in HSI classification. According to the feature utilized, HSI classification methods can be roughly divided into hand-crafted feature based methods and the deep learning feature based methods. A detailed review can be seen from Section 2. For hand-crafted feature based methods, HSI is often represented by the features designed manually [1–7]. However, due to their shallow structure, the representation ability of such features is limited, especially for HSIs which often exhibit high nonlinearity aroused by the high-dimensionality and mixture of pixels. On the contrary, deep learning feature based methods can automatically extract features from training data with deep architectures. It has been proved that those deep features perform well in representing the complicated nonlinearity of data, which has promoted the development of deep learning feature based HSI classification methods in recent years [8–12].

Since the convolutional kernels should be updated through the network training, traditional deep learning based methods exhaust much training time. To address this problem, an advanced CNNs architecture has been proposed recently, which adopts the kernels pre-learned from clustering the training data without updating them in the training process any more. One typical method is the K-means Net proposed in [13], where each CNNs kernel is first learned from a specific cluster obtained by conducting the K-means algorithm on training data. Nevertheless, the cluster number K (i.e., the number of kernels in CNNs) of K-means Net should be assigned empirically, which limits the representational power of CNNs. Specifically, a different number K of kernels designed manually in the convolutional layer will change the structure of CNNs and thus influence the output of CNNs. In addition, the number K is expected to be adaptive to different images and tasks. Therefore, how to data-adaptively choose a proper number of kernels is crucial for representing data characteristics with CNNs. However, most of the existing CNNs based HSI classification methods fail to pay sufficient consideration to this problem.

In this study, we propose a MCFSFDP based CNNs framework for HSI classification. First, inspired by clustering by fast search and find of peaks (CFSFDP) [14], a novel clustering method, named modified clustering by fast search and find of peaks (MCFSFDP), is proposed to data-adaptively learn a specific number of kernels from training data. The convolution kernels can be automatically determined by the center of each cluster and the inter-cluster margin, which guarantees the pre-learned kernels to be suitable for the data structure. Then, the CNNs framework with those pre-learned convolutional kernels is employed to classify each pixel in the HSI. Extensive experimental results demonstrate that the proposed method outperforms several state-of-the-art CNNs based methods in classification accuracy.

In summary, the proposed CNNs framework has two key advantages: (1) a specific number of convolutional kernels can be data-adaptively learned from training data, which can well represent the data characteristics; and (2) the MCFSFDP based CNNs framework is effective for HSI classification.

2. Related Work

Based on the feature adopted in classification of HSI, the HSI classification method can be roughly divided into two categories, including the hand-crafted feature based methods and the deep learning feature based methods.

2.1. Hand-Crafted Feature Based Methods

Linear features extracted by principal component analysis (PCA) [15] and partial least squares (PLS) [16] are applied to classify the HSI data. The kernel methods are further developed to exploit the nonlinear feature of HSI [17]. To depict the spatial texture of image, the wavelet transform (WT) methods [18,19] have been widely used, which often show different scales and perform effectively for classification in the high spatial resolution remotely sensed (HSRRS) data. Considering the complicated spatial correlation, some Gaussian Markov Random Field (GMRF) [20,21] methods are proposed to model such correlation within a graph structure. In [22], a spatial feature index that measured the gray similarity distance in every direction is used to describe the shape feature in local area that is surrounding a pixel in HSI. An adaptive mean-shift (MS) analysis framework [2] is proposed for object extraction and classification of HSI over urban areas, which is able to obtain an object-oriented representations of HSI data. Li et al. [3] integrate the spectral and spatial information in a Bayesian framework, which utilizes a Multinomial Logistic Regression (MLR) algorithm to learn the posterior probability distributions from the spectral information. In addition, this method uses subspace projection to better characterize noise, highly mixed pixels and contextual information. In [4], a mathematical morphology (MM) based method is utilized to process the HSI data. In this approach, opening and closing morphological transforms are used to isolate bright (opening) and dark (closing) structures in images, where bright/dark means brighter/darker than the surrounding features in the images. To model different kinds of structural information, morphological attribute profiles (APs) are adopted to provide a multi-level characterization for an image created by the sequential application of morphological attribute filters [23]. Based on Gray Level Co-occurrence matrix (GLCM), Zortea et al. attempt to extract the contextual information of images by concatenating the spectral features used for classification [1]. To improve the classification result of HSI, the Edge-Aware Filtering (EAF) and Edge-Preserving Filtering (EPF) methods are proposed in [24,25]. Based on the EPF method, a spectral-spatial classification framework was proposed in [25], which can significantly enhance the classification accuracy. Kang et al. propose combining a recursion with image fusion to enhance the image classification accuracy [26]. Recently, the Bag-of-Words (BOW) model has shown a promising way to handle the remote sensing imagery classification problem. In the BOW model, images can be represented by the frequency of visual words that are constructed by quantizing local features with a clustering method, such as K-means and so on [27,28]. Due to the capacity of extracting the handcrafted local features, such as local structural points, color histogram and texture features [29,30], BOW based methods present good performance. Manifold regularized kernel logistic regression (KLR) are proposed to solve multi-view image classification [31]. To integrate different levels of features for saliency detection, Wang et al. [32] propose a multiple-instance learning based framework that fuses the low-level, mid-level, and high-level features into a unified model. While effective, the trepresentation capacity of the manual feature extraction based methods is limited.

2.2. Deep Learning Feature Based Methods

Recently, with the development of deep learning technology, lots of methods based on deep learning have been developed for image classification, such as deep brief network (DBN) and stacked auto-encoder (SAE). The DBN and SAE are unsupervised learning methods that are also used for spectral-spatial classification of hyperspectral data without using the label information [9,33]. The concept of deep learning is introduced into the hyperspectral data classification for the first time [9]. The Canonical Correlation Analysis Network is useful for multi-view image classification [34]. With the development of convolutional neural networks (CNNs) [35], which has been widely applied to the image processing and achieved spectacular effects, more and more deep CNNs frameworks have emerged, such as AlexNet [36], VGGNet [37], GoogLeNet [38] and ResNet [39], which can provide results comparable with human beings in image classification and recognition tasks. Those methods can automatically learn features from the training data, which can replace the manually-engineered features, and have shown significant effects on HSI classification [8–10]. For example, Li et al. [40] applied 3D-CNNs for spectral-spatial feature extraction and classification, where 3D kernels were used to extract the feature from HSI cube without any preprocessing or post-processing. In [41], the transfer learning method for HRRS scene classification is used for transferring features from successfully pre-learned CNNs. Different from the CNNs methods, the convolutional kernels are updated in the training process, and the kernels in PCA-Net [42] and K-means Net are pre-learned before the network training and don't need to be updated in the network training. In addition, the kernels come from data directly. PCA-Net [42] adopts the principle components of training data as multistage filter banks, while K-means Net learns the kernels by clustering the training data. In this study, we mainly focus on the K-means Net. Although K-means Net can be directly applied to the classification and reduces the training time by employing the pre-learned kernels, it is difficult to determine the number of kernels that is crucial for the performance. To address this issue, we attempt to adaptively generate a specific number of kernels from the training data of CNNs framework.

3. MCFSFDP Based CNNs

The traditional CNNs framework contains the convolutional layer, fully connected layer and a classification layer. The convolution layer is updated through the error feedback process, which is different from the pre-learned convolutional kernels based CNNs framework.

The proposed MCFSFDP based CNNs method includes three major modules: (1) data pre-processing module, which extracts patches from block samples; (2) MCFSFDP based kernel learning module, which learns the convolutional kernels from those extracted patches; and (3) classification modules which utilize the learned convolution kernels.

The flowchart of our MCFSFDP based CNNs method is shown in Figure 1.

Figure 1. The flowchart of the MCFSFDP based CNNs method.

3.1. Data Pre-Processing

In this study, we follow the standard data pre-processing principle in K-means Net [13]. Specifically, a HSI used in this classification task is denoted by $\boldsymbol{R}$. Though HSI is 3D data, it also can be seen as a collection of 2D images (i.e., images from different bands). Here, we denote the HSI as 2D form. First, we randomly select M pixels from $\boldsymbol{R}$, and then extract M corresponding blocks $\{B_i\}_{i=1}^{M}$ with a size of $m \times m$ as samples, where each block is centered at each selected pixel. These extracted M samples are roughly divided into three parts, namely, training samples, validation samples and testing samples. The property of center block pixel is described by all the pixels in the block. Then, $\{B_i\}_{i=1}^{M}$ are put into the network and the center pixel labels of block B_i are used as the ground truth for training.

In addition, we randomly extract N patches $\{P_j\}_{j=1}^{N}$ with a size of $n \times n$ from M_T training samples, M_T denotes the number of training samples, where $M_T < M$ and $n < m$. The extracted N patches $\{P_j\}_{j=1}^{N}$ are used for learning the convolutional kernels with a size of $n \times n$ via MCFSFDP. The producing process of the block (sample) and patch is shown in Figure 2.

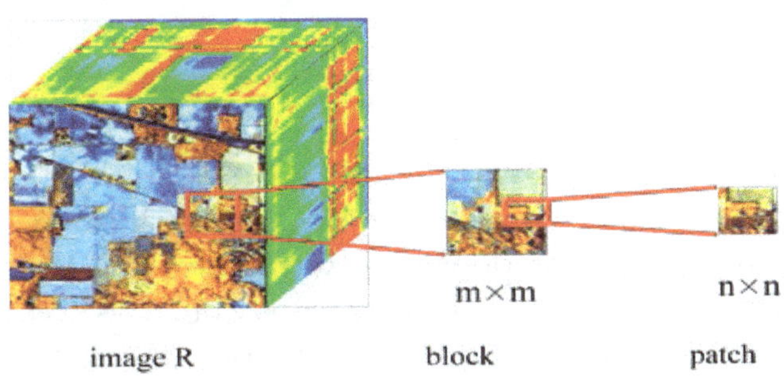

Figure 2. The block (sample) is extracted from image R and the patch is extracted from block, respectively.

3.2. MCFSFDP Based CNNs Kernels Learning

To obtain the kernels with those cropped patches, a suitable clustering method is necessary. Lots of clustering methods have been proposed, among which clustering by fast search and find of peaks (CFSFDP) [14], is a typical state-of-the-art method. The reason for partial success of CFSFDP on clustering is based on the idea that "cluster centers are characterized by a higher density than their neighbors and by a relatively large distance from points with higher densities" and the cluster centers can be determined through two thresholds of distance and density [14].

Though CFSFDP has shown its power for clustering, we find that when we apply it directly to generate the kernels for CNNs, the generated kernels are not always optimal for hyperspectral image classification tasks. This phenomena is observed from the experimental results (a similar conclusion also can be seen from the results in Section 4.3.1). In our opinion, we consider kernels (filters) as the standards for comparing the samples, which also show the evaluation standards for determining which cluster they belong to. Since the inter-cluster points are difficult to classify, we should also select several inter-cluster points with representations as the clusters (kernels). To address this problem, we propose a new clustering method based on CFSFDP, which only uses distance threshold to generate the kernel centers. The proposed method differs from the traditional CFSFDP in two aspects: (1) CFSFDP simultaneously uses the points with a large distance and high density to determine the cluster center, which easily excludes the outlier points into the generation of cluster centers; while the proposed MCFSFDP method only uses distance threshold to generate the cluster center, the cluster centers can be generated from either outlier points (with only large distance) or points of density; (2) the number of clusters via CFSFDP is determined 'semi-automatically', i.e., an extra frame needs to be introduced to help determine the number of clusters, while the number of clusters can be automatically determined through the proposed method. We give the details of the proposed method as follows.

The same as the CFSFDP algorithm in [14], we assume that the cluster centers are characterized by a higher density than their neighbors and by a relatively large distance from points with higher densities.

Following this idea, we firstly reshape each patch P_j into a column vector as a data point j with a size of $1 \times n^2$. For each point j, we compute two values: its local density ρ_j and its distance δ_j from the point with higher density, where, if the point j has the highest density, δ_j denotes the largest distance between j and other points.

Both of these values depend only on the Euclidean distances d_{jk} between any pair of data points j and k. The local density ρ_j of data j is defined as

$$\rho_j = \sum_k \chi(d_{jk} - d_c), \tag{1}$$

where $\chi(x) = 1$ if $x < 0$ and $\chi(x) = 0$ otherwise, and d_c is a cut-off distance. Basically, ρ_j is equal to the number of points that are closer than d_c to point j. δ_j is evaluated through computing the minimum distance between the point j and any other point with higher density in Equation (2):

$$\delta_j = \min_{k:\rho_k > \rho_j} (d_{jk}) \,. \tag{2}$$

For the point with the highest density, we usually take $\delta_j = \max_k(d_{jk})$. Note that δ_j is much larger than the typical nearest neighbor distance only for points that are local or global maxima in the density. Thus, the cluster centers are recognized as points for which the value of δ_j is anomalously large and the value of ρ_j is higher than a value density at the same time. To show the distance and density of each point intuitively, we give the decision graph of 10,000 patches with a size of 10×10 from the real Indian pines dataset in Figure 3.

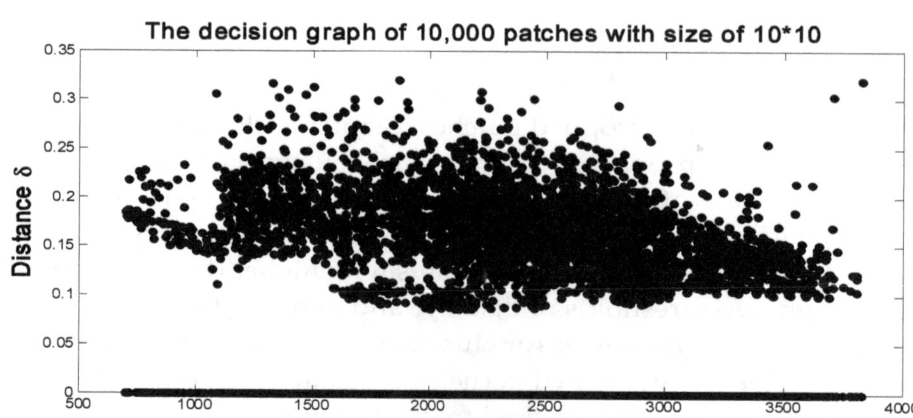

Figure 3. Decision graph of 10,000 patches with a size of 10×10 on the Indian pines dataset.

Different from choosing cluster centers in CFSFDP [14], we use the MCFSFDP algorithm to learn the kernels adaptively. Firstly, we choose the distance δ as the only threshold for choosing kernels from the decision graph in MCFSFDP.

To adapt the kernels and choose the number of kernels, we select the optimal distance threshold value δ_A as the following steps:

$$num_v = f(\delta_v), \tag{3}$$

$$con_v = [f(\delta_{v+1}) - f(\delta_v)]/(\delta_{v+1} - \delta_v), \tag{4}$$

$$quo_v = |con_v/con_{v+1}|. \tag{5}$$

where, in Equation (3), δ_v denotes the value of distance that contains points and $f(\delta_v)$ gives the mapping relationship of the number of points whose distances are equal or larger than δ_v, as shown as Figure 4a. In Equation (4), where $\delta_{v+1} \geq \delta_v$, con_v denotes the differential of $f(\delta_v)$, which is an intermediate result between Equations (3) and (5). Equation (5) denotes the variation quantity of the number of points with δ_v, shown as Figure 4b.

δ_A denotes the adaptive distance threshold, and the points whose distances are larger than δ_A are chosen as CNN kernels. δ_A is a critical point that must satisfy the number num_v and num_{v+1} of points are stable (in other words, they have a similar quantity), at the same time, the value $|con_v/con_{v+1}|$ is larger than the value $|con_{v+1}/con_{v+2}|$. In this time, δ_v is selected as the adaptive distance threshold δ_A.

In other words, to determine the adaptive distance threshold δ_A intuitively, from Figure 4a, we can find the value region δ_v (0.25–0.30) from curve 1 when num_v begins to approach to 0; as can be seen from Figure 4b, con_v with the distance value δ_v in region (0.25–0.30) has a local maxima at $\delta_v = 0.28$. The distance δ_v (0.28) that belongs to the region (0.25–0.30) is confirmed as the adaptive threshold distance as δ_A. In conclusion, by observing Figure 4, the adaptive distance threshold δ_A is determined as 0.28 on the Indian Pines dataset.

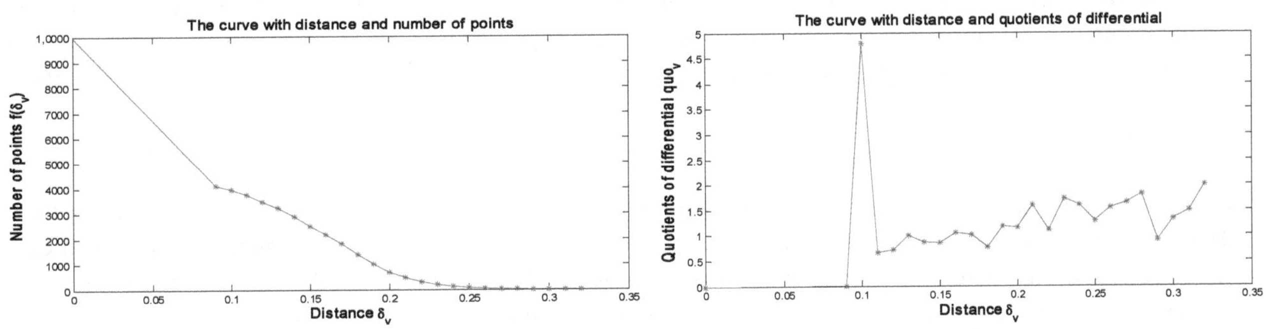

(a) (b)

Figure 4. The curve for determining the adaptive distance with patches with a size of 10 × 10 the on Indian pines dataset. (**a**) shows the curve of point-number over distance δ_v; (**b**) gives the curve of quotients of differential over distance δ_v.

Finally, the points j with the distance value $\delta_j > \delta_A$ are adaptively chosen as the kernels and thus the number of kernels is also adaptively determined through the threshold δ_A. Those chosen points are then reshaped to patches with a size of $n \times n$ as the convolutional kernels in the CNNs framework. The CNNs with the pre-learned adaptive kernels are called MCFSFDP Net. The pre-learned kernels are denoted as w^k in the following sections.

3.3. Convolutional Neural Networks

With the pre-learned kernels w^k, a convolutional neural network such as [13] is designed for per-pixel level HSI classification. This CNNs structure consists of an input layer, a convolutional layer, a pooling layer, a fully connected layer and a soft-max layer, as shown in Figure 5.

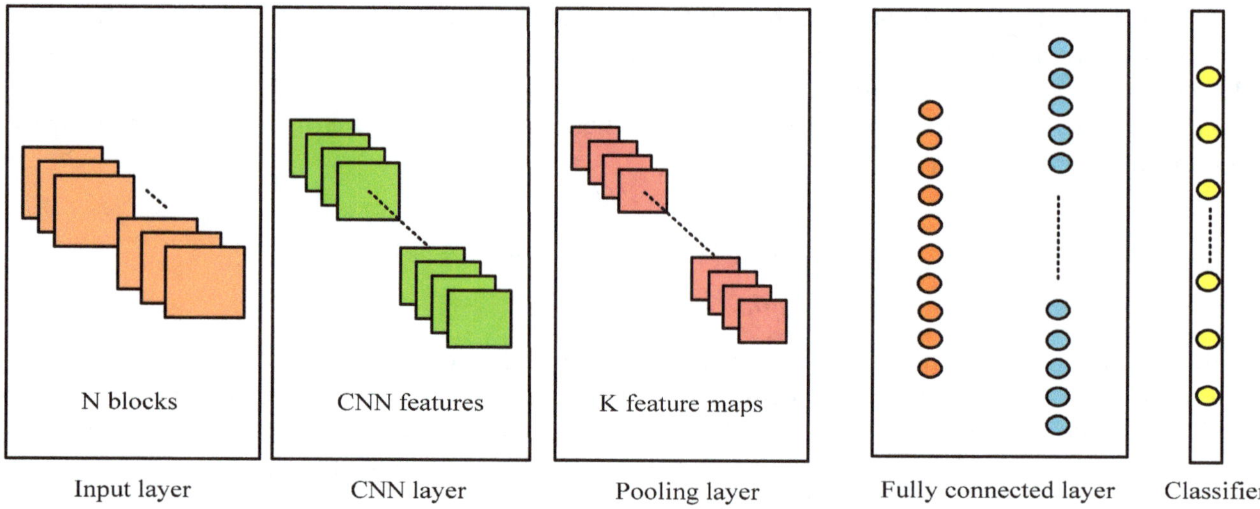

Figure 5. The structure of MCFSFDP based CNNs.

There are k kernels in the convolutional layer. Each feature map is calculated by taking the dot product between the k-th kernel w^k of size $n \times n$, $w \in R^{n \times n \times k}$, and local context area x of size $m \times m$ with c number of channels, $x \in R^{m \times m \times c}$. The feature map corresponding with the k-th filter $f \in R^{(m-n+1) \times (m-n+1)}$ is calculated as:

$$f_{ij}^k = \sigma(\sum_c \sum_{a=0}^{n-1} \sum_{b=0}^{n-1} w_{abc}^k x_{i+a,j+b}^c), \tag{6}$$

where σ is the rectified linear unit (ReLU). The kernels were pre-trained using the MCFSFDP algorithm.

The maximum pooling over a local non-overlapping spatial region is adopted to down-sample the convolutional layer. The pooling layer for the k-th filter, $g \in R^{(m-n+1)/p \times (m-n+1)/p}$, is calculated as:

$$g_{ij}^k = \max(f_{1+p(i-1),1+p(j-1)}^k, \ldots, f_{pi,1+p(j-1)}^k, \ldots, f_{1+p(i-1),pj}^k, \ldots, f_{1+pi,pj}^k). \tag{7}$$

The k feature maps are reshaped to the column vectors and all the column vectors are connected with a fully connected auto-encode unit. The autoencode unit is used to process the connected column vector and represented the feature of the column vector. The output results of the hide layer in the auto-encode unit were used to connect the classification layer.

The last CNNs step is a soft-max layer used for final classification.

4. Experiments and Analysis

Three datasets were utilized to validate the feasibility and effectiveness of the proposed CNNs based MCFSFDP method (named as MCFSFDP Net) in HSI classification. In the following sections, dataset and experimental settings are described firstly, and then the effectiveness and the superiority of the proposed method are tested.

4.1. Datasets

To find images with less categories and obvious discriminations between categories, we firstly select an image dataset with a size of 256×256. The image of this dataset has been manually labeled as three categories, including mountains, sky and roads. One hundred samples with a size of 25×25 from each category that were extracted from this image. We randomly choose 210 context area samples for training, 30 samples for validation and 60 other samples for testing. The details of selected image samples were given in Table 1.

Table 1. Ground truth classes and their respective sample numbers in Dataset 1.

Class	Samples		
	Training	Validation	Testing
Mountain	70	10	20
Sky	70	10	20
Road	70	10	20

In order to evaluate the proposed method on complex data, Dataset 2 includes the benchmark Indian Pines image, which is HSI data captured by the airborne visible imaging spectrometer (AVIRIS) sensor with a moderate spatial resolution of 20 m over the Indian Pines test site in northwestern Indiana in 1992. As shown in Figure 6, this image contains145 $\times$ 145 pixels and 224 spectral bands, whose wavelength ranges from 0.4 to 2.5 um. The number of bands of corrected data was reduced to 200 (extracted the 1–200 bands). In addition, 6476 image context area samples with a size of 19 $\times$ 19 were extracted. Among them, 3238, 647 and 2591 samples were used for training, validation and testing, respectively. The details of each category of image samples were given in Table 2.

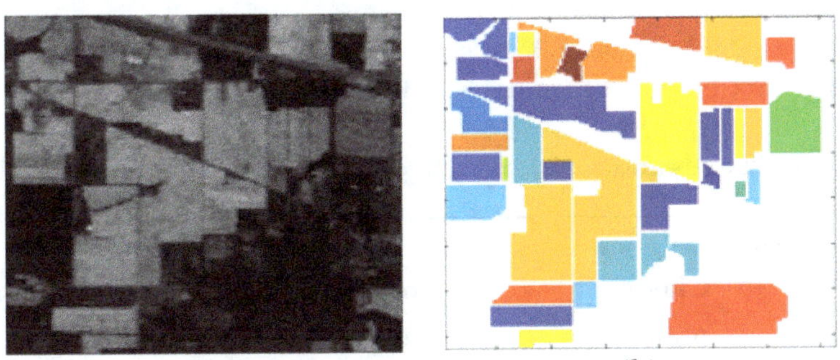

(a) (b)

Figure 6. The Indian Pines on Dataset 2. (**a**) shows the composite image; (**b**) shows the groundtruth of Indian Pines dataset, where the white area denotes the unlabeled pixels.

Table 2. Groundtruth of classes and their respective sample numbers on Indian Pines scene.

Class		Samples			
Number	Classes	Total	Training	Validation	Testing
1	Alfalfa	46	23	4	19
2	Corn-notill	1288	636	132	520
3	Corn-mintill	63	29	7	27
4	Corn	35	17	3	15
5	Grass-pasture	180	90	14	76
6	Grass-trees	730	342	84	304
7	Grass-pasture-mowed	28	16	1	11
8	Hay-windrowed	94	45	8	41
9	Oats	20	10	2	8
10	Soybean-notill	807	406	71	330
11	Soybean-mintill	2067	1019	215	833
12	Soybean-clean	227	124	22	81
13	Wheat	204	107	28	69
14	Woods	560	307	44	209
15	Buildings-Grass-Trees-Drives	73	38	9	26
16	Stone-Steel-Towers	54	29	3	22
	Total	**6476**	**3238**	**647**	**2591**

The third Dataset 3 includes the benchmark Pavia University image, which is HSI data captured by a ROSIS sensor with a moderate spatial resolution of 1.3 m over the flight campaign over Pavia, northern Italy. As shown in Figure 7, this image contains 610 × 610 pixels and 103 spectral bands. The number of bands was reduced to 100 (extracted the 1–100 bands). Furthermore, 34,400 image context area samples with a size of 11 × 11 were extracted. Among them, 17,200, 3440 and 13,760 samples were used for training, validation and testing, respectively. The details of each category of samples were given in Table 3.

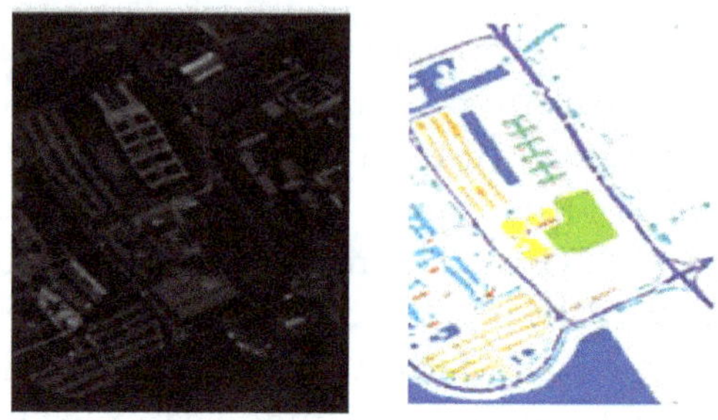

(a) (b)

Figure 7. The Pavia University in Dataset 3. (**a**) shows the composite image; (**b**) shows the groundtruth of the Pavia University dataset, white area denotes the unlabeled pixels.

Table 3. Groundtruth of classes and their respective sample numbers in the Pavia University scene.

Class		Samples			
Number	**Classes**	**Total**	**Training**	**Validation**	**Testing**
1	Asphalt	5446	2718	580	2148
2	Meadows	12,695	6307	1320	5068
3	Gravel	1314	674	126	514
4	Trees	2709	1329	241	1139
5	Painted metal sheets	1345	688	153	504
6	Bare Soil	5029	2517	453	2059
7	Bitumen	1330	686	120	524
8	Self-Blocking Bricks	3630	1810	362	1458
9	Shadows	902	471	85	346
	Total	**34,400**	**17,200**	**3440**	**13,760**

4.2. Experimental Parameter Settings

Ten thousand patches were randomly extracted from the training samples for learning kernels. For each dataset, the sample (blocks) size and the number of patches should be maintained consistently in different pre-learned CNNs frameworks.

The CNNs framework that is shown in Figure 5 uses one convolutional layer, one pooling layer, one auto-encode layer and a classifier. In our algorithm, the pooling layer adopted the non-overlap rule, the number of neurons in the hide layer of auto encode was set to 100 and the maximum iterations for training the classifier was 400. The learning rate is 0.0001 and momentum is 1. The batch sizes on the three datasets are chosen as 10, 50 and 200, respectively.

The codes are running on the computer with Intel Xeon E5-2678 V3 2.50 GHz × 2 (Intel, Santa Clara, CA, USA), NVIDIA Tesla (NVIDIA, Santa Clara, CA, USA) K40c GPU × 2, 128 GB RAM, 120 GB SSD and Matlab 2016a (MathWorks, Natick, MA, USA). The gradient is computed via batch gradient descent, which is not computed by GPU.

The average test accuracy is calculated on 10 independent Monte Carlo runs.

4.3. Experimental Results

4.3.1. Effectiveness of the Kernels Learned by MCFSFDP

The aim of this experiment is to validate the effectiveness of the kernels learned by MCFSFDP. To this end, we compared those kernels with those learned as the cluster center obtained by CFSFDP algorithm. Those two kinds of kernels were then integrated into the same CNNs framework for HSI classification on Dataset 1. To obtain fair comparison results, both of the numbers of kernels in those two methods were fixed at 49. The kernel size was set to 14×14 and the pooling size was designed as 4×4. The average testing classification accuracy of those two methods was shown in Table 4.

Table 4. The testing accuracy compared with learned 49 kernels via CFSFDP and MCFSFDP-M on Dataset 1.

Methods	CFSFDP Net	MCFSFDP Net
Accuracy (%)	81.67 ± 0.5904	**95.00 ± 0.5887**

It reveals that the kernels learned by the MCFSFDP are more effective than the kernels learned by the CFSFDP.

4.3.2. Effectiveness of the Kernels Number Determined by MCFSFDP

To demonstrate the effectiveness of the kernels number determined by MCFSFDP, we compared MCFSFDP with its variants for classification in each dataset. Those variants shared the same CNNs architecture and the kernel learning scheme excepted choosing the kernels number manually. Dataset 1, Dataset 2 and Dataset 3 were used in the experiment. For each dataset, the kernel size and the pooling size can be found in Table 5.

Table 5. The chosen block size, kernel size and pooling size of each dataset.

Dataset	Dataset 1	Dataset 2	Dataset 3
Block Size	25×25	19×19	11×11
Kernel Size	10×10	6×6	2×2
Pooling Size	4×4	7×7	2×2

We report the testing classification accuracy of all these methods on each dataset in Tables 6–8, respectively. Each variant is denoted as MCFSFDP-M Net followed with a specific number which indicates the kernel number chosen manually. Similarly, the number that followed MCFSFDP Net represents the kernel number automatically determined by the proposed method.

Table 6. The testing accuracy of MCFSFDP-M Net compared with MCFSFDP Net on Dataset 1.

Methods	MCFSFDP-M Net-20	MCFSFDP-M Net-25	MCFSFDP-M Net-41	MCFSFDP-M Net-55	MCFSFDP Net-35
Accuracy (%)	93.33 ± 0.5887	95.00 ± 0.5904	95.00 ± 0.5904	95.00 ± 0.5904	**96.67 ± 0.5887**
Distance threshold	0.19	0.18	0.16	0.15	**0.17**
Number of kernels	20	25	41	55	**35**

Table 7. The testing accuracy of MCFSFDP-M Net compared with MCFSFDP Net on Dataset 2.

Methods	MCFSFDP-M Net-14	MCFSFDP-M Net-24	MCFSFDP-M Net-31	MCFSFDP-M Net-83	MCFSFDP-M Net-151	MCFSFDP Net-50
Accuracy (%)	95.29 ± 0.0870	96.51 ± 0.4146	97.03 ± 0.1940	97.07 ± 0.3434	96.82 ± 0.1457	**97.84 ± 0.2249**
Distance threshold	0.27	0.26	0.25	0.23	0.22	**0.24**
Number of kernels	14	24	31	83	151	**50**

Table 8. The test accuracy of MCFSFDP-M Net compared with MCFSFDP Net on Dataset 3.

Methods	MCFSFDP-M Net-19	MCFSFDP-M Net-42	MCFSFDP-M Net-152	MCFSFDP Net-78
Accuracy (%)	88.98 ± 0.2651	89.32 ± 0.1908	89.54 ± 0.1002	**90.58 ± 0.1477**
Distance threshold	0.08	0.07	0.05	**0.06**
Number of kernels	19	42	152	**78**

In Table 6, the proposed method determines the kernel number as 35. The manually chosen kernel number in other variants are 20, 25, 41 and 55, respectively. The accuracy, distance threshold and the number of kernels for each method are shown in different rows. It can be seen that the proposed method shows the best classification accuracy. Similar phenomenon arises in Tables 7 and 8. Therefore, we can conclude that the proposed method is able to seek a good kernel number for different datasets.

4.3.3. Performance Evaluation of MCFSFDP Net

In this part, the proposed method was compared with three state-of-the-art pre-learned kernels based CNNs methods, including K-means Net [13], PCA-Net [42] and Random Net. For fair comparison, the same CNNs architecture was adopted by all comparison methods. The number of kernels for K-means Net, PCA-Net and Random Net was set to 50, while the proposed method determines the number of kernels automatically. For each dataset, the kernel size and the pooling size can be found in Table 9.

Table 9. The testing accuracy of different CNNs methods compared with MCFSFDP Net on Dataset 1.

Methods	K-Means Net-50	PCA Net-50	Random Net-50	MCFSFDP Net-35
Accuracy (%)	93.33 ± 0.5887	90.00 ± 1.8175	95.00 ± 1.8175	**96.67 ± 0.5887**

It reveals that the proposed algorithm can produce more accuracy for pixel classification than those three types of pre-learned kernels based CNNs methods on this dataset as shown in Table 9. Moreover, the proposed MCFSFDP Net with 35 kernels that has less computational complexity than comparison methods with 50 kernels in the training process.

The average testing classification accuracy of our proposed algorithm, K-means Net, PCA-Net and Random Net on Dataset 2 was given in Table 10. The results obviously show that the proposed MCFSFDP Net obtains better accuracy than those three types of pre-learned kernels based CNNs methods, which is consistent with the results obtained from Dataset 1.

Table 10. The testing accuracy of different CNNs methods compared with MCFSFDP Net on Dataset 2.

Methods	K-Means Net-50	PCA Net-50	Random Net-50	MCFSFDP Net-50
Accuracy (%)	95.02 ± 0.3343	97.30 ± 1.1916	97.12 ± 0.6195	**97.84 ± 0.2249**

The average classification accuracy of our proposed method compared with another three kernels pre-learned based CNNs on the Pavia University image was presented in Table 11. The results show that our proposed CNNs method is more accurate than those three types of pre-learned kernels based CNNs methods. Even if the proposed method needs more kernels number to perform the better classification result.

Table 11. The testing accuracy of different CNNs methods compared with MCFSFDP Net on Dataset 3.

Methods	K-Means Net-50	PCA Net-50	Random Net-50	MCFSFDP Net-78
Accuracy (%)	89.77 ± 0.3399	90.14 ± 0.2652	90.47 ± 0.5113	**90.58 ± 0.1477**

5. Discussion

5.1. Effect ofthe Number of Kernels

In the MCFSFDP-M Net, the number of kernels influences the pixel-level classification. Figure 8 shows the classification accuracy achieved with different numbers A_k that were manually selected via MCFSFDP on Dataset 1, Dataset 2 and Dataset 3.

Figure 8a shows the classification results with the variation of kernel numbers A_k on each kernel size $n \times n$ on Dataset 1. The accuracy of MCFSFDP-M Net computation cannot be enhanced when the kernel number A_k was increased. Figure 8b shows the highest accuracy on Dataset 2. While the kernel number is manually chosen via MCFSFDP, the accuracy can get a high point in the number range of the kernels, as the adaptive kernels learned through the MCFSFDP method. It demonstrates again that the accuracy cannot be enhanced with the increased kernel number on Dataset 3, as shown in Figure 8c.

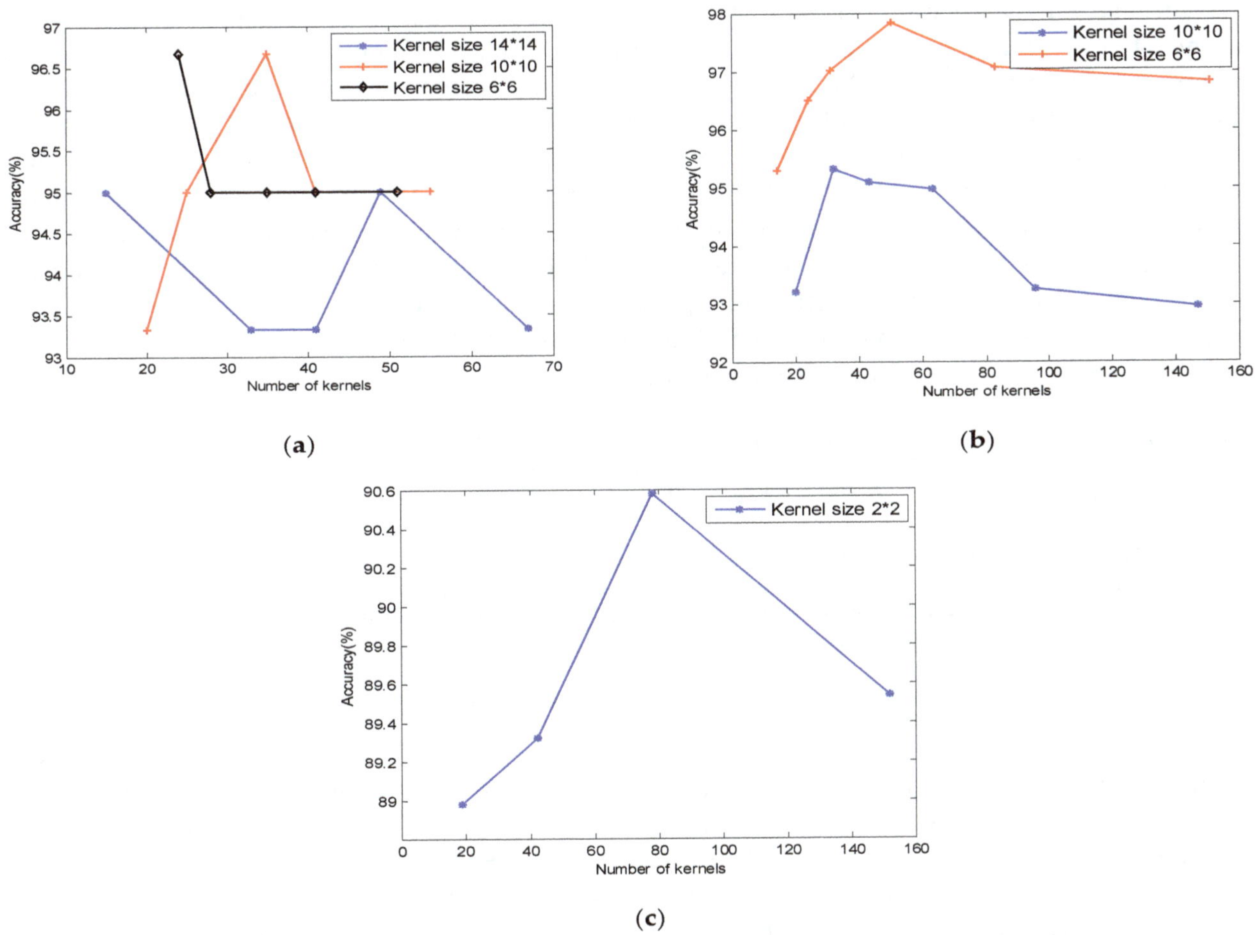

Figure 8. The classification accuracy influence with the number of kernels. (**a**) the classification accuracy with the increased number of kernels with different kernel size on Dataset 1; (**b**) the classification accuracy with the increased number of kernels with different kernel size on Dataset 2; (**c**) the classification accuracy with the increased number of kernels on Dataset 3.

5.2. Effect of the Kernel Size

In our proposed MCFSFSP based CNN method, the kernel size has a major impact on the pixel classification performance. Table 12 gives the average classification accuracy obtained by using different kernel size. It shows that the highest classification accuracy was achieved when kernel size was set to 10 × 10 and 6 × 6 on Dataset 1 and 6 × 6 on Dataset 2.

Table 12. The average classification accuracy obtained by using different kernel size.

Dataset		Dataset 1		Dataset 2	
Pooling Size	4 × 4	4 × 4	4 × 4	5 × 5	7 × 7
Kernel Size	14 × 14	10 × 10	6 × 6	10 × 10	6 × 6
Number of Kernels	15	35	24	32	50
Distance Value	0.22	0.17	0.17	0.28	0.24
Accuracy (%)	95	**96.67**	**96.67**	95.33	**97.84**

6. Conclusions

In this paper, we propose a novel CNNs classification framework for HSIs, which can data-adaptively learn a specific number of kernels from the training data. In particular, this model adopts the MCFSFDP algorithm to cluster the training data, and then the convolutional kernels can be determined automatically by the cluster center and inter-cluster margin. With those pre-learned kernels, a CNNs framework is developed for classifications. We have compared the proposed CNNs framework against three state-of-the-art deep learning methods with pre-trained kernels on three datasets. The experimental results demonstrate the superiority of the proposed CNNs framework in classification accuracy. Moreover, we validate that the proposed method is able to seek a good kernel number for a specific dataset. These adaptively learned kernels can help us understand the complexity of data and adjust the CNNs architecture for good feature extraction.

In terms of future research, we will exploit a multi-layer architecture via MCSFDP based CNNs to enhance the classification accuracy with less samples.

Acknowledgments: This work was supported by the Key Project of the National Natural Science Foundation of China (Grant No. 61231016), the National Natural Science Foundations of China (Grant No. 61471297, Grant No. 61671385 and Grant No. 61301192) and the China 863 Program (Grant No. 2015AA016402).

Author Contributions: All of the authors made significant contributions to this work. Chen Ding and Yanning Zhang devised the approach and analyzed the data; Yong Xia, Wei Wei, Lei Zhang and Ying Li helped design the remote sensing experiments and provided advice for the preparation and revision of the work; Chen Ding performed the experiments.

References

1. Zortea, M.; Martino, M.D.; Serpico, S. A SVM ensemble approach for spectral-contextual classification of optical high spatial resolution imagery. In Proceedings of the 2007 IEEE International Geoscience and Remote Sensing Symposium, Barcelona, Spain, 23–28 July 2007; pp. 1489–1492.
2. Huang, X.; Zhang, L. An Adaptive Mean-Shift Analysis Approach for Object Extraction and Classification From Urban Hyperspectral Imagery. *IEEE Trans. Geosci. Remote Sens.* **2008**, *46*, 4173–4185. [CrossRef]
3. Li, J.; Bioucas-Dias, J.M.; Plaza, A. Spectral-Spatial Hyperspectral Image Segmentation Using Subspace Multinomial Logistic Regression and Markov Random Fields. *IEEE Trans. Geosci. Remote Sens.* **2012**, *50*, 809–823. [CrossRef]
4. Benediktsson, J.A.; Palmason, J.A.; Sveinsson, J.R. Classification of hyperspectral data from urban areas based on extended morphological profiles. *IEEE Trans. Geosci. Remote Sens.* **2005**, *43*, 480–491. [CrossRef]
5. Wei, W.; Zhang, Y.; Tian, C. Latent subclass learning-based unsupervised ensemble feature extraction method for hyperspectral image classification. *Remote Sens. Lett.* **2015**, *6*, 257–266. [CrossRef]
6. Zhang, L.; Wei, W.; Tian, C.; Li, F.; Zhang, Y. Exploring Structured Sparsity by a Reweighted Laplace Prior for Hyperspectral Compressive Sensing. *IEEE Trans. Image Process.* **2016**, *25*, 4974–4988. [CrossRef]

7. Zhang, L.; Wei, W.; Zhang, Y.; Shen, C.; van den Hengel, A.; Shi, Q. Dictionary learning for promoting structured sparsity in hyperspectral compressive sensing. *IEEE Trans. Geosci. Remote Sens.* **2016**, *54*, 7223–7235. [CrossRef]
8. Zhang, L.; Zhang, L.; Du, B. Deep Learning for Remote Sensing Data: A Technical Tutorial on the State of the Art. *IEEE Geosci. Remote Sens. Mag.* **2016**, *4*, 22–40. [CrossRef]
9. Chen, Y.; Lin, Z.; Zhao, X.; Wang, G.; Gu, Y. Deep Learning-Based Classification of Hyperspectral Data. *IEEE J. Sel. Top. Appl. Earth Obs. Remote Sens.* **2014**, *7*, 2094–2107. [CrossRef]
10. Zhao, W.; Du, S. Spectral-Spatial Feature Extraction for Hyperspectral Image Classification: A Dimension Reduction and Deep Learning Approach. *IEEE Trans. Geosci. Remote Sens.* **2016**, *54*, 4544–4554. [CrossRef]
11. Wang, Q.; Lin, J.; Yuan, Y. Salient Band Selection for Hyperspectral Image Classification via Manifold Ranking. *IEEE Trans. Neural Netw. Learn. Syst.* **2016**, *27*, 1279. [CrossRef] [PubMed]
12. Wang, Q.; Yuan, Y.; Yan, P. Visual Saliency by Selective Contrast. *IEEE Trans. Circuits Syst. Video Technol.* **2013**, *23*, 1150–1155. [CrossRef]
13. Längkvist, M.; Kiselev, A.; Alirezaie, M.; Loutfi, A. Classification and Segmentation of Satellite Orthoimagery Using Convolutional Neural Networks. *Remote Sens.* **2016**, *8*, 329. [CrossRef]
14. Rodriguez, A.; Laio, A. Clustering by fast search and find of density peaks. *Science* **2014**, *344*, 1492–1496. [CrossRef] [PubMed]
15. Timmerman, M.E. Principal Component Analysis (2nd ed.) by I.T. Jolliffe. *J. Am. Stat. Assoc.* **2003**, *98*, 1082–1083. [CrossRef]
16. Rosipal, R.; Krämer, N. Overview and recent advances in partial least squares. In *Subspace, Latent Structure and Feature Selection, Proceedings of the Statistical and Optimization Perspectives Workshop (SLSFS 2005), Bohinj, Slovenia, 23–25 February 2005*; Springer: Berlin/Heidelber, Germany, 2006; pp. 34–51.
17. Camps-Valls, G.; Bruzzone, L. *Kernel Methods for Remote Sensing Data Analysis*; John Wiley & Sons: River Street Hoboken, NJ, USA, 2009.
18. Myint, S.W. Wavelets for Urban Spatial Feature Discrimination: Comparisons with Fractal, Spatial Autocorrelation, and Spatial Co-occurrence Approaches. *Photogramm. Eng. Remote Sens.* **2004**, *70*, 803–812. [CrossRef]
19. Zhu, C.; Yang, X. Study of remote sensing image texture analysis and classification using wavelet. *Int. J. Remote Sens.* **1998**, *19*, 3197–3203. [CrossRef]
20. Dong, Y.; Forester, B.C.; Milne, A.K. Segmentation of radar imagery using the Gaussian Markov random field model. *Int. J. Remote Sens.* **1999**, *20*, 1617–1639. [CrossRef]
21. Dong, Y.; Forster, B.C.; Milne, A.K. Comparison of radar image segmentation by Gaussian-and Gamma-Markov random field models. *Int. J. Remote Sens.* **2003**, *24*, 711–722. [CrossRef]
22. Zhang, L.; Huang, X.; Huang, B.; Li, P. A pixel shape index coupled with spectral information for classification of high spatial resolution remotely sensed imagery. *IEEE Trans. Geosci. Remote Sens.* **2006**, *44*, 2950–2961. [CrossRef]
23. Dalla Mura, M.; Benediktsson, J.A.; Waske, B.; Bruzzone, L. Morphological Attribute Profiles for the Analysis of Very High Resolution Images. *IEEE Trans. Geosci. Remote Sens.* **2010**, *48*, 3747–3762. [CrossRef]
24. He, K.; Sun, J.; Tang, X. Guided image filtering. *IEEE Trans. Pattern Anal. Mach. Intell.* **2011**, *35*, 1397.
25. Kang, X.; Li, S.; Benediktsson, J.A. Spectral-Spatial Hyperspectral Image Classification With Edge-Preserving Filtering. *IEEE Trans. Geosci. Remote Sens.* **2014**, *52*, 2666–2677.
26. Kang, X.; Li, S.; Benediktsson, J.A. Feature Extraction of Hyperspectral Images With Image Fusion and Recursive Filtering. *IEEE Trans. Geosci. Remote Sens.* **2014**, *52*, 3742–3752.
27. Sivic, J.; Zisserman, A. Video Google: A Text Retrieval Approach to Object Matching in Videos. In Proceedings of the Ninth IEEE International Conference on Computer Vision (ICCV), Nice, France, 13–16 October 2003; pp. 1470–1477.
28. Lazebnik, S.; Schmid, C.; Ponce, J. Beyond bags of features: Spatial pyramid matching for recognizing natural scene categories. In Proceedings of the 2006 IEEE Computer Society Conference on Computer Vision and Pattern Recognition, New York, NY, USA, 17–22 June 2006; pp. 2169–2178.
29. Xia, G.S.; Delon, J.; Gousseau, Y. Accurate Junction Detection and Characterization in Natural Images. *Int. J. Comput. Vis.* **2014**, *106*, 31–56.
30. Xia, G.S.; Delon, J.; Gousseau, Y. Shape-based Invariant Texture Indexing. *Int. J. Comput. Vis.* **2010**, *88*, 382–403.

31. Liu, W.; Liu, H.; Tao, D.; Wang, Y.; Lu, K. Manifold regularized kernel logistic regression for web image annotation. *Neurocomputing* **2016**, *172*, 3–8.
32. Wang, Q.; Yuan, Y.; Yan, P.; Li, X. Saliency Detection by Multiple-Instance Learning. *IEEE Trans. Cybern.* **2013**, *43*, 660–672. [PubMed]
33. Chen, Y.; Zhao, X.; Jia, X. Spectral-Spatial Classification of Hyperspectral Data Based on Deep Belief Network. *IEEE J. Sel. Top. Appl. Earth Obs. Remote Sens.* **2015**, *8*, 2381–2392.
34. Yang, X.; Liu, W.; Tao, D.; Cheng, J. Canonical Correlation Analysis Networks for Two-view Image Recognition. *Inf. Sci.* **2017**, *385–386*, 338–352.
35. LeCun, Y.; Bottou, L.; Bengio, Y.; Haffner, P. Gradient-based learning applied to document recognition. *Proc. IEEE* **1998**, *86*, 2278–2324.
36. Krizhevsky, A.; Sutskever, I.; Hinton, G.E. ImageNet classification with deep convolutional neural networks. In *Advances in Neural Information Processing Systems 25, Proceedings of the Neural Information Processing Systems, Stateline, NV, USA, 3–8 December 2012*; Neural Information Processing Systems Foundation, Inc.: La Jolla, CA, USA, 2012; pp. 1097–2013.
37. Simonyan, K.; Zisserman, A. Very Deep Convolutional Networks for Large-Scale Image Recognition. *arXiv*, **2014**, arXiv:1409.1556.
38. Szegedy, C.; Liu, W.; Jia, Y.; Sermanet, P.; Reed, S.; Anguelov, D.; Rabinovich, A. Going deeper with convolutions. In Proceedings of the IEEE Conference on Computer Vision and Pattern Recognition, Boston, MA, USA, 8–10 June 2015; pp. 1–9.
39. He, K.; Zhang, X.; Ren, S.; Sun, J. Deep Residual Learning for Image Recognition. In Proceedings of the IEEE Conference on Computer Vision and Pattern Recognition, Las Vegas, NV, USA, 26 June–1 July 2016; pp. 770–778.
40. Li, Y.; Zhang, H.; Shen, Q. Spectral-Spatial Classification of Hyperspectral Imagery with 3D Convolutional Neural Network. *Remote Sens.* **2017**, *9*, 67. [CrossRef]
41. Hu, F.; Xia, G.S.; Hu, J.; Zhang, L. Transferring Deep Convolutional Neural Networks for the Scene Classification of High-Resolution Remote Sensing Imagery. *Remote Sens.* **2015**, *7*, 14680–14707. [CrossRef]
42. Chan, T.H.; Jia, K.; Gao, S.; Lu, J.; Zeng, Z.; Ma, Y. PCANet: A Simple Deep Learning Baseline for Image Classification? *IEEE Trans. Image Process.* **2015**, *24*, 5017–5032. [CrossRef] [PubMed]

10

Maritime Semantic Labeling of Optical Remote Sensing Images with Multi-Scale Fully Convolutional Network

Haoning Lin [1,2,3], Zhenwei Shi [1,2,3,*] and Zhengxia Zou [1,2,3]

1 Image Processing Center, School of Astronautics, Beihang University, Beijing 100191, China; harrylin@buaa.edu.cn (H.L.); zhengxiazou@buaa.edu.cn (Z.Z.)
2 Beijing Key Laboratory of Digital Media, Beihang University, Beijing 100191 China
3 State Key Laboratory of Virtual Reality Technology and Systems, School of Astronautics, Beihang University, Beijing 100191, China
* Correspondence: shizhenwei@buaa.edu.cn.

Academic Editors: Qi Wang, Nicolas H. Younan, Carlos López-Martínez, Xiaofeng Li and Prasad S. Thenkabail

Abstract: In current remote sensing literature, the problems of sea-land segmentation and ship detection (including in-dock ships) are investigated separately despite the high correlation between them. This inhibits joint optimization and makes the implementation of the methods highly complicated. In this paper, we propose a novel fully convolutional network to accomplish the two tasks simultaneously, in a semantic labeling fashion, i.e., to label every pixel of the image into 3 classes, sea, land and ships. A multi-scale structure for the network is proposed to address the huge scale gap between different classes of targets, i.e., sea/land and ships. Conventional multi-scale structure utilizes shortcuts to connect low level, fine scale feature maps to high level ones to increase the network's ability to produce finer results. In contrast, our proposed multi-scale structure focuses on increasing the receptive field of the network while maintaining the ability towards fine scale details. The multi-scale convolution network accommodates the huge scale difference between sea-land and ships and provides comprehensive features, and is able to accomplish the tasks in an end-to-end manner that is easy for implementation and feasible for joint optimization. In the network, the input forks into fine-scale and coarse-scale paths, which share the same convolution layers to minimize network parameter increase, and then are joined together to produce the final result. The experiments show that the network tackles the semantic labeling problem with improved performance.

Keywords: semantic labeling; convolution neural network; fully convolutional network; sea-land segmentation; ship detection

---

1. Introduction

Remote sensing imagery is one important solution to maritime surveillance, because of its wide field of view, satisfying spatial resolution and update frequency. Remote sensing imagery includes various kinds, ranging from hyper-spectral imagery [1], synthetic aperture radar (SAR) imagery [2], to optical imagery. These kinds of imaging technology serve varying purposes according to their different characteristics, and optical imagery is applied widely for its rich presentation and similar reception frequency to that of human eyes.

There has been a considerate amount of research in optical imagery understanding focusing on detection of different types of objects, such as roads [3,4], buildings [5,6], oil tanks [7,8], vehicles [9–11] and airplanes [12–14]. Aside from detecting scattered objects, the classification of scenes also receives a lot of attention recently, such as in [15–17], where the objective is to classify image patches into different classes, such as buildings, forest, harbor, etc.

Two of the most important tasks in understanding remote sensing images that is maritime-related, would be sea-land segmentation and ship detection. Research on ship detection originally focuses on off-shore ships with relatively simple background, majorly on SAR imagery [2,18,19]. In recent literature, both sea-land segmentation [20,21] and ship detection [22–27] tasks are addressed with complex frameworks, which consists of cascaded procedures and have to be designed and fine-tuned with expert knowledge. That is, when the source sensor of the images is changed, the carefully designed framework always has to be re-calibrated or even re-designed by experts. The complex steps that constitute the framework also make the implementation difficult. Furthermore, ship detection, especially when including in-dock ships, are highly dependent of the performance of sea-land segmentation, making it less robust in precarious sea-land situation. Furthermore, to tackle the two problems separately, also inhibits the joint optimization of the designed algorithm.

The recent advancement in the deep learning community motivates us to address these problems with deep neural networks. Deep learning, as a subcategory of soft computing [28–34], is seeing great attention. In our previous work, we focus on the detection of objects instance-wise, i.e., acquiring the location and bounding box of the objects in interest. In this paper, we propose to address the sea-land segmentation and ship detection at the same time, with a deep neural network, in a semantic labeling perspective. The network allows us to cope with these problems in an end-to-end fashion, without complex procedures, and without handcrafted features.

The semantic labeling of everyday images recently receives increasing attention [35–38] and is regarded as a more challenging task compared to object classification and detection of images. Semantic labeling provides a pixel-to-pixel label map corresponding to the input image, as opposed to only a single label in classification task. It also, from another point of view, provides the boundaries of the detected object, as opposed to only bounding boxes in detection task. This is similar to the saliency detection methods [39,40], with the difference that saliency detection is more general and pays less attention to object boundaries. One general approach of semantic labeling is to first process images into over-segmented areas and then classify each area with its extracted features [35]. Yet with the fast-paced development of deep learning, it also proves to be able to achieve state-of-the-art semantic labeling in everyday images [36–38]. Moreover, since recent research shows that neural networks based on everyday object knowledge can have satisfactory performance on remote sensing imagery [41], the application of deep learning in remote sensing imagery is promising.

In both sea-land segmentation and ship detection tasks, semantic labeling using deep networks shows great potential. First, deep network is able to learn high level features, as opposed to that in other methods, features has to be handcrafted and are complex to implement. Second, semantic labeling's pixel-labeling nature allows it to be independent of bounding boxes and are relatively indifference to objects' size and shape. This helps because sea and land are of arbitrary sizes and shapes, and the bounding boxes of in-dock ships are hard to acquire.

However, the deep network, when used to address these remote sensing problems, is faced with one critical problem, to balance between the hardware requirement and the network's efficacy. In remote sensing images, the area of interest can be of arbitrary size. This lead to the need for a network with extra-large receptive field (which will be extensively discussed in Section 3.2), which requires increased amount of weights for the network layers, which then leads to excessive graphics processing unit (GPU) memory requirement from the network and increased computation in training and testing.

To ameliorate the trade-off between the network's receptive field and the GPU memory requirement, we introduce a novel multi-scale structure for the semantic labeling network, which greatly increases the receptive field of the network, with only a small number of parameter increase.

The main idea of our multi-scale structure is different than those of the conventional ones, where shortcuts are created between convolution layers of different levels to utilize the finer feature maps in order to produce finer outputs. Our structure focuses on enlarging the receptive field of the network to incorporate information from larger scale, which is important for understanding remote sensing images.

In the network, the input data is processed in two separate layers, crop layer and resize layer, into different scales of data, fine-scale and coarse-scale, respectively. The fine-scale path with crop layer keeps the fine details in the data, but with small sized receptive field, while the coarse-scale path with resize layer down-samples the data, omitting high-resolution textures in exchange for large sized receptive field. In this paper, the coarse-scale path is more suitable for discriminating between sea and land, for their large proportion in area and usually obscure boundaries (near beaches and other natural shore-lines, for example). The fine-scale path is suitable for ship detection, for exactly the opposite reasons.

The main contribution of the paper is listed as follows,

1. Joint sea-land segmentation & in-dock ship detection. The information extracted by the network is used both for sea-land segmentation and ship detection. The sharing of the information can lead to better performance, especially in in-dock ship detection, since it is no longer dependent on other separated sea-land segmentation methods and can be trained jointly.
2. A different perspective into multi-scale structure for remote sensing images with small parameter number increase. The conventional multi-scale structures connect different feature maps from different layers that represents different level of semantics, aiming to fully utilize fine-scale features. The proposed structure aims to widen the receptive field of the network, designed specifically for remote sensing images. With the multi-scale structure, the network is able to achieve tasks that require different scales, while maintaining relatively small number of parameter and low calculation complexity. An extensive experiment is conducted to compare our proposed structure to several variants to show its superiority in learning speed and performance.

The following content is structured as follows. In Section 2, a brief introduction to fully convolutional network is given. In Section 3, the proposed multi-scale structure is described in detail and the receptive field of the network is analyzed. In Section 4, the given framework and other methods are experimented on two remote sensing datasets. Finally, Section 6 concludes this paper.

2. Fully Convolutional Network

In this section, we will provide a brief introduction to the fully convolutional network (FCN) upon which we construct our semantic labeling framework.

CNN proves to be extremely effective in image related tasks, such as object detection and classification [42,43]. Based on CNN, Fully Convolutional Networks (FCN) are designed to predict a label map rather than a single label for an input image, by replacing fully connected layers in CNN with small sized convolution layers [44]. FCN's pixel-to-pixel label map output naturally suits the need of semantic labeling. Figure 1, modified from in [44], shows a typical FCN structure, when used to semantic label a remote sensing image.

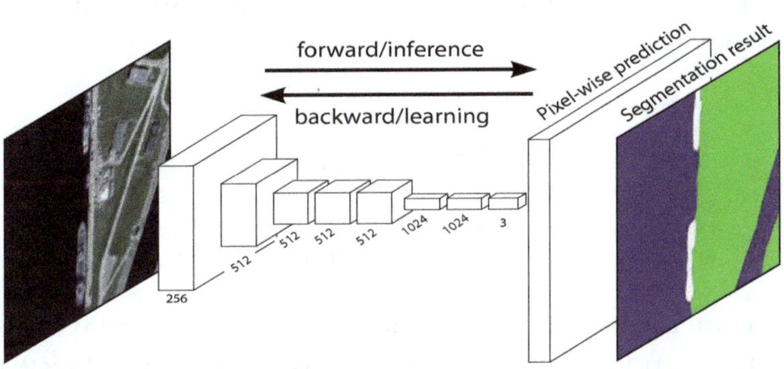

Figure 1. A typical FCN, each cuboid indicating an output matrix of a convolution layer. The numbers indicate the size of the 3rd dimension of each cuboid, or equally, the number of kernels of the corresponding layer.

A typical FCN consists of convolution layers, pooling layers and activation layers [44]. There are also softmax layers for output and loss layers for training. An input data matrix (say, an RGB image $X \in R^{h \times w \times 3}$ or a grayscale image $X \in R^{h \times w \times 1}$) is processed through each layer in sequence in a neural network.

A convolution layer consists of an array of kernel matrices, with which the input data are convoluted. In a convolution layer, the data is processed with the following calculation,

$$Y_l = f_l * X \tag{1}$$

where $*$ is a 3D (3 Dimensional) convolution operator, X is the input matrix of the layer, f_l is the lth kernel of the layer and Y_l is the output matrix correspond to the lth kernel. Here it is mandatory that f_l and X are of the same size in 3rd dimension, so that the size of dimension 3 of Y_l is necessarily 1. Finally, the Y_ls of a layer are concatenated in 3rd dimension, resulting in

$$Y(x, y, l) = Y_l(x, y) \tag{2}$$

where Y is the complete output matrix of the convolution layer. x, y and l are indexes for Y of dimension 1, 2 and 3, respectively. Convolution layers are designed to capture local features and are translation invariant, and the output matrix of a convolution layer is usually called a feature map, since the output represents the extracted features of each single pixels of the input image, with pixel-to-pixel correspondence.

Activate functions are often added after convolution layers to provide non-linear properties for a network to enhance the expressive ability of the features. In an activation function, an element-wise operation is conducted,

$$Y(x, y, z) = f(X(x, y, z)) \tag{3}$$

where x, y, z are indexes of 3 dimensions of a matrix and X, Y are input and output matrices, respectively. f is the function of the layer. In a simple but rather popular activation layer, Relu [45],

$$f(\cdot) = max(0, \cdot) \tag{4}$$

A pooling layer, acting like a down-sampling filter, is often inserted among other layers. It is designed to progressively reduce the size of transferred data to reduce the amount of parameters and enhance the generalization of a network. The most common form of a pooling layer uses max operation to produce results for each local area of the input,

$$Y(x, y, z) = \max_{(i,j) \in \Omega} (X(i, j, z)) \tag{5}$$

where x, y, z, i and j are indexes of their according dimensions of the matrices and X, Y are input and output matrices, respectively. In addition,

$$(i, j) \in \Omega \iff \begin{cases} i \geq x \times step \\ i < x \times step + kernel\_size \\ j \geq y \times step \\ j < y \times step + kernel\_size \end{cases} \tag{6}$$

where *step* and *kernel_size* are the two hyperparameters of the pooling layer, determining the stride of the output according to input and the size of Ω, respectively. For simplicity, the indexes here follow the convention in programming and start from zero.

In an FCN, there are no fully connected layers, which connect all the elements in the input matrix and output results that ignore all spatial information. Convolution layers with kernels of size

1×1 are implemented instead, producing an output matrix of corresponding spatial dimensions [44]. Because some of the convolution layers at the top of the network still act as a role of traditional fully connected layers, in our paper we still distinguish these layers and symbolize them as "fc" layers following the notation in [36].

A softmax layer is a layer for output, it takes in matrices of arbitrary-scaled elements and outputs a matrix of probabilities, with the formula of

$$Y(x, y, j) = \frac{e^{X(x,y,j)}}{\sum_{k=1}^{K} e^{X(x,y,k)}} \tag{7}$$

where X and Y are input and output matrix of the softmax layer, respectively. x, y, j and k are indexes of their corresponding dimensions. K is the size of dimension 3 of X. In a semantic labeling network, the softmax layer outputs the probabilities of every pixel belonging to every category.

A loss layer takes in data both from outputs of previous layers and from ground truth labels. The gradients are firstly calculated from the difference of both sides, and then are back-propagated to previous layers. The kernels of each layer are then updated according to the gradients. This process is gone through iteratively and the network will be trained.

3. Multi-Scale Network for Semantic Labeling

The semantic labeling of maritime scenes calls for multi-scale features because of the tremendous size difference between the sea, land and ships. Sea-land identification demands wide spatial range of input for richer context and comprehensive understanding, whereas small targets, such as ships, demand context of smaller scale but more detailed information from local area. The feature of multi-scale has been extensively utilized in neural networks. Liang et al. connects output of the first few layers to the last layer for attention on fine-resolution layers [36]. Paisitkriangkrai et al. trains several CNNs with different resolution of input images [46]. Eigen et al. concatenates layers that are designed for different scales into a whole [37]. These networks either are trained separately on every scale, resulting in far more parameters to train, or leave the layers trained without the knowledge of its corresponding scale information. Here we present a multi-scale FCN specifically designed for remote sensing imagery. This framework enlarges the receptive field of the network, while preserving the ability to take in fine details, with only a small increase in the number of parameters.

3.1. Network Structure

To implement the multi-scale structure in the network, we introduce two layers, crop layer and resize layer, which is illustrated in Figure 2. In a resize layer, the input is down-sampled, and in the crop layer, the input is center-cropped. The input are separated into what we call fine and coarse scale, respectively, after these two layers.

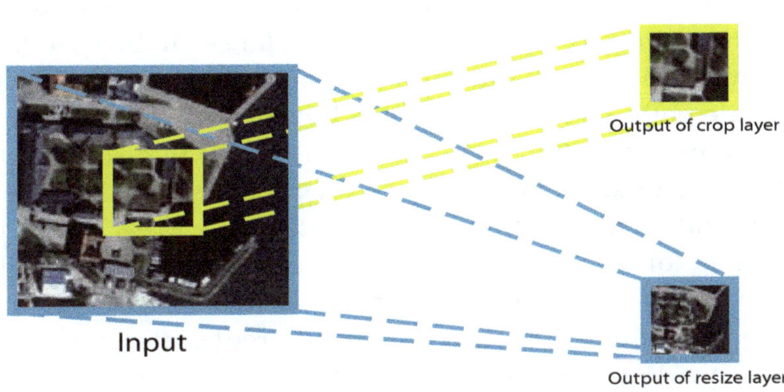

Figure 2. An illustration of the input/output of crop layer and the resize layer.

With the two layers, we have two separate paths of images as well as ground truth labels going through the network at the same time, each representing a different scale.

In Figure 3, the whole network is illustrated. First, an input image patch is duplicated and preprocessed in 2 separate ways, fine-scale and coarse-scale, using aforementioned crop layer and resize layer and are fed into following convolution layers. The outputs of the two preprocessing layers, although cover different areas (Area Yellow vs. Area Blue), are of the same size and thus can be fed into the same layer configuration with the same weights. The convolution layers are configured as DeepLab-LargeFOV [36], with its first 13 convolution layers and are interlaced with Relu and pooling layers. We implement this convolution configuration because it proves to have state-of-the-art semantic labeling performance in everyday images.

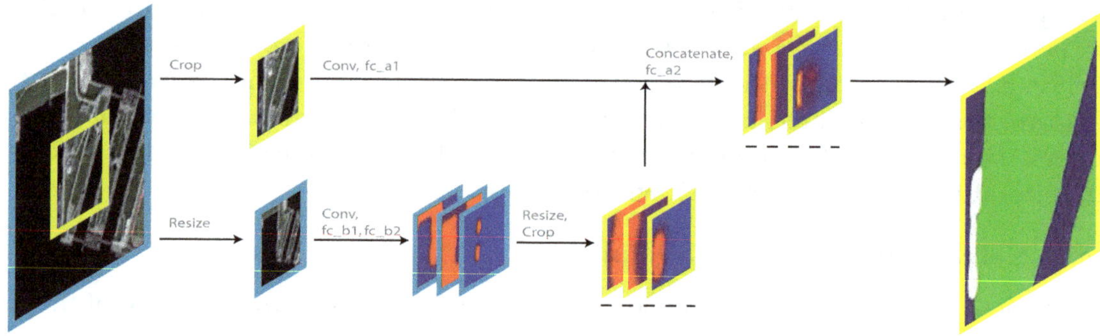

Figure 3. An illustration of the proposed network. The texts on the arrowed lines specify the layers the data go through to produce the displayed results. The color of the outline (blue and yellow) of each result marks the corresponding input area it represents for the sake of clarity. The results that are directly connected to loss layers are underlined with dashed lines.

In Figure 3, some of the results of certain layers are represented as a group of 3 slices, each slice representing the probability map of a corresponding class. We name these groups as score maps for convenience. A score map, in essence, is the same as a feature map, where both maps are the output of a certain convolution layer, but the position and the configuration of the convolution layer define the semantics that the layer is to learn to output the scores of each pixel to belong to a certain category (In fact, a score map is a direct output of a softmax layer, which is placed after a convolution layer. But the function of a softmax layer is relatively trivial compared to the other layers, so the softmax layers are not mentioned either in the figure or in the text).

The network also utilizes two loss layers, each to train the layers in different scales. In Figure 3, the score maps that are connected to a loss layers are underlined with dashed lines. As for the ground truth labels that are needed by the loss layers, they are acquired in the same way as the input patches. The original labels go through crop layer and resize layer separately, and then are fed into loss layers in fine-scale loss layer and coarse-scale loss layer, respectively. With the two loss layers, the convolution layers and fc_** layers learn to produce score maps in accordance. To be specific, the coarse-scale score map is predicted purely from coarse-scale data and then are modified with resize and crop layers, and finally are fed into fine-scale path (Layer fc_a2) to produce fine-scale score map jointly.

Although in Figure 3 the convolution layers are divided into fine-scale and coarse-scale, it is only for clarity. In the practical implementation the fine-scale and coarse-scale data are concatenated first and fed into the same convolution layers, and then separated back to each scale before producing score maps. The weights are shared for convolution layers on the same level and also between fc_*1 layers to minimize the number of parameters. Here the feature extraction mechanism of the convolution layers are not scale specific. The feature maps extracted from different scales may share different semantics, but they are equally effective. We presume that just as deep features can generalize from everyday objects to remote sensing domains [41], deep features can also generalize between different scales of scenes, hence the sharing weights between scales.

The crop layer we design has a simple forward function, which center crops the input with a single parameter, $scale\_factor$. The crop function only work on spatial dimensions, which means only the sizes of the first 2 dimension will change. It also has no backward function, meaning no gradient is transferred back through this layer, for simplicity. The simplification is plausible because (a) it is located in special positions in the network, twice after data layers and once after convolution layers and (b) the convolution layers before it have shared weights and can already learn from both scales and (c) coarse-scale layers should focus on sea-land classification and need to take little knowledge from fine-scale losses.

It is also worth mentioning that in [36,47], a 'hole algorithm' is introduced into the deep network for convolution layers, to increase the receptive field of the layers while keeping the number of weights unchanged. A simple explanation would be to put 'holes' in the layer kernels to enlarge the kernels spatially, while maintaining the number of parameters in a kernel,

$$K_{hole}(x,y,z) = \begin{cases} K(x/hole, j/hole, z), \\ \qquad \text{if } x \textbf{ mod } hole = 0, y \textbf{ mod } hole = 0 \\ 0, \\ \qquad \text{otherwise} \end{cases} \tag{8}$$

where x, y, z are indexes of their according dimensions of the matrices, **mod** means to calculate the remainder of division and K, K_{hole} are original and modified kernel of the convolution layer with 'hole algorithm', respectively. $hole$ is a hyperparameter of the layer, specifying how large the 'hole' you want to insert into the kernels. In this paper, we keep the 'hole algorithm' as is implemented in [36], but with tuned-down size of the hole, in order to acquire more subtle details for the label result.

Table 1 lists the structure setup of the network. The layer names are either self-explanatory or mentioned in the text, so the layer types are omitted. Apart from the Relu layers after each convolution layer, the layers from conv1_1 to pool5b are listed as the exact setup order.

Table 1. The setup of the network.

<table>
<tr><th>Layer Name</th><th>Kernel Size</th><th>Kernel Num.</th><th>Remarks</th></tr>
<tr><td>conv1_1, conv1_2</td><td rowspan="12">3</td><td>64</td><td>-</td></tr>
<tr><td>pool1</td><td>-</td><td>step: 2 type: max</td></tr>
<tr><td>conv2_1, conv2_2</td><td>128</td><td>-</td></tr>
<tr><td>pool2</td><td>-</td><td>step: 2 type: max</td></tr>
<tr><td>conv3_1, conv3_2, conv3_3</td><td>256</td><td>-</td></tr>
<tr><td>pool3</td><td>-</td><td>step: 2 type: max</td></tr>
<tr><td>conv4_1, conv4_2, conv4_3</td><td>512</td><td>hole: 2</td></tr>
<tr><td>pool4</td><td>-</td><td>step: 1 type: max</td></tr>
<tr><td>conv5_1, conv5_2, conv5_3</td><td>512</td><td>hole: 2</td></tr>
<tr><td>pool5a</td><td>-</td><td>step: 1 type: max</td></tr>
<tr><td>pool5b</td><td>-</td><td>step: 1 type: average</td></tr>
<tr><td>fc_a1, fc_b1</td><td>512</td><td>hole: 2</td></tr>
<tr><td>fc_a2, fc_b2</td><td>1</td><td>3</td><td>-</td></tr>
</table>

3.2. Receptive Field Analysis

The receptive field is a vital concept that can affect a network's performance. It is a biologically-inspired term from animals' visual cortex. In a network, it describes the spatial range of input pixels that can contribute to the calculation of a single element in the output. With larger receptive field, each layer can take in more context and represent more abstract meanings. For a network to determine if a pixel belongs to a ship, it is important that the network can determine if the pixel belongs to the forecastle deck or the side of a ship. For ship detection, the receptive field is best to be large enough to cover the space of ship and its context, and for sea-land segmentation, extensively larger receptive field is needed.

The crop layer and the resize layer introduced in the former section is introduced into the network to specifically enlarge the receptive field of one path, while also maintaining the detail feature in another path. The resize layer acts as a downsampling filter, which shrinks the spatial size of the input at the cost of losing detail information, while allowing enlarging the receptive field of the following network path (coarse-scale) without any modification to the existing convolutional layers. The crop layer ensures the input image is cropped to the same spatial area as the desired and maintains the detail information (for the fine-scale). In the training procedure, the parameters of the network is jointly optimized to decrease the loss to the ground truth label. With the different scale of the input data that is given to each path, the network can automatically learn the optimized task for each path, as shown in Section 4.4.

Figure 4 also illustrates the relation between the kernel size and the receptive field of each convolution layer. The kernel size determine the area of the input data to calculate one single element in the output. We can see that with the layers going deeper, or with larger kernel size, the receptive field of the the output layer will increase. Apart from that, the pooling layer also can increase the receptive field, with the possible draw-back of lowering the layer resolution.

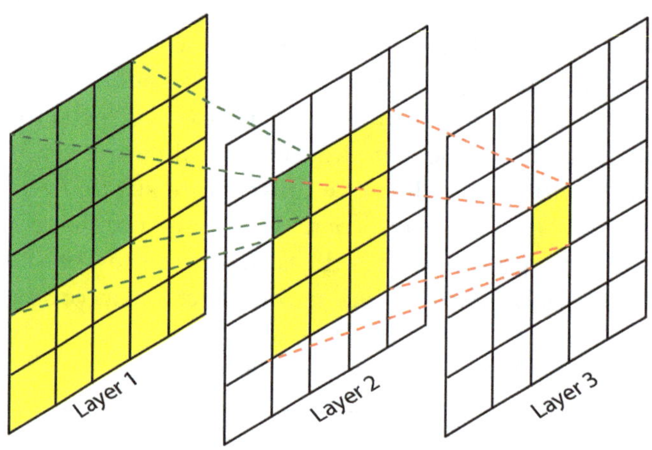

Figure 4. The receptive field of each convolution layer with a 3×3 kernel. The green area marks the receptive field of one pixel in Layer 2, and the yellow area marks the receptive field of one pixel in Layer 3.

The whole network's receptive field can calculated by stacking up each layer's receptive radius,

$$R = \sum_{L} r_l \tag{9}$$

where R is the radius of the network receptive field ($2 \times R + 1 = receptive\_field\_size$), l is the index of the layers and affects receptive field, including pooling layers and convolution layers. L is the total number of the above layers, and r is the radius of the layer's receptive field when considering the subsampling effect of its previous pooling layers. r is defined by

$$r = r_{kernel} \times \prod_{N} step_n \tag{10}$$

where r_{kernel} is the radius of the kernel of the current layer ($2 \times r_{kernel} + 1 = kernel\_size$), N is the number of pooling layers between the input layer and the current layer, and $step_n$ is the step size of the pooling layer n.

Apart from the crop layers and resize layers, all the layers that affect receptive field are listed in Table 1. All the layers without *hole* hyperparameter have an *r_kernel* of 1, and those with $hole = 2$ have an *r_kernel* of 2. With 13 convolution layers, 5 pooling layers and 2 fc_** layers, our non-multi-scale network has a receptive field of 259×259 pixels. While the ships in our datasets average about 25×150 in size, the receptive field is suitable only for ship detection tasks. To enlarge a network's receptive field for sea-land segmentation, traditional methods such as increasing kernel size or increasing the depth of the network all lead to huge amount of increase in weight number, which in turn leads to harder network training, higher hardware requirement and more computation complexity. The resize layer in the multi-scale structure has a similar effect on receptive fields as the pooling layer, therefore the receptive field of network is enlarged as if a pooling layer is put at the very beginning of the network. With $scale\_factor$ of both resize layer and crop layer set to 3, the receptive field is roughly scaled up by 9 times in area, into 775×775, with only two additional layers (fc_b1 and fc_b2) with weights. The convolution layers which hold large proportion of the weights stay unchanged, and the network is still able to be fine-tuned from the pre-trained original one.

3.3. Data Preprocessing

Different from most of the previous works that focus on open datasets only containing extracted small image patches, we focus our framework on large, relatively complete images that general remote sensing images are distributed as. The network training and testing on large images brings in new problems, such as how to effectively extract samples or image patches for the network.

Because of the limitation of GPU memory, both training and testing images have to be cut into relatively small patches before being fed into the neural network. For training, we select samples from training images randomly. To be specific, for each original image in the training set, we randomly generate N triplets (x, y, θ), with each symbol indicating pixel coordinate x, coordinate y, and the rotating angle. For each triplet one training sample is selected according to (x, y, θ), with (x, y) being the coordinate of the selected patch in the original image. Finally each sample is rotated by angle θ. For testing samples, the patches are extracted in a sliding-window manner, with a stride the same as the size of the fine-scale input patch, so that the fine-scale inputs has no overlaying on each other. The experiment shows that the network we train performs well on patch borders, especially with the help of the multi-scale scheme. When put back together, the label maps connects to each other well, with no obvious artifacts.

In the literature of semantic labeling, the balancing of samples is barely mentioned, because its application background is mainly on daily images in well-prepared dataset and the problem of unbalanced samples does not exist. In the remote sensing dataset, it is crucial to balance the samples (in this context, to balance the number of pixels of different categories) first for the network to learn equally from different classes. Without the balanced samples, the network will lean towards better performance on sea-land classification, neglecting the accuracy of ship category. In this experiment, firstly, we limit the number of samples that do not contain ships, secondly, we utilize one of the functions of loss layer in DeepLab's Caffe implementation [36], the ability to ignore the loss on the pixels that are labeled to a special class, *ignore*. We randomly set the ground truth label of sea and land pixels to *ignore*, so that when calculating the loss value of the network, the actual functioning ground truth pixels are category-balanced. In our experiment, without balancing the sample, the accuracy of ship detection would decrease dramatically (by 10%).

The samples that contain ships are rotated several times because a ship presented on an remote sensing image can be of arbitrary possible orientation. We also control the number of samples extracted from an image, so that the image is covered roughly twice by the training samples. Although convolution layers have the property of translation invariance, we sample the images more times to counter-act the border effect of the convolution (the borders of a input matrix has to be padded by zero before convolution to maintain output matrix size, thus compromising the effectiveness of features close to the borders).

4. Experiments

This network is implemented with Caffe [48], on Ubuntu 14.04, with one Titan X. The network is trained with mini-batched Stochastic Gradient Descent (SGD) with momentum and step learning rate. The batch size is set to 14, base learning rate 0.001, which drops by a factor of 10 every 2000 iterations. We use momentum 0.9, weight decay 5 and doubled learning rate for biases following the implementation in [36,44]. The network is first initialized with pre-trained weights from ImageNet dataset and then fine-tuned with remote sensing data, to compensate the limited amount of training images. With the pre-trained weights, the network converges to a satisfactory extent at only 4000 iterations.

The selection of the Caffe framework and the training scheme follows the common acknowledgment in the deep learning community [36,49]. Although there are plenty of selection of deep learning framework to use (such as TensorFlow, Torch), the accuracy-wise performance has only a very limited variation [49]. The choice of training scheme has also undergone extensive investigation [50] and we follow [36] because of the similar network architecture. We also experiment other modified version of SGD [50] but yield inferior results.

We experiment our proposed method on two different datasets. The first dataset we use consists of 6 panchromatic (grayscale) images from GaoFen-1 satellite each with above $18,000 \times 18,000$ pixels and a resolution of 2.5 m/pixel. The second dataset has 21 images (RGB) from Google Map, each with above 5000×5000 pixels and a resolution of 1 m/pixel. Both datasets focuses on areas with harbors, where both ships and various types of terrain exist. Although datasets (such as SPOT-4) that has lower resolution can provide competitive results for sea/land segmentation [51], we select the high resolution imagery to meet the requirement for in-shore ship detection, as proposed in [23,25].

Remote sensing datasets from Google Map has received extensive research in the recent days and are recognized as a valid source for remote sensing research [52]. Although imagery from Google Map may be enhanced to different extents, we qualitatively find that the imagery are not too varied to the degree that human cannot distinguish the objects in the imagery in the way on daily life objects, i.e., objects in Google Earth still are faithful to real colors and textures. Nevertheless, we here provide the coordinates and the sensor of the images we use for the experiment. All of the images from Google Map are produced by Digital Mapping Camera (DMC) collected from United States Geological Survey (USGS) High Resolution Orthoimagery and the coordinates of the most north-west pixel of the images are listed in Table 2.

Table 2. Coordinates of maps used in Google Map dataset (excerpt).

Map No.	Longitude	Latitude	Map No.	Longitude	Latitude	Map No.	Longitude	Latitude
1	129.687E	33.122N	2	127.645E	26.214N	3	21.958W	64.132N
4	132.520E	34.199N	5	79.926W	9.233N	6	21.936W	64.140N
7	139.627E	35.267N	8	129.837E	32.702N	9	15.580E	56.128N
10	129.687E	33.122N	11	12.590E	55.662N	12	10.160E	54.293N
13	8.126E	53.504N	14	30.720E	46.450N	15	4.197W	50.355N
16	1.113E	50.774N	17	3.1884E	51.312N	18	27.886E	43.155N
19	4.773E	52.927N	20	8.306W	51.801N	21	12.094E	54.147N

For the Google dataset, we select 7 images as test data and the other 14 as training data. For GaoFen-1 dataset, we find that to augment the training data by including Google images can improve the performance, so we convert the 14 RGB training images to grayscale images and join them to 2 of the GaoFen-1 images as training data, and choose the remaining 4 as test data. Note that the distribution of the ships also varies across the dataset, where the dataset from Google Map has far more in-dock ships. In the test data, Google Map dataset has 55 ships in total, including 50 in-dock ships, while GaoFen-1 dataset has 160 ships in total, including only 20 in-dock ships.

As a novel effort to implement deep learning semantic labeling into the maritime area, we focus our detection target on large navy ships/oil tankers to limit the scale of the target. The length of the target ships vary from 80 m to under 200 m in Google Map dataset and around 300 m in GaoFen-1 dataset. This lead to a similar scale for ships on both datasets because of the different resolution. The ships vary from 80 to 200 pixels in length.

For performance evaluation, we follow the method that is widely used in segmentation/semantic labeling tasks [36,46] . We count the number of pixels that are correctly labeled and those that are not, and compute the confusion matrix and Intersection-Over-Unions (IOU) for each task. Values in a confusion matrix indicate the percentage of the pixels labeled to the column class in the pixels belonging to the row class, meaning a row of a confusion matrix sums to one. Whereas IOU is calculated as

$$\mathrm{IOU} = \frac{true\ positives}{true\ positives + false\ positives + false\ negatives} \tag{11}$$

4.1. Benefit of Multi-Class Classification

Previous in-shore ship detection methods rely heavily on the acquirement of shore-line as a 1st step. This step contributes to locating possible areas with in-shore ships and eliminating complex inland areas that could produce huge number of false alarms. Traditionally there are two options to acquire shore-line information, (a) manually labeled shore-line database, which has two problems, the need for constant update and the need for accurate registration between database and image; And (b) a separate algorithm for the detection of shore-line, which is time consuming and requires tedious optimization (possibly hand-tuned) iterated between shore-line and ship detection algorithm.

In our framework the two problems are tackled at the same time and are jointly optimized to achieve better performance. We experiment our network on GaoFen-1 dataset on two different scenarios to show the benefit of multi-class classification of our framework, (1) the network is tasked to classify only 2 classes, Non-ship and Ship and (2) the network is tasked to classify 3 classes, Sea, Land and Ship. Table 3 shows with 3-class task, the network's accuracy on Ship is greatly improved, the network's learning time is also decreased. This is because with the 3-class task, the network in training is given extra information to comprehend the context of the task and by jointly classify multiple classes, the network learns the spatial relationship between the different classes (the Ship have minimal probability to appear in the middle of Land but maximal probability at the brink between Land and Sea). The 3-class problem also provides a more balanced sample pool so the network is easier to train with larger learning rate and faster converging speed.

Table 3. The comparison of Accuracy/Recall on Ship and training time on 2-Class/3-Class problems.

Problem	Accuracy	Recall	Trained Epoch
2-Class	85.3%	83.9%	160
3-Class	94.1%	83.4%	80

Although in the remote sensing imagery the land area features most complex objects, we find the classification of these objects will not contribute to the performance of the task. This is because ship detection is majorly focused at the brink between land and sea. The classification of objects enclosed by land will not provide any additional information for ship-detection, while also unnecessarily taking up the capacity of the network.

4.2. Comparison between Different Realization of Multi-Scale Structure

The multi-scale structure which ensembles paths of different receptive fields has various ways of implementation. In this section we list a few different multi-scale structures and compare the convergent speeds and parameter numbers of the various structures to show the superiority of our choice. Note that although there are already many multi-scale structures proposed in literature, by creating shortcuts between layers to ensemble feature maps of different level of semantics, we here focus on multi-scale structures that use input of different scales.

Figure 5 depicts the some of the common structures feasible for multi-scale implementation that we experiment in the comparison. Network A (Figure 5a) is the most basic multi-scale structure which simply averages the results of different scales. This network is similar to basic scale augmentation of training samples. The network does not learn the relationship between different scales. Network B (Figure 5b) concatenate the results after the convolution layers. Here the fc_** layers start to learn the weights of different paths to classify different objects. In Network C (Figure 5c), the concatenation takes place after the first fc_** layers. The proposed network is similar to Network C except the the network has two loss layers, each on the top of either path. In contrast, Network A, B, C only have one loss layer at the very top of the network.

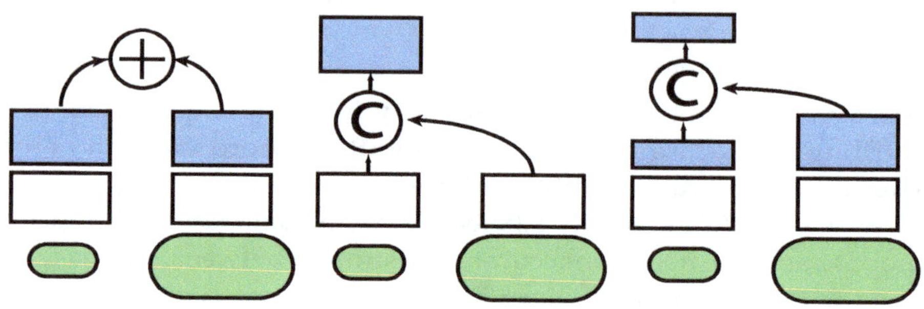

(**a**) (**b**) (**c**)

Figure 5. Different multi-scale structures in our experiment, (**a**) the features are summed up element-wise at the very end, (**b**) features are concatenated before fc_** layers and (**c**) features are concatenated between fc_** layers. Here green blocks indicate inputs of different scales, white indicates convolution layers, blue indicates fc_** layers, circle with a plus indicates element-wise addition operation of feature maps, circle with a C indicates concatenation operation. The loss layers are placed on the very top of each network.

We experiment the different networks on Google Map images. Figure 7 shows the training average Accuracy/epoch, Recall/epoch and IOU/epoch curve of these networks. It shows that the proposed method, with 2 loss layers at different path, has overall faster learning speed and higher performance.

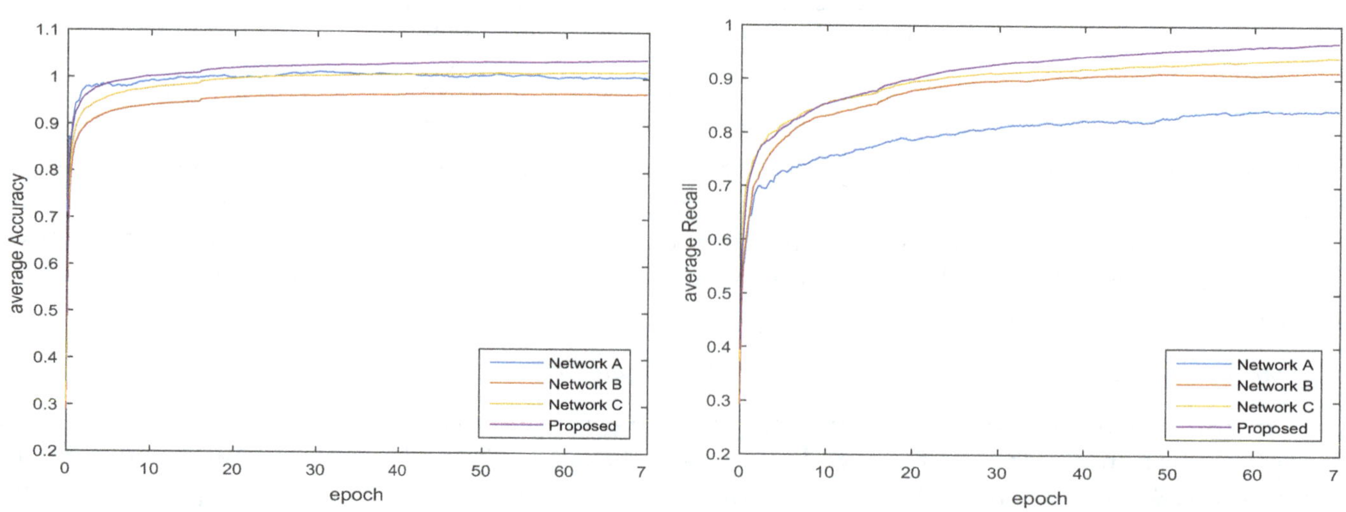

(**a**) (**b**)

Figure 6. *Cont.*

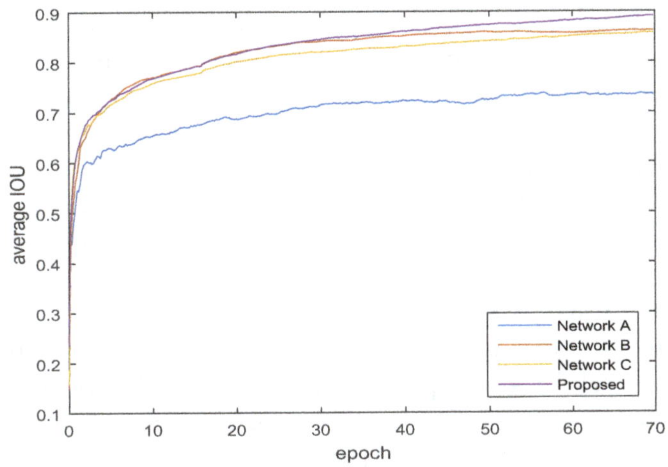

(c)

Figure 7. Training average Accuracy/epoch, Recall/epoch and IOU/epoch curve of networks with different multi-scale structure in Figure 5, with (**a**-**c**) labeled accordingly.

In Table 4 we show the the numbers of parameters and the computation time of a single forward/backward routine on a 961 × 961 training patch of these networks. Since Network C differs from the proposed one only in that it has one fewer loss layer, its data is omitted. The table shows that our proposed multi-scale structure has relatively small number of parameters and considerate fewer computation time compared to Network B.

Table 4. The numbers of parameters of the networks and computation time of a single forward/backward routine per patch.

Results	Network A	Network B	Proposed
# of Parameters	15.2 m	15.7 m	15.2 m
Computation time	0.1525 s	0.2825 s	0.1425 s

We also include an experiment to show the performance comparison between the usage of hole algorithm and different convolution architecture. In one of the experiment we cancel the hole implementation and in the other experiment we use Resnet [53] as the convolution structure. The IOU is shown in Table 5. As shown in the table, the hole implementation and the convolution structure both slightly increase the performance compared to the counterpart. The implementation with Resnet has the worst performance despite it is the more recent architecture. This is because Resnet introduce heavy pooling and the small details are further neglected. This can be seen in the 4% drop IOU in Ship performance.

Table 5. IOUs of the proposed method/without hole/with Resnet.

IOU (%)	Sea	Land	Ship
Proposed	98.2	98.7	68.3
W/O hole	98.0	98.5	68.1
W/Resnet	96.9	97.2	64.4

4.3. Comparison with Other Methods

For a performance baseline, we also experiment the SLIC (Simple Linear Iterative Clustering) method [35] and DenseCRF (Dense Conditional Random Field) [54] two of the most widely used semantic labeling methods other than deep learning networks, to approach the same problem. SLIC is a widely approved way of producing superpixels as a preprocessing step for other process such as object localization and semantic labeling. We first break the large images into small, irregular segments called superpixels and then learn to classify each superpixel into different categories as described in [35], i.e., to extract color, shape and texture info of each superpixel as features and train an adaboost classifier [55] for classification. DenseCRF is a widely used multi-class image segmentation method based on fully connected random field. This model accounts for unary and pairwise potentials among pixels at the same time. The pairwise potential can address the difference between pixels in arbitrary feature spaces and the unary potentials are computed independently on each pixel. The unary potentials are treated as the initial guess of each pixel's category, and the pairwise potentials are to rectify the results. The solution to this model is yielded in an iterative fashion and leads to a refined classification results of each pixel. In this experiment, we follow the implementation in [54], in which the unary potentials are acquired using TextonBoost [56]. Although for the unary potentials there are multiple selection such as convolutional network, we use TextonBoost for the consistency to the original paper.

Table 6 shows the comparison results. We notice that because of the nature of the categories in this problem, pixels that belongs to sea or land takes an extremely great proportion (over 99%), affecting the statistics in the evaluation. So we randomly ignore pixels belonging to sea or land in the evaluation, to ensure the numbers of pixels in different categories are of similar order of magnitude (the ratio between the areas of land, sea and ships is balanced roughly to 4:4:1). We also list the IOUs without the balanced evaluation in Table 7 for completeness, but for future results, we will only show the ones with balanced evaluation.

Table 6. Accuracy of segmentation of different methods. Confusion matrix with percentages row-normalized and IOU of each class.

a SLIC/DenseCRF/proposed network on GaoFen-1 images.

% of Total	Sea	Land	Ship	IOU
Sea	96.1/95.3/**99.5**	3.7/4.6/0.5	0.2/0.1/0.0	93.4/71.9/**99.5**
Land	2.7/8.2/1.4	94.8/91.4/**98.6**	2.4/0.4/0.0	/47.1/**98.6**
Ship	0/27.4/12.8	53.9/61.1/3.8	46.0/11.5/**83.4**	44.8/11.5/**83.4**

b SLIC/DenseCRF/proposed network on Google Map images.

% of Total	Sea	Land	Ship	IOU
Sea	91.4/95.1/**98.2**	8.1/4.8/1.7	0.5/0.1 0.0	78.1/72.0/**98.2**
Land	14.1/8.6/1.2	64.9/91.0/**98.7**	21.1/0.5/0.0	40.4/43.9/**98.7**
Ship	3.1/24.5/6.8	51.9/59.8/24.9	45.0/15.7/**68.3**	37.0/15.6/**68.3**

Table 7. IOUs of the proposed method without balanced evaluation.

IOU (%)	Sea	Land	Ship
GaoFen-1	99.3	95.8	59.0
Google Map	96.9	97.2	40.5

We notice that the SLIC method performs poorly in this problem because a) the superpixels produced are of bad accuracy even with carefully tuned parameters (initial region size and spatial regularizer) and b) at the classification stage, the features extracted are not rich enough to distinguish each category. The DenseCRF's iteration method relies greatly on its initial result, the unary potentials from TextonBoost, which is initially designed for everyday image circumstances. The experiment shows that TextonBoost is, however, not suitable for remote sensing images. We presume the failure of DenseCRF and SLIC is generally due to the fact that remote sensing images have scarce (if any) color information and objects are of much smaller size compared to those in everyday images.

4.4. Experiments on Multi-Scale Structure

Table 8 shows the performance comparison between the network with and without multi-scale structure. The multi-scale structure enhances the network's ability to discriminate categories in different scales, with accuracy on sea and land greatly improved.

Table 8. Accuracy of segmentation with or without multi-scale. Confusion matrix with percentages row-normalized and IOU for each class.

a network without/with multi-scale on GaoFen-1 images.

% of Total	Sea	Land	Ship	IOU
Sea	**99.6**/99.5	0.3/0.5	0.0/0.0	97.7/**99.5**
Land	8.2/1.4	91.8/**98.6**	0.0/0.0	89.2/**98.6**
Ship	14.5/12.8	2.8/3.8	82.7/**83.4**	82.6/**83.4**

b network without/with multi-scale on Google Map images.

% of Total	Sea	Land	Ship	IOU
Sea	97.9 / **98.2**	2.1 / 1.7	0.0 / 0.0	94.9 / **98.2**
Land	2.3 / 1.2	97.6 / **98.7**	0.0 / 0.0	86.4 / **98.7**
Ship	1.3 / 6.8	33.1 / 24.9	65.6 / **68.3**	65.4 / **68.3**

As is shown in Figure 8, after the training of the network, we extracted the weights of Layer fc_a2, which is used to combine the information from fine-scale feature maps and coarse-scale score maps. Only 20 weights of each kernel are shown for clarity. The layer learns that sea and land score maps from coarse-scale have greater weights and ship score map have relatively lesser weight (as it should, intuitively, since coarse-scale network are more reliable for sea-land segmentation).

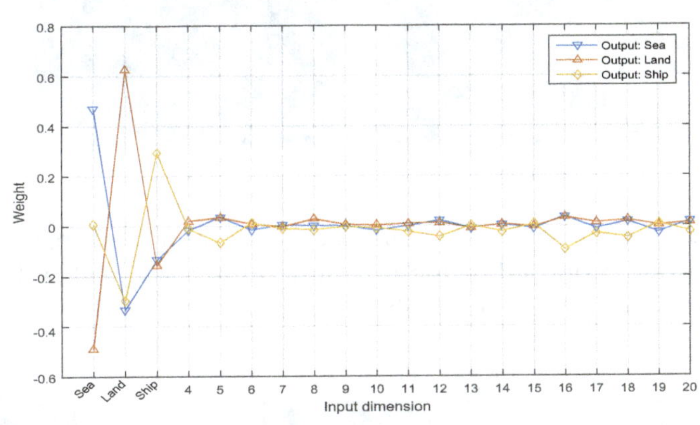

Figure 8. First 20 weights of Layer fc_a2 plotted as lines. Each line represents the weights corresponding to a specific output category (sea, land and ship) as listed in the legend. Each dot on the line represent a weight corresponding to an input dimension. The first 3 input dimensions corresponds to coarse-scale score slices of sea, land and ship, respectively, and the other dimensions corresponds to feature maps from fine-scale Layer fc_a1.

4.5. Qualitative Experiments

The qualitative performance is shown in Figures 9 and 10. Figure 9 features the comparison between our proposed method with and without multi-scale structure. The result with multi-scale structure tends to be more accurate and continuous, especially on GaoFen-1 dataset. Also note that images from GaoFen-1 dataset have more ships off-shore, which can be relatively easy for the network and add to better quantitative performance on GaoFen-1 dataset in Table 6. It is also noticed that the segmentation boundaries are not quite accurate with respect to the original images. This is in accordance with the initial results of DeepLab network [36], which, later implements DenseCRF as a post-process to acquire better segmentation boundaries. However, DenseCRF does not yield satisfactory results in our experiments, due to the fact that the objects in our dataset lack color differentiation and clear boundaries, especially in GaoFen-1 images.

Qualitative comparisons between our proposed method and DenseCRF, SLIC are shown in Figure 11. The compared methods presents inferior results because of two aspects, the classification and the segmentation. DenseCRF and SLIC both have worse performance when compared to our deep network and can not fully identify the ship body. In addition, when the shadows on the ship is evident, these two methods often classify these shadows into sea category.

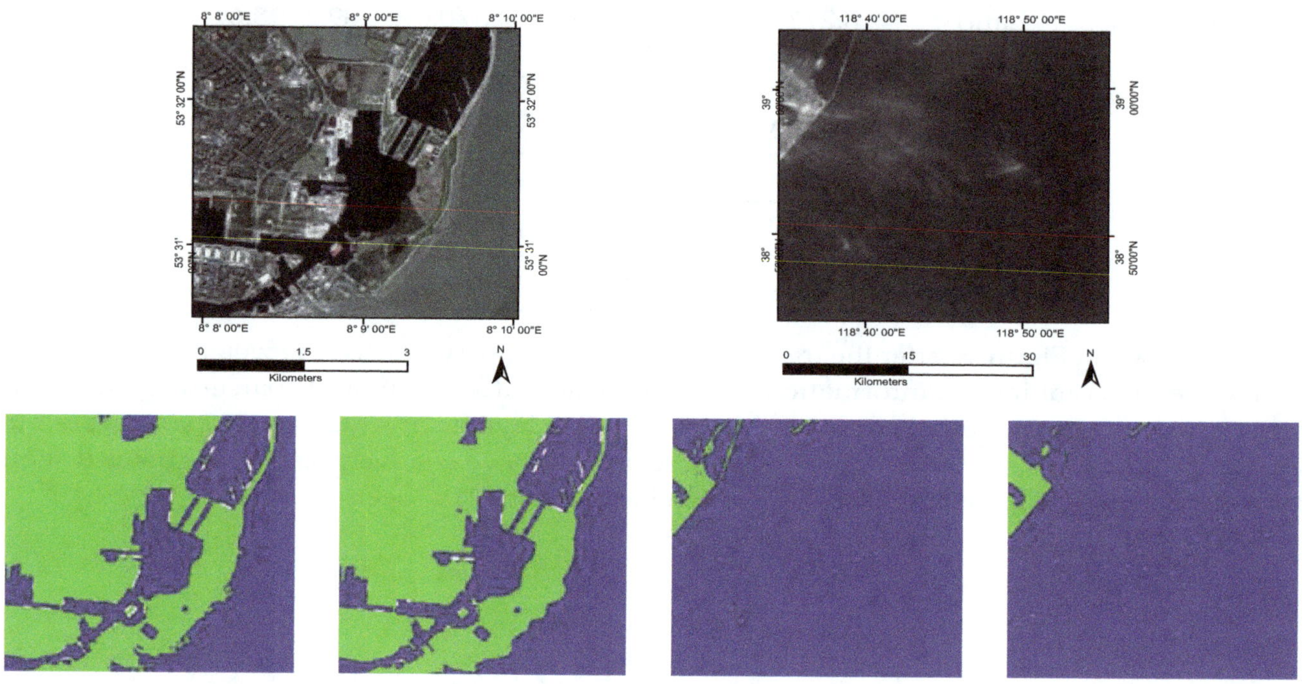

(a) (b) (c) (d)

Figure 9. Semantic labeling results on Google Map images (**a**,**b**) and GaoFen-1 images (**c**,**d**). The images are arranged as original (**top**), proposed method without multi-scale (**a**,**c**) and proposed method with multi-scale (**b**,**d**). Here sea, land and ship are labeled as blue, green and white, respectively.

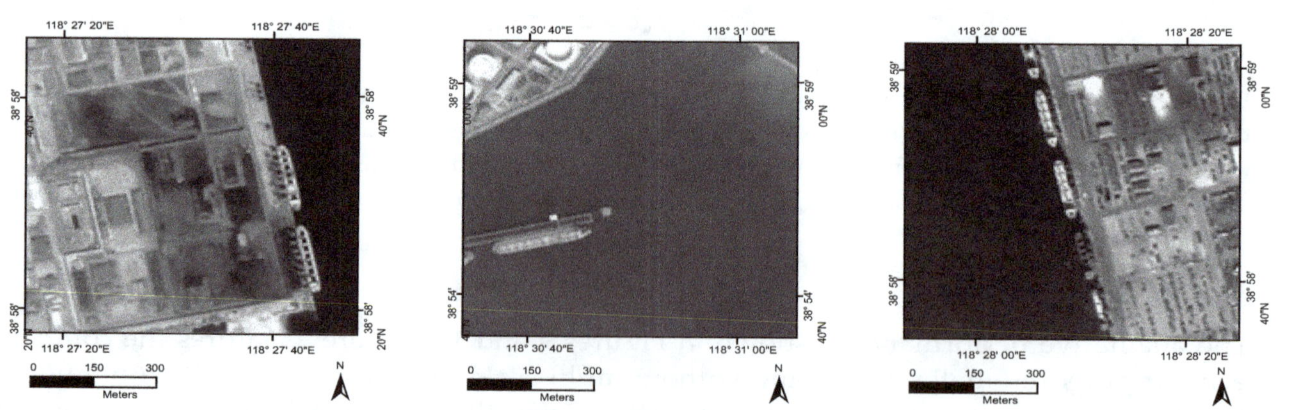

Figure 10. *Cont.*

(a) (b) (c)

Figure 10. Zoomed in semantic labeling results (**bottom**) on GaoFen-1 images (**a-c**), presented with The original image (**top**). Here sea, land and ship are labeled as blue, green and white, respectively.

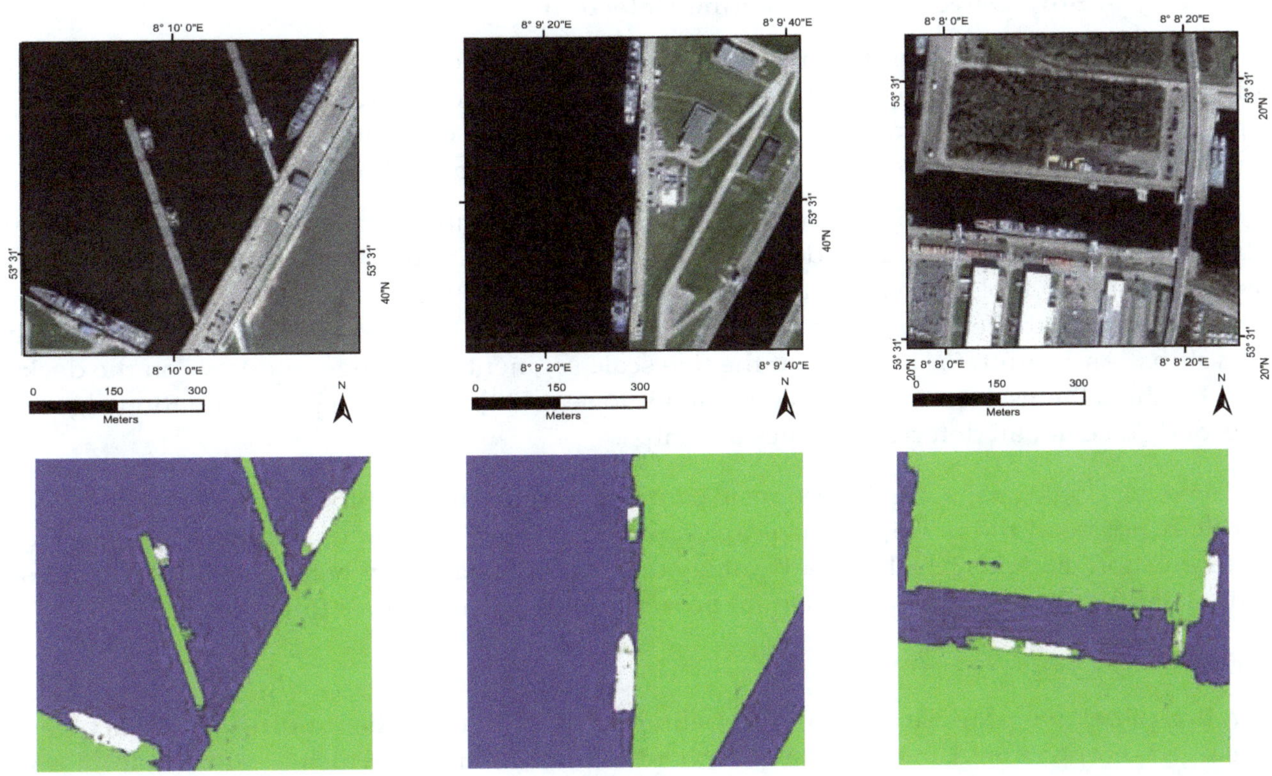

(a) (b) (c)

(d) (e) (f)

Figure 11. *Cont.*

(g) (h) (i)

Figure 11. Zoomed in semantic labeling results of DenseCRF (**a-c**), SLIC (**d-f**) and proposed method (**g-i**) with original images (**top**). Here sea, land and ship are labeled as blue, green and white, respectively.

5. Feasibility of Ship Detection via Coastline Detection

Coastline detection has undergone extensive research over the last decades [51,57,58] and it is possible to consider it as an approach towards ship detection to regard dynamic ships as a temporal change in multi-temporal images. The automatic coastline detection can facilitate autonomous navigation, coastal resource management and coastal environmental protection.

Although the accuracy of coastline detection is constantly increasing, it is still not enough for direct implementation for ship detection [59]. Coastline detection methods commonly utilize image segmentation tools such as watershed transformation [59] or graph-based discrimination [60], which are based on the features of textures and intensities and have no knowledge to holistic objects such as ships. As a result, for instance, at the fine scale segmentation stage, shadows on the decks that are cast by ships themselves are often segmented into seas [23]. Besides, a post-validation algorithm is still needed since not all detected changes are ships.

Moreover, single image ship detection, in contrast to multi-image ship detection, has the advantage that it does not need the multiple image registration and the storing of template images. Besides, change detection methods has the disadvantage that it is not accurate when image contrast has severe variation and that it needs constant manual power to update latest coast line.

6. Conclusions

In this paper, we propose a semantic labeling network with unified multi-scale structure which has enlarged receptive field and minimal parameter number increase, which is different from tradition multi-scale schemes that focus on utilizing finer-scale feature maps. The large receptive field is designed specifically for maritime remote sensing images and the experiments show that with the multi-scale semantic labeling scheme, an improved performance is achieved in the problem of sea-land segmentation and ship detection on both GaoFen-1 and Google Map images, under the circumstances that the ship targets are limited to large navy ships and oil tankers. In the future work, we will extend this work for more diversified ship targets such as yachts and fishing boats.

Acknowledgments: The work was supported by the National Natural Science Foundation of China under the Grant 61671037, the Beijing Natural Science Foundation under the Grant 4152031, the funding project of State Key Laboratory of Virtual Reality Technology and Systems, Beihang University under the Grant BUAA-VR-16ZZ-03. *(Corresponding author: Zhenwei Shi)*

Author Contributions: Haoning Lin designed the proposed model and implemented the experiments. Haoning Lin drafted the manuscript. Zhengxia Zou contributed to the improvement of the proposed model and edited the manuscript. Zhenwei Shi provided overall guidance to the project, reviewed and edited the manuscript and obtained funding to support this research.

References

1. Wang, Q.; Lin, J.; Yuan, Y. Salient band selection for hyperspectral image classification via manifold ranking. *IEEE Trans. Neural Netw. Learn. Syst.* **2016**, *27*, 1279–1289.
2. Tello, M.; López-Martínez, C.; Mallorqui, J.J. A novel algorithm for ship detection in SAR imagery based on the wavelet transform. *IEEE Geosci. Remote Sens. Lett.* **2005**, *2*, 201–205.
3. Mnih, V.; Hinton, G.E. Learning to detect roads in high-resolution aerial images. In Proceedings of the European Conference on Computer Vision, Crete, Greece, 5–11 September 2010; pp. 210–223.
4. Wang, J.; Song, J.; Chen, M.; Yang, Z. Road network extraction: A neural-dynamic framework based on deep learning and a finite state machine. *Int. J. Remote Sens.* **2015**, *36*, 3144–3169.
5. Sirmacek, B.; Unsalan, C. A probabilistic framework to detect buildings in aerial and satellite images. *IEEE Trans. Geosci. Remote Sens.* **2011**, *49*, 211–221.
6. Stankov, K.; He, D.C. Detection of buildings in multispectral very high spatial resolution images using the percentage occupancy hit-or-miss transform. *IEEE J. Sel. Top. Appl. Earth Obs. Remote Sens.* **2014**, *7*, 4069–4080.
7. Ok, A.O.; Başeski, E. Circular oil tank detection from panchromatic satellite images: A new automated approach. *IEEE Geosci. Remote Sens. Lett.* **2015**, *12*, 1347–1351.
8. Zhang, L.; Shi, Z.; Wu, J. A Hierarchical Oil Tank Detector with Deep Surrounding Features for High-Resolution Optical Satellite Imagery. *IEEE J. Sel. Top. Appl. Earth Obs. Remote Sens.* **2015**, *8*, 4895–4909.
9. Wen, X.; Shao, L.; Fang, W.; Xue, Y. Efficient feature selection and classification for vehicle detection. *IEEE Trans. Circuits Syst. Video Technol.* **2015**, *25*, 508–517.
10. Yang, L.; Bi, G.; Xing, M.; Zhang, L. Airborne sar moving target signatures and imagery based on LVD. *IEEE Trans. Geosci. Remote Sens.* **2015**, *53*, 5958–5971.
11. Yu, X.; Shi, Z. Vehicle detection in remote sensing imagery based on salient information and local shape feature. *Opt.-Int. J. Light Electron Opt.* **2015**, *126*, 2485–2490.
12. Cai, H.; Su, Y. Airplane detection in remote sensing image with a circle-frequency filter. In Proceedings of the International Conference on Space Information Technology, Beijing, China, 19–20 November 2005; p. 59852T.
13. Bo, S.; Jing, Y. Region-based airplane detection in remotely sensed imagery. In Proceedings of the 2010 3rd International Congress on Image and Signal Processing (CISP), Yantai, China, 16–18 October 2010; Volume 4, pp. 1923–1926.
14. An, Z.; Shi, Z.; Teng, X.; Yu, X.; Tang, W. An automated airplane detection system for large panchromatic image with high spatial resolution. *Opt.-Int. J. Light Electron Opt.* **2014**, *125*, 2768–2775.
15. Cheriyadat, A.M. Unsupervised feature learning for aerial scene classification. *IEEE Trans. Geosci. Remote Sens.* **2014**, *52*, 439–451.
16. Zhang, F.; Du, B.; Zhang, L. Saliency-guided unsupervised feature learning for scene classification. *IEEE Trans. Geosci. Remote Sens.* **2015**, *53*, 2175–2184.
17. Cheng, G.; Han, J.; Zhou, P.; Guo, L. Multi-class geospatial object detection and geographic image classification based on collection of part detectors. *ISPRS J. Photogramm. Remote Sens.* **2014**, *98*, 119–132.
18. Crisp, D.J. *The State-of-the-Art in Ship Detection in Synthetic Aperture Radar Imagery*; Technical Report, DTIC Document; DTIC: Fort Belvoir, VA, USA, 2004.
19. Vachon, P.; Campbell, J.; Bjerkelund, C.; Dobson, F.; Rey, M. Ship detection by the RADARSAT SAR: Validation of detection model predictions. *Can. J. Remote Sens.* **1997**, *23*, 48–59.
20. Zhang, T.; Yang, X.; Hu, S.; Su, F. Extraction of coastline in aquaculture coast from multispectral remote sensing images: Object-based region growing integrating edge detection. *Remote Sens.* **2013**, *5*, 4470–4487.
21. Cheng, D.; Meng, G.; Xiang, S.; Pan, C. Efficient sea–land segmentation using seeds learning and edge directed graph cut. *Neurocomputing* **2016**, *207*, 36–47.
22. Zhu, C.; Zhou, H.; Wang, R.; Guo, J. A novel hierarchical method of ship detection from spaceborne optical image based on shape and texture features. *IEEE Trans. Geosci. Remote Sens.* **2010**, *48*, 3446–3456.
23. Liu, G.; Zhang, Y.; Zheng, X.; Sun, X.; Fu, K.; Wang, H. A new method on inshore ship detection in high-resolution satellite images using shape and context information. *IEEE Geosci. Remote Sens. Lett.* **2014**, *11*, 617–621.

24. Shi, Z.; Yu, X.; Jiang, Z.; Li, B. Ship detection in high-resolution optical imagery based on anomaly detector and local shape feature. *IEEE Trans. Geosci. Remote Sens.* **2014**, *52*, 4511–4523.
25. You, H.M.; Pi, S.H. Ship detection at the dock using a polygon approximation method. In Proceedings of the 2015 IEEE International Conference on Grey Systems and Intelligent Services (GSIS), Leicester, UK, 18–20 August 2015; pp. 500–504.
26. Tang, J.; Deng, C.; Huang, G.B.; Zhao, B. Compressed-domain ship detection on spaceborne optical image using deep neural network and extreme learning machine. *IEEE Trans. Geosci. Remote Sens.* **2015**, *53*, 1174–1185.
27. Zou, Z.; Shi, Z. Ship Detection in Spaceborne Optical Image with SVD Networks. *IEEE Trans. Geosci. Remote Sens.* **2016**, *54*, 5832–5845.
28. Taormina, R.; Chau, K.W. Data-driven input variable selection for rainfall–runoff modeling using binary-coded particle swarm optimization and Extreme Learning Machines. *J. Hydrol.* **2015**, *529*, 1617–1632.
29. Liang, Z.; Shan, S.; Liu, X.; Wen, Y. Fuzzy prediction of AWJ turbulence characteristics by using typical multi-phase flow models. *Eng. Appl. Comput. Fluid Mech.* **2017**, *11*, 225–257.
30. Bellary, S.A.I.; Adhav, R.; Siddique, M.H.; Chon, B.H.; Kenyery, F.; Samad, A. Application of computational fluid dynamics and surrogate-coupled evolutionary computing to enhance centrifugal-pump performance. *Eng. Appl. Comput. Fluid Mech.* **2016**, *10*, 171–181.
31. Zhang, J.; Chau, K.W. Multilayer Ensemble Pruning via Novel Multi-sub-swarm Particle Swarm Optimization. *J. Univ. Comput. Sci.* **2009**, *15*, 840–858.
32. Wang, W.C.; Chau, K.W.; Xu, D.M.; Chen, X.Y. Improving forecasting accuracy of annual runoff time series using ARIMA based on EEMD decomposition. *Water Resour. Manag.* **2015**, *29*, 2655–2675.
33. Zhang, S.; Chau, K.W. Dimension reduction using semi-supervised locally linear embedding for plant leaf classification. *Emerg. Intell. Comput. Technol. Appl.* **2009**, 948–955.
34. Wu, C.; Chau, K.; Fan, C. Prediction of rainfall time series using modular artificial neural networks coupled with data-preprocessing techniques. *J. Hydrol.* **2010**, *389*, 146–167.
35. Achanta, R.; Shaji, A.; Smith, K.; Lucchi, A.; Fua, P.; Süsstrunk, S. SLIC superpixels compared to state-of-the-art superpixel methods. *IEEE Trans. Pattern Anal. Mach. Intell.* **2012**, *34*, 2274–2282.
36. Liang-Chieh, C.; Papandreou, G.; Kokkinos, I.; Murphy, K.; Yuille, A. Semantic Image Segmentation with Deep Convolutional Nets and Fully Connected CRFs. In Proceedings of the International Conference on Learning Representations, San Diego, CA, USA, 7–9 May 2015.
37. Eigen, D.; Fergus, R. Predicting depth, surface normals and semantic labels with a common multi-scale convolutional architecture. In Proceedings of the IEEE International Conference on Computer Vision, Santiago, Chile, 13–16 December 2015; pp. 2650–2658.
38. Noh, H.; Hong, S.; Han, B. Learning deconvolution network for semantic segmentation. In Proceedings of the IEEE International Conference on Computer Vision, Santiago, Chile, 13–16 December 2015; pp. 1520–1528.
39. Wang, Q.; Yuan, Y.; Yan, P.; Li, X. Saliency detection by multiple-instance learning. *IEEE Trans. Cybern.* **2013**, *43*, 660–672.
40. Wang, Q.; Yuan, Y.; Yan, P. Visual saliency by selective contrast. *IEEE Trans. Circuits Syst. Video Technol.* **2013**, *23*, 1150–1155.
41. Penatti, O.A.; Nogueira, K.; dos Santos, J.A. Do deep features generalize from everyday objects to remote sensing and aerial scenes domains? In Proceedings of the IEEE Conference on Computer Vision and Pattern Recognition Workshops, Boston, MA, USA, 7–12 January 2015; pp. 44–51.
42. Krizhevsky, A.; Sutskever, I.; Hinton, G.E. Imagenet classification with deep convolutional neural networks. In Proceedings of the Advances in Neural Information Processing Systems, Stateline, NC, USA, 3–8 December 2012; pp. 1097–1105.
43. Szegedy, C.; Liu, W.; Jia, Y.; Sermanet, P.; Reed, S.; Anguelov, D.; Erhan, D.; Vanhoucke, V.; Rabinovich, A. Going deeper with convolutions. In Proceedings of the IEEE Conference on Computer Vision and Pattern Recognition, Boston, MA, USA, 8–10 June 2015; pp. 1–9.
44. Long, J.; Shelhamer, E.; Darrell, T. Fully convolutional networks for semantic segmentation. In Proceedings of the IEEE Conference on Computer Vision and Pattern Recognition, Boston, MA, USA, 8–10 June 2015; pp. 3431–3440.
45. Nair, V.; Hinton, G.E. Rectified linear units improve restricted boltzmann machines. In Proceedings of the 27th International Conference on Machine Learning (ICML-10), Haifa, Israel, 21–24 June 2010; pp. 807–814.

46. Paisitkriangkrai, S.; Sherrah, J.; Janney, P.; Hengel, V.D. Effective semantic pixel labelling with convolutional networks and conditional random fields. In Proceedings of the IEEE Conference on Computer Vision and Pattern Recognition Workshops, Boston, MA, USA, 7–12 June 2015; pp. 36–43.
47. Papandreou, G.; Kokkinos, I.; Savalle, P.A. Untangling local and global deformations in deep convolutional networks for image classification and sliding window detection. *arXiv preprint* **2014**, arXiv:1412.0296.
48. Jia, Y.; Shelhamer, E.; Donahue, J.; Karayev, S.; Long, J.; Girshick, R.; Guadarrama, S.; Darrell, T. Caffe: Convolutional Architecture for Fast Feature Embedding. *arXiv preprint* **2014**, arXiv:1408.5093.
49. Shi, S.; Wang, Q.; Xu, P.; Chu, X. Benchmarking State-of-the-Art Deep Learning Software Tools. *arXiv preprint* **2016**, arXiv:1608.07249.
50. Zeiler, M. ADADELTA: An Adaptive Learning Rate Method. *arXiv* **2012**, arXiv:1212.5701.
51. Bayram, B.; Bayraktar, H.; Helvaci, C.; Acar, U. Coastline change detection using CORONA, SPOT and IRS 1D images. *Int. Arch. Photogramm. Remote Sens.* **2004**, *35*, 437–441.
52. Hu, Q.; Wu, W.; Xia, T.; Yu, Q.; Yang, P.; Li, Z.; Song, Q. Exploring the use of Google Earth imagery and object-based methods in land use/cover mapping. *Remote Sens.* **2013**, *5*, 6026–6042.
53. He, K.; Zhang, X.; Ren, S.; Sun, J. Deep residual learning for image recognition. In Proceedings of the IEEE Conference on Computer Vision and Pattern Recognition, Las Vegas, NV, USA, 26 June–1 July 2016; pp. 770–778.
54. Koltun, V. Efficient inference in fully connected crfs with gaussian edge potentials. *Adv. Neural Inf. Process. Syst.* **2011**, 109–117.
55. Freund, Y.; Schapire, R.E. A desicion-theoretic generalization of on-line learning and an application to boosting. In Proceedings of the European Conference on Computational Learning Theory, Barcelona, Spain, 13–15 March 1995; Springer: Berlin, Germany; pp. 23–37.
56. Shotton, J.; Winn, J.; Rother, C.; Criminisi, A. Textonboost: Joint appearance, shape and context modeling for multi-class object recognition and segmentation. In Proceedings of the European Conference on Computer Vision, Amsterdam, The Netherlands, 8–16 October 2006; Springer: Berlin, Germany; pp. 1–15.
57. Li, X.; Damen, M.C. Coastline change detection with satellite remote sensing for environmental management of the Pearl River Estuary, China. *J. Mar. Syst.* **2010**, *82*, S54–S61.
58. Ji, R.; Lu, Y.; Zuo, L. Coastline change detection of the Bohai Bay using satellite remote sensing. In Proceedings of the 2011 International Conference on Remote Sensing, Environment and Transportation Engineering (RSETE), Nanjing, China, 24–26 June 2011; pp. 168–171.
59. Sheng, G.; Yang, W.; Deng, X.; He, C.; Cao, Y.; Sun, H. Coastline detection in synthetic aperture radar (SAR) images by integrating watershed transformation and controllable gradient vector flow (GVF) snake model. *IEEE J. Ocean. Eng.* **2012**, *37*, 375–383.
60. Ding, X.; Zou, X.; Yu, T. Coastline detection in SAR images using discriminant cuts segmentation. In Proceedings of the IOP Conference Series: Earth and Environmental Science, Beijing, China, 7–8 July 2016; Volume 46, p. 012035.

11

Optimized Kernel Minimum Noise Fraction Transformation for Hyperspectral Image Classification

Lianru Gao [1], Bin Zhao [1,2], Xiuping Jia [3], Wenzhi Liao [4] and Bing Zhang [1,2,*]

1 Key Laboratory of Digital Earth Science, Institute of Remote Sensing and Digital Earth, Chinese Academy of Sciences, Beijing 100094, China; gaolr@radi.ac.cn (L.G.); zhaobin@radi.ac.cn (B.Z.)
2 University of Chinese Academy of Sciences, Beijing 100049, China
3 School of Engineering and Information Technology, The University of New South Wales, Canberra Campus, Bruce ACT 2006, Australia; x.jia@adfa.edu.au
4 Department of Telecommunications and Information Processing, Ghent University, Ghent 9000, Belgium; wenzhi.liao@telin.ugent.be
* Correspondence: zb@radi.ac.cn.

Academic Editors: Qi Wang, Nicolas H. Younan, Carlos López-Martínez and Prasad S. Thenkabail

Abstract: This paper presents an optimized kernel minimum noise fraction transformation (OKMNF) for feature extraction of hyperspectral imagery. The proposed approach is based on the kernel minimum noise fraction (KMNF) transformation, which is a nonlinear dimensionality reduction method. KMNF can map the original data into a higher dimensional feature space and provide a small number of quality features for classification and some other post processing. Noise estimation is an important component in KMNF. It is often estimated based on a strong relationship between adjacent pixels. However, hyperspectral images have limited spatial resolution and usually have a large number of mixed pixels, which make the spatial information less reliable for noise estimation. It is the main reason that KMNF generally shows unstable performance in feature extraction for classification. To overcome this problem, this paper exploits the use of a more accurate noise estimation method to improve KMNF. We propose two new noise estimation methods accurately. Moreover, we also propose a framework to improve noise estimation, where both spectral and spatial de-correlation are exploited. Experimental results, conducted using a variety of hyperspectral images, indicate that the proposed OKMNF is superior to some other related dimensionality reduction methods in most cases. Compared to the conventional KMNF, the proposed OKMNF benefits significant improvements in overall classification accuracy.

Keywords: hyperspectral image; feature extraction; dimensionality reduction; optimized kernel minimum noise fraction (OKMNF)

1. Introduction

Hyperspectral images provide very rich spectral information of earth objects [1,2]. In general, a hyperspectral image contains hundreds of spectral bands with high spectral resolution. However, the high dimensionality reduces the efficiency of hyperspectral data processing. Moreover, in hyperspectral image classification, another problem is known as the curse of dimensionality or the Hughes phenomenon [3]. Namely, the more spectral bands the image has, the more training samples are needed in order to achieve an acceptable classification accuracy. Obviously, it is not easy to be satisfied to the hyperspectral case [4]. Dimensionality reduction is a very effective technique to solve this problem [5,6]. Dimensionality reduced data should well represent the original data, and can be considered as the extracted features for classification [7–9]. When the data dimensionality is lower, the computing time will be reduced, and the number of training samples required will become less demanding [10–13]. Therefore, dimensionality reduction is a very critical pre-processing step for hyperspectral image classification [14–16]. Typically, several approaches exist for dimensionality reduction in hyperspectral data that can be split into two major groups. The first group includes band selection approaches. Such methods aim at selecting a subset of relevant data from the original information. This group includes not only a supervised method such as Bhattacharyya distance, Jeffries–Matusita distance, divergence, kernel dependence, mutual information, and spectral angle mapper, but also unsupervised methods such as geometric-based representative bands, dissimilar bands based on linear projection, manifold ranking [17] and dual clustering [18,19], which have proven to be valuable to achieve superior classification results. The second group relates to feature extraction approaches. Feature extraction methods transform original hyperspectral data into an optimized feature space by mathematical transformation, and then achieve dimensionality reduction through feature selection. A number of techniques have been developed for feature extraction. These techniques can be categorized as two major classes. The first class includes supervised feature extraction methods such as linear discriminant analysis (LDA) [20], nonparametric weighted feature extraction (NWFE) [21], sparse graph based feature extraction and their extensions [22–24]. The second class relates to unsupervised feature extraction approaches such as principal component analysis (PCA) [25] and minimum noise fraction (MNF) [26], sparse-graph learning-based dimensionality reduction method [27], which do not need priori knowledge on label information. PCA and MNF are two of the widely adopted methods for dimensionality reduction of hyperspectral images. As we all know, the performance of PCA highly relies on noise characteristics [26,28]. When the noise is not uniformly distributed across all of the spectral bands or when the noise variance is larger than the signal variance in one band, PCA cannot guarantee that the first few principal components have the highest image quality [26]. MNF generates new components ordered by image quality and provides better spectral features in the major components than PCA, no matter how the spectral noise is distributed [28]. Original MNF is a linear dimensionality reduction method. It is simple in processing and can be applied in most conditions. However, it is not easy for this method to handle the nonlinear characteristics within the data. The nonlinear characteristics of hyperspectral data is often due to the nonlinear nature of scattering as described in the bidirectional reflectance distribution function, multiple scattering within a pixel, and the heterogeneity of subpixel constituents [29,30]. The Kernel MNF (KMNF) method is developed to overcome this weakness in MNF [31–33]. KMNF is a nonlinear dimensionality reduction method, which introduces the use of kernel functions [34] to model the nonlinear characteristics within the data. The nonlinear transformation based on a kernel function can transform the original data into a higher dimensional feature space, and then a linear analysis can be followed in this space, as the complex nonlinear characteristics in the original input space have become simpler linear characteristics in the new feature space [35–39]. Using a similar theory of the kernel methods such as KMNF, kernel PCA (KPCA) was also proposed for nonlinear dimensionality reduction of hyperspectral images [40].

While MNF is a valuable dimensionality reduction method for hyperspectral image classification, it is found that the traditional version of MNF cannot provide desired results in real applications. From the theoretical and experimental analysis, it has been reported that noise estimation is the key factor leading to this problem [41–43]. In the traditional MNF, it is assumed that spatial neighboring pixels have very high correlation and the differences between these pixels can be considered as the noise. It works when the image has very high spatial resolution. Due to the limitation of hyperspectral sensors, hyperspectral images are often unable to offer high spatial resolution, and mixed pixels are very common in a hyperspectral image [44]. Thus, spatial information adopted in the traditional MNF is less reliable for estimating noise for a hyperspectral image. Obviously, the spectral resolution of hyperspectral images is very high, which means that hyperspectral images have strong spectral correlation between bands [45]. It has been found that the combination of the spatial and the spectral information is much more appropriate to estimate noise in hyperspectral images than only using single spatial information [46,47]. Optimized MNF (OMNF) utilized spectral and spatial de-correlation (SSDC) [48–50] to improve noise estimation [51]. However, existing SSDC combines the spectral information with only one spatial neighbor for noise estimation [48–50], leading to imperfect exploitation of spatial information. KMNF is a kernel version of MNF, and can well treat nonlinear characteristics within the data. However, the classification results using the features extracted by KMNF are often disappointing, and sometimes even worse than using MNF. The fundamental reason of this problem mainly also lies in the fact that the original KMNF adopts only spatial information to estimate noise that has a lot of errors and is not stable.

To overcome the above limitations, we propose a new framework to optimize KMNF (OKMNF) for feature extraction of hyperspectral data. Instead of only relying on single spatial information for noise estimation, the proposed OKNMF estimates noises by taking into account both spectral and spatial correlations through multiple linear regression. We also propose a more general method than SSDC [51–53] for noise estimation, where more spatial neighbors are exploited. Moreover, the proposed OKMNF can well treat nonlinear characteristics within the data, which cannot be effectively processed by linear OMNF and MNF. Therefore, OKMNF is much more stable and accurate than KMNF on the noise estimation, and enables better performances on both dimensionality reduction and its post application to classification. Last but not least, the proposed framework can be extended to a general model, when some other accurate noise estimation methods are available.

The remainder of this paper is organized as follows. In Section 2, the OKMNF method will be introduced in detail. Section 3 validates the proposed approach and reports experimental results, comparing them to several state-of-the-art alternatives. Section 4 discusses the performance of noise estimation algorithms and dimensionality reduction methods. Section 5 states the conclusions.

2. Proposed OKMNF Method

Let us consider a hyperspectral image data set with n pixels and b spectral bands organized as a matrix $\mathbf{X}$ with n rows and b columns. Hyperspectral images inevitably contain noises due to the sensor error and other environmental factors' influence. Normally, we can consider the original hyperspectral image $\mathbf{X}$ as a sum of a signal part and a noise part [26,54–56]:

$$\mathbf{x}(p) = \mathbf{x}_S(p) + \mathbf{x}_N(p), \tag{1}$$

where $\mathbf{x}(p)$ is the pixel vector in position p, $\mathbf{x}_N(p)$ and $\mathbf{x}_S(p)$ are noise and signal contained in $\mathbf{x}(p)$, respectively. In optical images, noises and signals are often considered to be independent. Thus, the covariance matrix $\mathbf{S}$ of image $\mathbf{X}$ could be written as a sum of the noise covariance matrix $\mathbf{S}_N$ and signal covariance matrix $\mathbf{S}_S$,

$$\mathbf{S} = \mathbf{S}_N + \mathbf{S}_S. \tag{2}$$

Let us consider $\widetilde{x}_k$ as the average of the kth band, and we can get the matrix $\mathbf{X}_{mean}$ with n rows b columns:

$$\mathbf{X}_{mean} = \begin{bmatrix} \widetilde{x}_1 & \widetilde{x}_2 & \cdots & \widetilde{x}_b \\ \widetilde{x}_1 & \widetilde{x}_2 & \cdots & \widetilde{x}_b \\ \vdots & \vdots & \vdots & \vdots \\ \widetilde{x}_1 & \widetilde{x}_2 & \cdots & \widetilde{x}_b \end{bmatrix}, \tag{3}$$

$\mathbf{Z}$ as the center matrix of $\mathbf{X}$, is given by

$$\mathbf{Z} = \mathbf{X} - \mathbf{X}_{mean}. \tag{4}$$

The covariance matrix $\mathbf{S}$ of images $\mathbf{X}$ could be written as

$$\mathbf{S} = \mathbf{Z}^{\mathrm{T}}\mathbf{Z}/(n-1). \tag{5}$$

Let us consider $\widetilde{x}_{Nk}$ as the average of the noise in kth band, and we can get the matrix $\mathbf{X}_{Nmean}$ with n rows and b columns:

$$\mathbf{X}_{Nmean} = \begin{bmatrix} \widetilde{x}_{N1} & \widetilde{x}_{N2} & \cdots & \widetilde{x}_{Nb} \\ \widetilde{x}_{N1} & \widetilde{x}_{N2} & \cdots & \widetilde{x}_{Nb} \\ \vdots & \vdots & \vdots & \vdots \\ \widetilde{x}_{N1} & \widetilde{x}_{N2} & \cdots & \widetilde{x}_{Nb} \end{bmatrix}. \quad (6)$$

$\mathbf{Z}_N$, as the center matrix of the noise matrix $\mathbf{X}_N$, can be computed as

$$\mathbf{Z}_N = \mathbf{X}_N - \mathbf{X}_{Nmean}. \quad (7)$$

The covariance matrix $\mathbf{S}_N$ of $\mathbf{X}_N$ could be expressed as

$$\mathbf{S}_N = \mathbf{Z}_N^{\mathrm{T}} \mathbf{Z}_N / (n-1). \quad (8)$$

The noise fraction **NF** could be defined as the ratio of the noise variance to the total variance, so for a linear combinations, $\mathbf{a}^T \mathbf{z}(p)$ [26,31], we get

$$\mathbf{NF} = \mathbf{a}^T \mathbf{S}_N \mathbf{a} / \mathbf{a}^T \mathbf{S} \mathbf{a} = \mathbf{a}^T \mathbf{Z}_N^T \mathbf{Z}_N \mathbf{a} / \mathbf{a}^T \mathbf{Z}^T \mathbf{Z} \mathbf{a}, \quad (9)$$

where **a** is the eigenmatrix of **NF**. In **NF**, it is significant that the noise is estimated reliably. The original KMNF method [31] mainly adopts the spatial neighborhood (3 by 3) feature of a hyperspectral image to estimate noise $\mathbf{Z}_N$ [57], as shown below:

$$\begin{aligned} n_{i,j,k} &= z_{i,j,k} - \hat{z}_{i,j,k} \\ &= z_{i,j,k} - (-z_{i-1,j-1,k} + 2z_{i,j-1,k} - z_{i+1,j-1,k} + 2z_{i-1,j,k} + , \\ &5z_{i,j,k} + 2z_{i+1,j,k} - z_{i-1,j+1,k} + 2z_{i,j+1,k} - z_{i+1,j+1,k})/9 \end{aligned} \quad (10)$$

where $z_{i,j,k}$ is the value of pixel located at line i, column j, and band k of the original hyperspectral image $\mathbf{Z}$, $\hat{z}_{i,j,k}$ is the estimated value of this pixel, and $n_{i,j,k}$ is the estimated noise value of $z_{i,j,k}$.

However, noise estimation based on spatial information alone can be unstable and data-selective [25,51,53]. It is because hyperspectral images do not always have very high spatial resolution, and the difference between pixels may contain a significant signal instead of pure noise. In contrast, in hyperspectral images, correlation between bands generally is very high. Therefore, we can incorporate the high correlations between bands for noise estimation, such as SSDC, which is a useful method for hyperspectral image noise estimation. In SSDC, the spatial and spectral correlations are removed through a multiple linear regression model, and the remaining residuals are the estimates of noise [49,50,58]. Recent works show that SSDC can offer reliable results for noise estimation when there are different land cover types in the hyperspectral images [50].

2.1. Noise Estimation

In noise estimation based on spectral and spatial de-correlation, an image is uniformly divided into non-overlapping small sub-blocks $\mathbf{X}_{sub}$ with $w \times h$ pixels, in order to reduce the influence of the variations in ground cover types. In SSDC, a multiple linear regression formula is adopted as follows for each pixel [49,50]:

$$x_{i,j,k} = a + bx_{i,j,k-1} + cx_{i,j,k+1} + dx_{p,k}, \quad (11)$$

$$x_{p,k} = \begin{cases} x_{i-1,j,k}; & i > 1, j = 1 \\ x_{i,j-1,k}; & j > 1 \end{cases}, \quad (12)$$

where $1 \le i \le w$, $1 \le j \le h$, and $(i,j) \ne (1,1)$, a, b, c, and d are the coefficients need to be determined. For each sub-block $\mathbf{X}_{sub}$, the multiple linear regression models could be written as

$$\mathbf{X}_{sub} = \mathbf{B}\boldsymbol{\mu} + \boldsymbol{\varepsilon}, \tag{13}$$

$$\mathbf{X}_{sub} = \begin{bmatrix} x_{1,2,k} \\ x_{1,3,k} \\ \vdots \\ x_{w,h,k} \end{bmatrix}, \mathbf{B} = \begin{bmatrix} 1 & x_{1,2,k-1} & x_{1,2,k+1} & x_{1,1,k} \\ 1 & x_{1,3,k-1} & x_{1,3,k+1} & x_{1,2,k} \\ \vdots & \vdots & \vdots & \vdots \\ 1 & x_{w,h,k-1} & x_{w,h,k+1} & x_{w,h-1,k} \end{bmatrix}, \boldsymbol{\mu} = \begin{bmatrix} a \\ b \\ c \\ d \end{bmatrix}, \tag{14}$$

where $\mathbf{X}_{sub}$ is sub-block matrix, $\mathbf{B}$ is the spectral-spatial neighborhoods matrix, $\boldsymbol{\mu}$ is the coefficients matrix, and $\boldsymbol{\varepsilon}$ is residual value.

However, SSDC integrates spectral information and one spatial neighbor in multiple linear regression for noise estimation. This way the spatial information might not be well exploited to estimate noise. To solve this problem, we propose two methods to improve the SSDC, named SSDC_1 and SSDC_2, where more spatial neighbors are incorporated into multiple linear regression for noise estimation.

We define SSDC_1 in the same multiple linear regression (same as Equation (11)) framework, but adopts the spatial neighbor parts $x_{p,k}$ as follows:

$$x_{p,k} = \begin{cases} (x_{i-1,j,k} + x_{i+1,j,k})/2; & i > 1, j = 1 \\ (x_{i,j-1,k} + x_{i,j+1,k})/2; & j > 1 \end{cases}, \tag{15}$$

where $\mathbf{X}_{sub}$ and $\boldsymbol{\mu}$ are the same as SSDC, but $\mathbf{B}$ is different from it, and can be defined as follows:

$$\mathbf{B} = \begin{bmatrix} 1 & x_{1,2,k-1} & x_{1,2,k+1} & (x_{1,1,k} + x_{1,3,k})/2 \\ 1 & x_{1,3,k-1} & x_{1,3,k+1} & (x_{1,2,k} + x_{1,4,k})/2 \\ \vdots & \vdots & \vdots & \vdots \\ 1 & x_{w,h,k-1} & x_{w,h,k+1} & (x_{w,h-1,k} + x_{,w,h+1,k})/2 \end{bmatrix}. \tag{16}$$

We can also improve multiple linear regression, which we define as SSDC_2:

$$x_{i,j,k} = a + bx_{i,j,k-1} + cx_{i,j,k+1} + dx_{i,j-1,k} + ex_{i,j+1,k}, \tag{17}$$

where $\mathbf{X}_{sub}$ is the same as SSDC, but $\mathbf{B}$ and $\boldsymbol{\mu}$ are defined as follows:

$$\mathbf{B} = \begin{bmatrix} 1 & x_{1,2,k-1} & x_{1,2,k+1} & x_{1,1,k} & x_{1,3,k} \\ 1 & x_{1,3,k-1} & x_{1,3,k+1} & x_{1,2,k} & x_{1,4,k} \\ \vdots & \vdots & \vdots & \vdots & \vdots \\ 1 & x_{w,h,k-1} & x_{w,h,k+1} & x_{w,h-1,k} & x_{w,h+1,k} \end{bmatrix}, \boldsymbol{\mu} = \begin{bmatrix} a \\ b \\ c \\ d \\ e \end{bmatrix}. \tag{18}$$

$\boldsymbol{\mu}$ could be estimated by

$$\hat{\boldsymbol{\mu}} = (B^T B)^{-1} B^T \mathbf{X}_{sub}. \tag{19}$$

Signal value could be estimated through

$$\hat{\mathbf{X}}_{sub} = \mathbf{B}\hat{\boldsymbol{\mu}}. \tag{20}$$

Finally, the noise value $\mathbf{N}_{sub}$ can be obtained by

$$\mathbf{N}_{sub} = \mathbf{X}_{sub} - \hat{\mathbf{X}}_{sub}. \tag{21}$$

The procedure of noise estimation is summarized in Algorithm 1.

Algorithm 1. Noise Estimation.

Input: hyperspectral image **X**, sub-block width $w \times h$.
Step 1: compute the coefficients a, b, c, d and e of the multiple linear regression models for each sub-block using Equation (11) or Equation (17); then:
$x_{i,j,k} = a + bx_{i,j,k-1} + cx_{i,j,k+1} + dx_{p,k}$, or
$x_{i,j,k} = a + bx_{i,j,k-1} + cx_{i,j,k+1} + dx_{i,j-1,k} + ex_{i,j+1,k}$
Step 2: estimate noise: $n_{i,j,k} = x_{i,j,k} - \hat{x}_{i,j,k}$
Output: noise data **N**.

We analyze the influences of sub-block size by using hyperspectral image as shown in Figure 1a. From the experiments, we found that, when the sub-block size is 4×4, or 5×5, some sub-blocks are homogeneous and have similar DN values in certain bands; thus, it makes the matrix inversion in multiple linear regression infeasible. When the sub-block size is too large, such as 15×15 and 30×30, some sub-blocks contain multiple types of earth surface features, and the results of noise estimation become inaccurate and instable. When the sub-block size is 6×6, as shown in Figures 2 and 3, the results of noise estimation are reliable and stable. Therefore, we set the sub-block size to 6×6 for SSDC, $SSDC_1$ and $SSDC_2$. The width and height of each sub-block are set as $w = 6$, $h = 6$.

2.2. *Kernelization and Regularization*

After noise is estimated through SSDC, $SSDC_1$ or $SSDC_2$, it will be included in KMNF. In KMNF, in order to get the new components ordered by image quality after dimensionality reduction, we should minimize the **NF**. For the convenience of mathematics, we can maximize the $1/\mathbf{NF}$, which can be presented as

$$1/\mathbf{NF} = \mathbf{a}^T\mathbf{S}\mathbf{a}/\mathbf{a}^T\mathbf{S}_N\mathbf{a} = \mathbf{a}^T\mathbf{Z}^T\mathbf{Z}\mathbf{a}/\mathbf{a}^T\mathbf{Z}_N^T\mathbf{Z}_N\mathbf{a}. \tag{22}$$

We can get to the dual formulation by reparametrizing and setting $\mathbf{a} \propto \mathbf{Z}^T\mathbf{b}$ [31,34]:

$$1/\mathbf{NF} = \mathbf{b}^T\mathbf{Z}\mathbf{Z}^T\mathbf{Z}\mathbf{Z}^T\mathbf{b}/\mathbf{b}^T\mathbf{Z}\mathbf{Z}_N^T\mathbf{Z}_N\mathbf{Z}^T\mathbf{b}. \tag{23}$$

For the kernelization of $1/\mathbf{NF}$, we will consider an embedding map

$$\Phi : x \to \Phi(x), \tag{24}$$

where $x \in R^n$, $\Phi(x) \in R^N$, $N > n$, and nonlinear mapping $\Phi(x)$ can transform the original data x into higher dimensional feature space F [34].

After mapping $\Phi(x)$, the kernelized $1/\mathbf{NF}$ can be expressed as

$$1/\mathbf{NF} = \mathbf{b}^T\Phi(\mathbf{Z})\Phi(\mathbf{Z})^T\Phi(\mathbf{Z})\Phi(\mathbf{Z})^T\mathbf{b}/\mathbf{b}^T\Phi(\mathbf{Z})\Phi(\mathbf{Z}_N)^T\Phi(\mathbf{Z}_N)\Phi(\mathbf{Z})^T\mathbf{b}. \tag{25}$$

Traditionally, the inner products $\langle\Phi(x), \Phi(y)\rangle$ $(x, y \in R^n)$ sometimes can be computed more efficiently as a direct function of the input features, without explicitly computing the mapping $\Phi(x)$ [34]. This function is called the kernel function κ, which can be expressed as

$$\kappa(x, y) = \langle\Phi(x), \Phi(y)\rangle. \tag{26}$$

Therefore, Equation (25) could be written as

$$1/\mathbf{NF} = \mathbf{b}^T\kappa^2\mathbf{b}/\mathbf{b}^T\kappa_N\kappa_N^T\mathbf{b}, \tag{27}$$

where $\kappa = \Phi()\Phi()^T$ with elements $\kappa(z_i, z_j)$, and $\kappa_N = \Phi(\mathbf{Z})\Phi(\mathbf{Z}_N)^T$ with elements $\kappa(z_i, z_{Nj})$. To ensure the uniqueness of the result in Equation (27), we regulate the 1/NF by introducing a regulator

r, similarly to what the other kernel methods (e.g., KMNF, KPCA [28,31]) have done. This way, we get a version which is regulated as

$$1/\mathbf{NF} = \mathbf{b}^T[(1-r)\kappa^2 + r\kappa]\mathbf{b}/\mathbf{b}^T\kappa_N\kappa_N^T\mathbf{b}. \tag{28}$$

2.3. OKMNF Transformation

The regulated version described above is a symmetric generalized eigenvalue problem, which could be solved by maximizing the Rayleigh quotient in Equation (28). Therefore, this problem can be written as

$$[(1-r)\kappa^2 + r\kappa]\mathbf{b} = \lambda\kappa_N\kappa_N^T\mathbf{b}, \tag{29}$$

$$[(1-r)\kappa^2 + r\kappa]\mathbf{b} = \lambda(\kappa_N\kappa_N^T)^{1/2}(\kappa_N\kappa_N^T)^{1/2}\mathbf{b}, \tag{30}$$

$$(\kappa_N\kappa_N^T)^{-1/2}[(1-r)\kappa^2 + r\kappa](\kappa_N\kappa_N^T)^{-1/2}[(\kappa_N\kappa_N^T)^{1/2}\mathbf{b}] = \lambda[(\kappa_N\kappa_N^T)^{1/2}\mathbf{b}], \tag{31}$$

where λ and $(\kappa_N\kappa_N^T)^{1/2}\mathbf{b}$ are eigenvalues and eigenvectors of $(\kappa_N\kappa_N^T)^{-1/2}[(1-r)\kappa^2 + r\kappa](\kappa_N\kappa_N^T)^{-1/2}$, respectively. $\mathbf{a} \propto \mathbf{Z}^T\mathbf{b}$, after mapping $\Phi(x)$, $\mathbf{Z}^T\mathbf{b}$ transforms to $\Phi(\mathbf{Z})^T\mathbf{b}$. Thus, we can get the value of $\mathbf{b}$, and the feature extraction result $\mathbf{Y}$ can be obtained by:

$$\begin{aligned}\mathbf{Y} &= \Phi(\mathbf{Z})\mathbf{a} \\ &= \Phi(\mathbf{Z})\Phi(\mathbf{Z})^T\mathbf{b} \\ &= \kappa\mathbf{b}\end{aligned}. \tag{32}$$

From the above analysis, we can see that noise estimation is a very critical step in the OKMNF method. Firstly, in the original data space, based on original hyperspectral data $\mathbf{Z}$, we get the estimated data $\hat{\mathbf{Z}}$ calculated by multiple linear regression models. Then, we transform the original real hyperspectral data $\mathbf{Z}$ and the estimated data $\hat{\mathbf{Z}}$ to the kernel space. In this space, we get the results of noise estimation through calculating the difference of kernel $\mathbf{Z}$ and kernel $\hat{\mathbf{Z}}$. It means that the noise is estimated in the kernel space. Finally, we get the transformation matrix by maximizing regulated $1/\mathbf{NF}$ and achieve the dimensionality reduction. A good noise estimation is important for effective dimensionality reduction.

In many real applications, a hyperspectral image typically has a huge amount of pixels. Then, the kernel matrix could be very large (for example, the matrix sizes of κ and κ_N are n by n, and n is the number of pixels). In this case, even in conventional hyperspectral remote sensing images, the kernel matrix will exceed the memory capacity of an ordinary personal computer. For example, a hyperspectral image of n = 512 × 512 pixels, the size of the kernel matrix is $n \times n$ = (512 × 512) × (512 × 512) elements. To reduce memory cost and computational complexity, we can randomly subsample the image and perform the kernel eigenvalue analysis only on these selected samples (suppose m), which can be used as training samples. We can generate a transformed version of the entire image by mapping all pixels onto the primal eigenvectors obtained from the subset samples. The procedure of OKMNF is summarized in Algorithm 2.

Algorithm 2. The Proposed OKMNF.

Input: hyperspectral image $\mathbf{X}$, and m training samples.
Step 1: compute the residuals (noises) of training samples: $n_{i,j,k} = x_{i,j,k} - \hat{x}_{i,j,k}$.
Step 2: dual transformation, kernelization and regularization of $1/\mathbf{NF}$ using Equation (22).
Step 3: compute the eigenvectors of $(\kappa_N\kappa_N^T)^{-1/2}[(1-r)\kappa^2 + r\kappa](\kappa_N\kappa_N^T)^{-1/2}$.
Step 4: mapping all pixels onto the primal eigenvectors.
Output: feature extraction result $\mathbf{Y}$.

3. Experiments and Results

This section designs three experiments to evaluate the performances of a few noise estimation algorithms and dimensionality reduction methods. The first experiment using real images with different land covers is to assess the robustness of noise estimation algorithms adopted in OKMNF, and the results are shown in Figure 4. The other two experiments are to validate the performances of dimensionality reduction methods in terms of maximum likelihood-based classification (ML) on two real hyperspectral images. The experimental results of Indian Pines image (as shown in Figure 5) are shown in Figures 6–9. The experimental results of Minamimaki scene image (as shown in Figure 10) are shown in Figures 11–13.

3.1. Parameter Tuning

In Equation (28), we introduced a parameter r to guarantee the uniqueness of the eigenvectors. Figures 7a and 12a show the sensitivity of kernel dimensionality reduction methods (KPCA, KMNF, and OKMNF) with respect to r. We can see that the values of parameter r have little effect on kernel dimensionality reduction methods, and OKMNF gets overall better or comparable accuracy than KMNF and KPCA. To fairly compare different dimensionality reduction methods, we adopt the optimal value of parameter r within the range of requirements when the classification accuracy of hyperspectral images achieves the maximum value. According to our empirical study, in the Indian Pines scene, r of OKMNF, KMNF, and KPCA are all set to 0.0025, and in the Minamimaki Scene, r of KMNF is set to 0.1, and r of OKMNF and KPCA are both set to 0.005.

Another important parameter is the number of subsamples (pixels), m. They were used to derive eigenvectors for data transformation. Figures 7b and 12b show the sensitivity of kernel dimensionality reduction methods (KPCA, KMNF, and OKMNF) with respect to m. We can see that the values of parameter m have little effect on KPCA. To the Indian Pines scene and the Minamimaki scene, the classification accuracy of OKMNF and KMNF both evidently descend when the value of parameter m is greater than 100. However, OKMNF shows lower sensitivity on parameter m than KMNF, and is even better or comparable to KPCA when the value of parameter m is less than 80. We fix the number of the extracted features to see the impact of subsample size on classification. We see the performance decrease, as the number of subsample increases. The reason is that when m increases, more extracted features are required. To reduce the computational time and memory use, we will adopt a small number of subsamples. It is an important empirical rule that can be considered in the applications of OKMNF. Here, we also adopt the optimal value of parameter m within the range of requirements when the classification accuracy of hyperspectral images achieves the maximum value. According to our empirical study, in the Indian Pines scene, m of OKMNF and KPCA are both set to 63, and m of KMNF is set to 42. In the Minamimaki Scene, m of KMNF and KPCA are both set to 30, and m of OKMNF is set to 25.

In this paper, the employed kernel function is the Gaussian radial basis function, which is the same as KPCA, KMNF, and OKMNF [59] The Gaussian radial basis function is defined as

$$\kappa(\mathbf{x}_i, \mathbf{x}_j) = \exp[-||\mathbf{x}_i - \mathbf{x}_j||^2/(2\sigma^2)], \tag{33}$$

where $\mathbf{x}_i$ and $\mathbf{x}_j$ are vectors of observations, $\sigma = s\sigma_0$, σ_0 is the mean distance between the observations in feature space and s is a scale factor [33,37]. Figures 7c and 12c show the sensitivity of KPCA, KMNF, and OKMNF with respect to s. We can see that both OKMNF and KPCA show better performance than KMNF. In the Indian Pines scene, OKMNF performs better than KPCA. Just like above, we adopt the optimal value of parameter s within the range of requirements when the classification accuracy of hyperspectral images achieves the maximum value. According to our empirical study, s of KPCA, KMNF, and OKMNF are set to 35, 1, and 15 for the Indian Pines scene, respectively. Then, for the Minamimaki scene, s of OKMNF is set to 25, and s of KPCA and KMNF are both set to 10.

3.2. Experiments on Noise Estimation Algorithms in KMNF and OKMNF

To assess the performance of noise estimation algorithms adopted in KMNF and OKMNF, six real Airborne Visible/Infrared Imaging Spectrometer (AVIRIS) radiance images with very different land cover types were used in this experiment. These images are shown in Figure 1. Each of them contains 300 × 300 pixels, and covers spectral wavelengths from 400 nm to 2500 nm. Normally, the random noise in AVIRIS sensor images is mainly additive and uncorrelated with the signal [60]. More detailed descriptions are shown in Table 1.

We assess the performance of noise estimation algorithms by computing noise standard deviation, after we get noise data through Algorithm 1. The local standard deviation (LSD) of each sub-block is estimated by

$$LSD = [\frac{1}{w \times h - 4} \sum_{i=1}^{w} \sum_{j=1}^{h} n_{i,j,k}^2]^{\frac{1}{2}} \tag{34}$$

where $w \times h - 4$ means that four parameters are used in the multiple linear regression model and that the degree of freedom is $w \times h - 4$. The LSD of each sub-block is calculated as the noise estimate of that region. The mean value of these LSD is considered as the best estimate of the band noise.

The AVIRIS hyperspectral imageries in Figure 1 were acquired from July 1996 to June 1997. Figure 1a–f are cut from the same image, respectively. Therefore, their noise level should be the same [50].

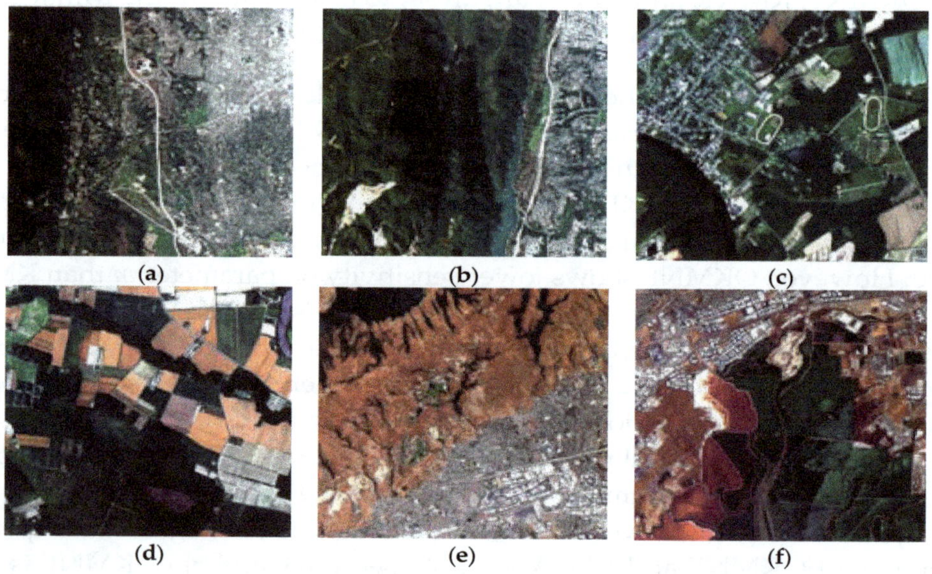

Figure 1. Airborne Visible/Infrared Imaging Spectrometer radiance images used for noise estimation, where (**a**) is the first subimage of Jasper Ridge; (**b**) is the second subimage of Jasper Ridge; (**c**) is the first subimage of Low Altitude; (**d**) is the second subimage of Low Altitude; (**e**) is the first subimage of Moffett Field; and (**f**) is the second subimage of Moffett Field.

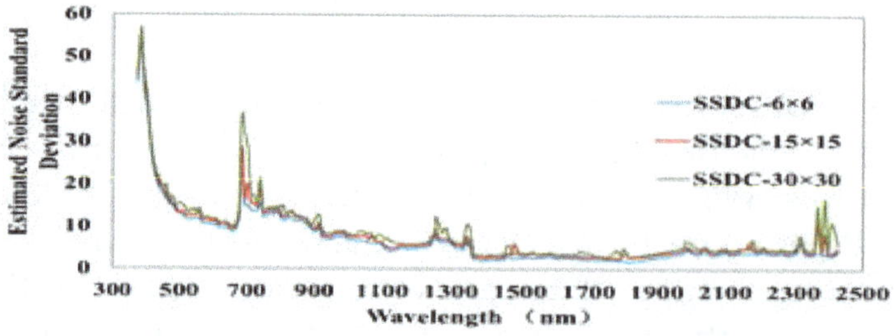

Figure 2. Noise estimation results of spectral and spatial de-correlation (SSDC) of Figure 1a in a different size of sub-block.

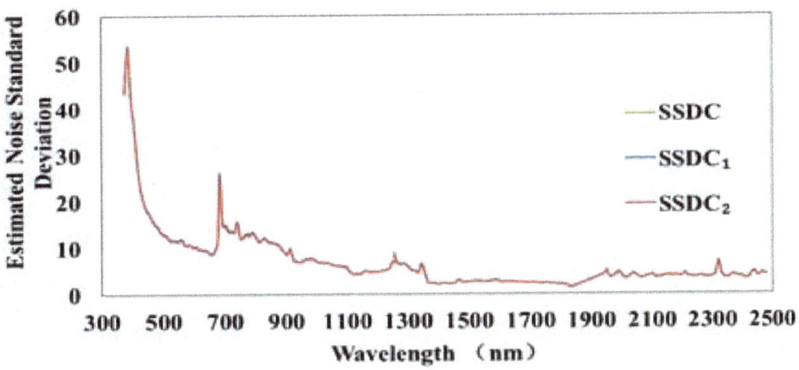

Figure 3. Noise estimation results of SSDC, $SSDC_1$, and $SSDC_2$ of Figure 1a in the 6 × 6 size of sub-block.

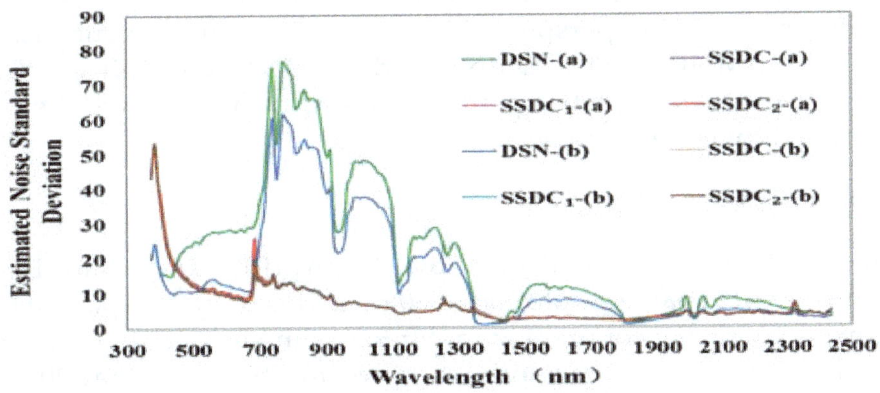

(**a**)

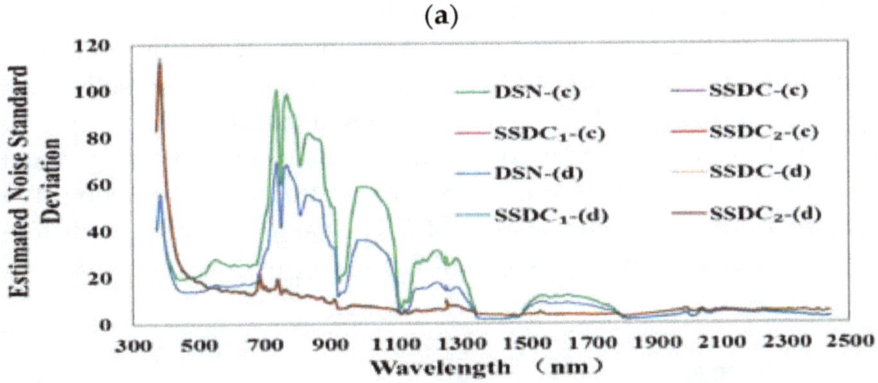

(**b**)

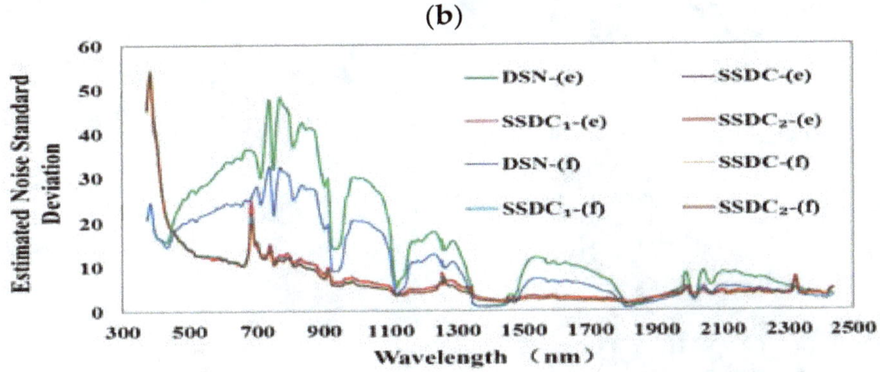

(**c**)

Figure 4. Noise estimation results of (**a**) Figure 1a,b; (**b**) Figure 1c,d; (**c**) Figure 1e,f, through the difference of spatial neighborhood (DSN) used in kernel minimum noise fraction (KMNF), and the SSDC, $SSDC_1$, and $SSDC_2$ used in optimize KMNF (OKMNF).

Table 1. Detailed description of Airborne Visible/Infrared Imaging Spectrometer images shown in Figure 1.

	Spatial Resolution	Acquired Site	Acquired Time	Image Description
(a) (b)	20 m	Jasper Ridge	3 April 1997	Dominated by a heterogeneous city area Dominated by a homogeneous vegetation area
(c) (d)	3.4 m	Low Altitude	5 July 1996	Dominated by a heterogeneous city area Homogeneous farmland
(e) (f)	20 m	Moffett Field	20 June 1997	A mix of a heterogeneous city area and a homogeneous bare soil Dominated by a homogeneous water

3.3. Experiments on Dimensionality Reduction Methods

In these experiments, the dimensionality reduction performance of OKMNF is evaluated in terms of classification results on two real hyperspectral images. Classification accuracies using the features extracted by PCA, KPCA, MNF, KMNF, OMNF, and OKMNF (OKMNF-SSDC, OKMNF-$SSDC_1$, and OKMNF-$SSDC_2$) are compared. Each experiment was run ten times, and the average of these ten experiments was reported for comparisons.

3.3.1. Experiments on the Indian Pines Image

The experimental dataset was collected by the AVIRIS at Indian Pines. The image contains 145 $\times$ 145 pixels with spatial resolution of 20 m, and is with 220 spectral bands from 400 nm to 2500 nm. In this experiment, we compare with different dimensionality reduction methods based on original image including all the 220 bands. It is worth observing that 20 bands covering the region of water absorption are really noisy, thus allowing us to analyze the robustness of the different dimensionality reduction methods to real noise. As shown in Figures 5 and 9, large classes are considered in this experiment. In addition, 25% of samples are randomly selected for training and the others 75% are employed for testing [61,62]. The numbers of training and testing samples are listed in Table 2. The first three features extracted by different dimensionality reduction methods are shown in Figure 8. The overall accuracies of ML classification after different dimensionality reduction methods are shown in Table 3 and Figure 6. The results of ML classification after different dimensionality reduction (number of features = 5) methods are shown in Figure 9.

(a) (b)

Figure 5. (**a**) original Indian Pines image; (**b**) ground reference map containing nine land-cover classes.

Table 2. Training and testing samples used in Indian Pines image.

Classes	Training	Testing
Corn-no till	359	1075
Corn-min till	209	625
Grass/Pasture	124	373
Grass/Trees	187	560
Hay-windrowed	122	367
Soybean-no till	242	726
Soybean-min till	617	1851
Soybean-clean till	154	460
Woods	324	970
Total	2338	7007

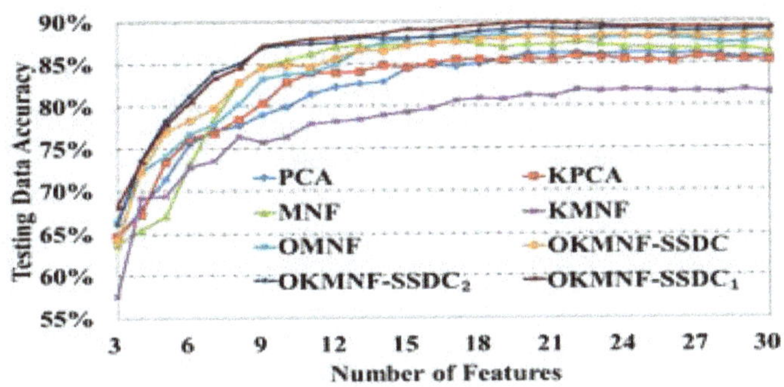

Figure 6. Comparison of accuracies of maximum likelihood-based classification (ML) classification after different dimensionality reduction methods.

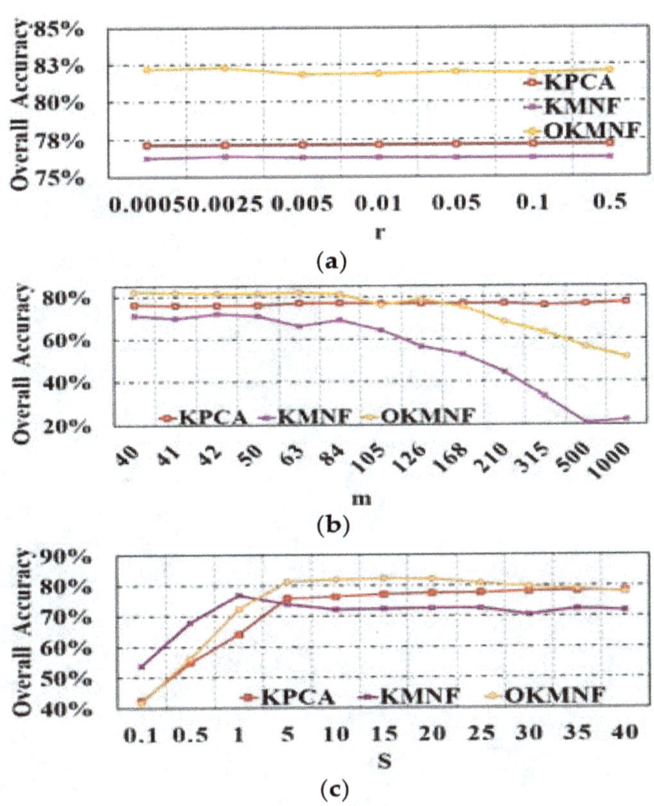

Figure 7. Parameter tuning in the experiments using the Indian Pines dataset for ML classification after different feature extraction methods (number of features = 8), where (**a**) is r versus accuracies; (**b**) is m versus accuracies; (**c**) is s versus accuracies.

Table 3. The overall accuracies of maximum likelihood-based classification (ML) classification after different dimensionality reduction methods.

Number of Features	PCA	KPCA	MNF	KMNF	OMNF	OKMNF-SSDC	OKMNF-SSDC$_1$	OKMNF-SSDC$_2$
3	64.57%	64.84%	63.75%	57.63%	66.50%	64.21%	**68.19%**	66.19%
4	67.90%	67.33%	65.44%	69.14%	72.25%	72.23%	73.34%	**73.60%**
5	71.35%	73.41%	67.14%	69.36%	74.15%	76.93%	77.74%	**78.29%**
6	75.60%	76.23%	73.09%	72.76%	76.69%	78.31%	80.48%	**81.22%**
7	77.08%	76.81%	78.43%	73.53%	77.88%	79.73%	83.35%	**84.07%**
8	77.65%	78.45%	82.76%	76.39%	80.21%	82.86%	84.56%	**85.03%**
9	79.01%	80.32%	84.74%	75.71%	83.27%	84.59%	**87.21%**	86.93%
10	79.92%	82.82%	85.43%	76.35%	83.84%	84.87%	**87.56%**	87.26%
11	81.40%	83.96%	86.16%	77.88%	83.97%	84.69%	**87.94%**	87.44%
12	82.27%	83.96%	86.93%	78.18%	84.96%	85.66%	**88.17%**	87.60%
13	82.67%	84.10%	87.15%	78.42%	86.61%	86.63%	**88.33%**	88.13%
14	82.90%	84.84%	87.08%	78.94%	87.57%	86.50%	**88.63%**	88.04%
15	84.49%	84.54%	87.33%	79.31%	87.95%	87.17%	**89.10%**	88.04%
16	84.87%	85.03%	87.48%	79.72%	88.30%	87.43%	**89.04%**	88.17%
17	84.72%	85.50%	87.55%	80.66%	88.05%	87.64%	**89.37%**	88.41%
18	85.02%	85.50%	87.34%	80.89%	88.28%	87.91%	**89.51%**	88.84%
19	85.50%	85.37%	86.91%	80.78%	88.47%	87.98%	**89.68%**	89.00%
20	86.16%	85.59%	87.27%	81.25%	88.25%	88.20%	**89.82%**	89.30%
21	86.21%	85.41%	87.21%	81.13%	88.37%	88.24%	**89.77%**	89.14%
22	86.23%	85.89%	87.57%	81.88%	88.10%	88.01%	**89.64%**	89.03%
23	86.00%	85.76%	87.28%	81.76%	88.00%	88.30%	**89.55%**	89.15%
24	86.24%	85.49%	86.97%	81.88%	88.17%	88.28%	**89.35%**	89.14%
25	86.27%	85.40%	86.87%	81.82%	88.08%	88.23%	**89.42%**	89.14%
26	86.06%	85.30%	86.74%	81.66%	88.11%	88.34%	**89.34%**	88.89%
27	86.27%	85.84%	86.76%	81.72%	87.85%	88.20%	**89.28%**	88.82%
28	85.96%	85.59%	86.84%	81.60%	87.43%	88.27%	**89.27%**	88.88%
29	85.71%	85.50%	86.80%	81.92%	87.50%	88.25%	**89.28%**	88.91%
30	85.89%	85.39%	86.31%	81.60%	87.87%	88.24%	**89.23%**	88.97%

PCA: principal component analysis; KPCA: kernel PCA; MNF: minimum noise fraction; KMNF: kernel minimum noise fraction; OMNF: optimized MNF; OKMNF: optimized kernel minimum noise fraction.

KPCA KMNF OKMNF-SSDC OKMNF-SSDC$_1$ OKMNF-SSDC$_2$

Figure 8. The first three features (from up to bottom) of kernel PCA (KPCA), KMNF, OKMNF-SSDC, OKMNF-SSDC$_1$, and OKMNF-SSDC$_2$.

Figure 9. The results of ML classification after different dimensionality reduction methods (number of features = 5).

3.3.2. Experiments on the Minamimaki Scene

This scene was collected by the Pushbroom Hyperspectral Imager (PHI) sensor over Minamimaki, Japan. The PHI sensor was developed by the Shanghai Institute of Technical Physics of the Chinese Academy of Sciences, China. The data has 200 × 200 pixels with a spatial resolution of 3 m and 80 spectral bands from 400 nm to 850 nm. As shown in Figure 10, this image has six classes. About 10% of samples per class were randomly selected for training and the other 90% were employed for testing. The numbers of training and testing samples are listed in Table 4. The overall accuracies of ML classification after different dimensionality reduction methods are shown in Table 5 and Figure 11. The results of ML classification (number of features = 3) are shown in Figure 13.

Figure 10. (**a**) true color image of the Minamimaki scene; (**b**) ground reference map with 6 classes.

Table 4. Training and testing samples used in the Minamimaki scene.

Classes	Training	Testing
Bare soil	1238	11,150
plastic	33	300
Chinese cabbage	29	245
forest	111	1000
Japanese cabbage	425	3830
pasture	20	153
Total	1856	16,678

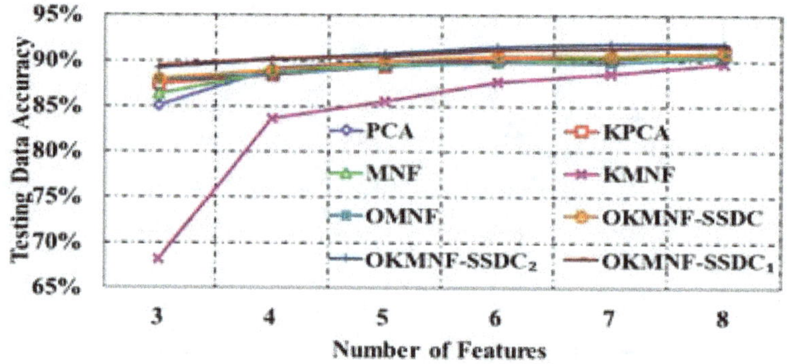

Figure 11. Comparison of accuracies of ML classification after different dimensionality reduction methods.

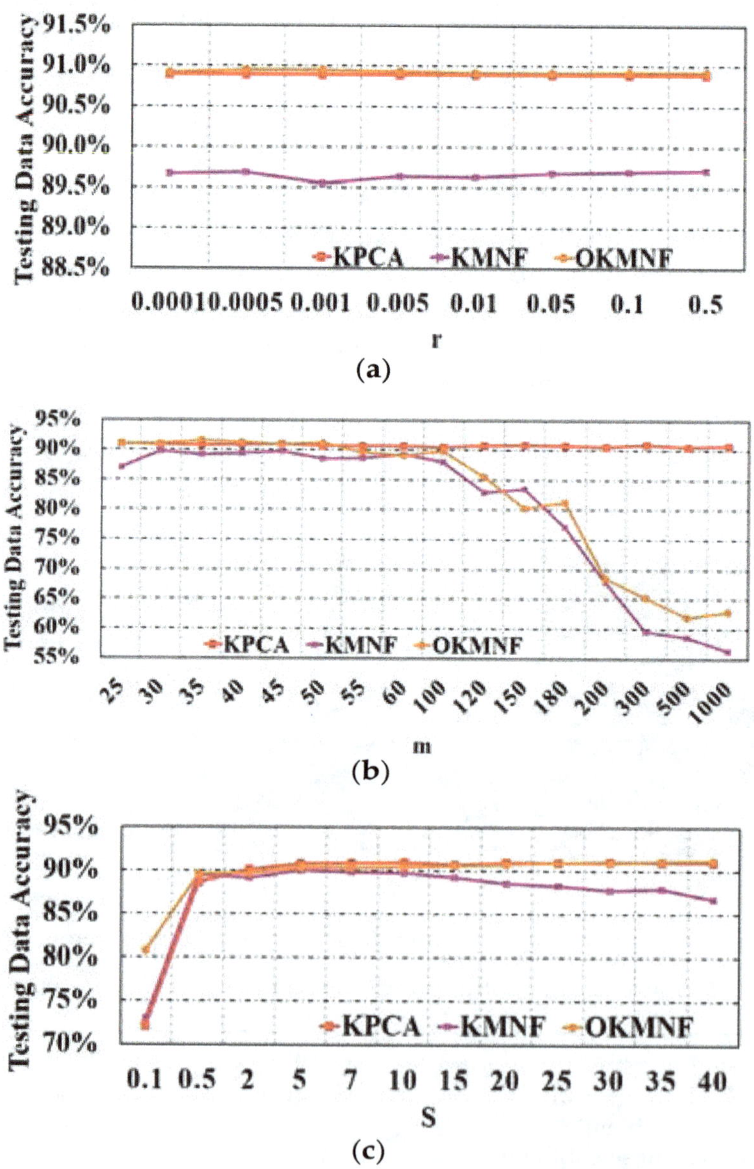

Figure 12. Parameter tuning in experiments using the Minamimaki dataset for ML classification after different dimensionality methods (number of features = 8), where (**a**) is r versus accuracies; (**b**) is m versus accuracies; (**c**) i s s versus accuracies.

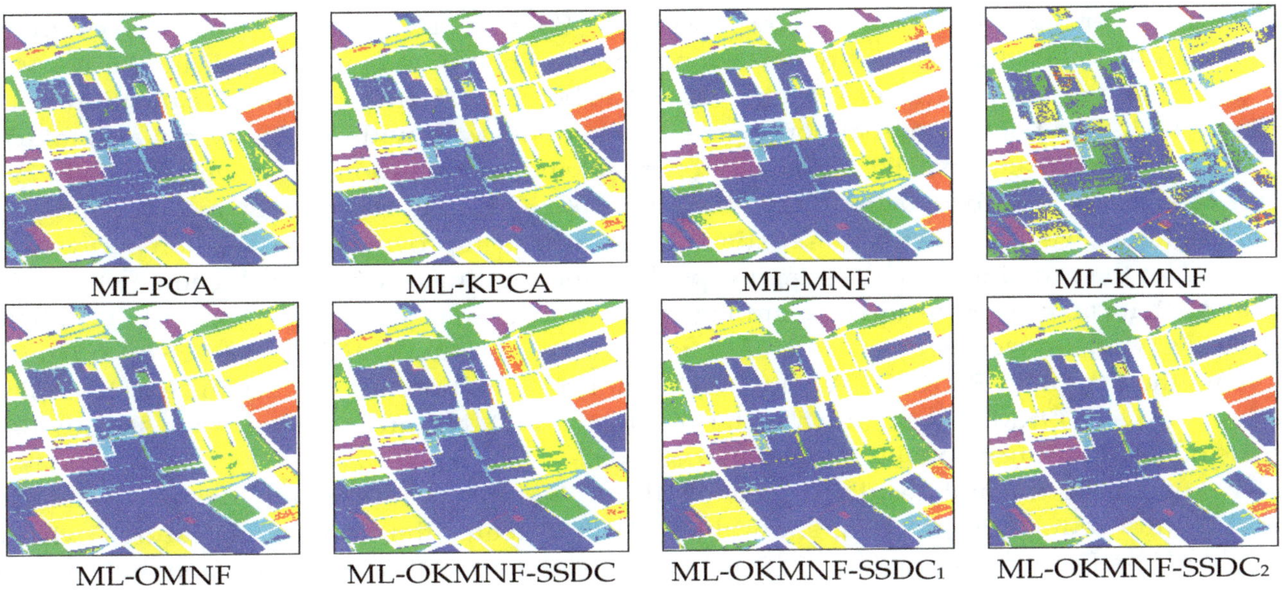

Figure 13. The results of ML classification after different dimensionality reduction methods (number of features = 3).

Table 5. The overall accuracies of ML classification after different dimensionality reduction methods.

Method	**Number of Features**					
	3	**4**	**5**	**6**	**7**	**8**
PCA	85.14%	89.13%	89.86%	90.17%	90.44%	90.75%
KPCA	87.43%	88.41%	89.46%	90.19%	90.22%	90.87%
MNF	86.34%	88.73%	89.48%	89.69%	90.32%	90.59%
KMNF	68.30%	83.81%	86.02%	87.69%	88.61%	89.66%
OMNF	87.82%	88.51%	89.32%	90.10%	89.88%	90.51%
OKMNF-SSDC	88.10%	88.94%	89.98%	90.60%	90.63%	90.97%
OKMNF-SSDC$_1$	**89.46%**	**90.18%**	90.44%	91.19%	91.39%	91.68%
OKMNF-SSDC$_2$	89.24%	90.17%	**90.78%**	**91.56%**	**91.88%**	**91.89%**

4. Discussion

This section discusses the performances of noise estimation algorithms, and these results are shown in Section 3.2. In addition, the results of the dimensionality reduction methods are shown in Section 3.3.

Based on the experiment of assessing the performance of noise estimation algorithms adopted in KMNF and OKMNF, it can be seen in Figure 4 that the estimated noise curves through the difference of spatial neighborhood used in KMNF show a strong relationship with land cover types in the scene, and the noise levels are not the same for the two subimages from the same image. There are no such problems when the noise is estimated by OKMNF through SSDC, SSDC$_1$ and SSDC$_2$. We can see that SSDC, SSDC$_1$ and SSDC$_2$ are more reliable noise estimation methods than that used in KMNF. Thus, we can adopt SSDC, SSDC$_1$ and SSDC$_2$ to estimate noise for OKMNF.

Based on the experiment of assessing the performance of dimensionality reduction methods from Section 3.3.1, it can be seen in Figure 8 that the feature quality of KMNF is worse than other dimensionality reduction methods. OKMNF, by considering SSDC, SSDC$_1$ or SSDC$_2$ for noise estimation, outperforms the other dimensionality reduction methods. It can be seen in Table 3, and Figures 6 and 9 that the classification results using transformed data by MNF are not always better than those of PCA on low dimension space. KMNF performs worse than KPCA. By considering the spectral and spatial de-correlation for noise estimation, linear OMNF always performs better than PCA and mostly better than MNF. OKMNF, by considering SSDC, $SSDC_1$ or $SSDC_2$ for noise estimation, outperforms the other dimensionality reduction methods (including linear OMNF and kernel MNF), with less sensitivity for parameter settings, as well as better performances for classification. This is because OKMNF not only can treat nonlinear characteristics well within the data but also take into account both spectral and spatial correlations for reliable noise estimation. Moreover, OKMNF-$SSDC_1$ and OKMNF-$SSDC_2$ perform better than OKMNF-SSDC. This indicates that, by incorporating more spatial neighbors, we enable better noise estimation, as well as improve the classification performances.

Based on the experiment of assessing the performance of dimensionality reduction methods from Section 3.3.2, it can be seen in Table 5, and Figures 11 and 13 that the performances of PCA, KPCA, MNF, and OMNF are very similar, and all of them are better than KMNF. When we optimized the KMNF method through SSDC, $SSDC_1$ and $SSDC_2$ noise estimation, the performance of KMNF was greatly improved. OKMNF gets much better results than KMNF, and also performs slightly better than the other four dimensionality reduction methods.

The two experimental results, based on the experiment of assessing the performance of dimensionality reduction methods, show that: (1) the greater the number of features extracted, the higher classification accuracy is; (2) it is better not to use KMNF for dimensionality reduction in many cases, the overall accuracies of ML classification after KMNF are lower than MNF and other dimensionality reduction methods; (3) our proposed OKMNF, OKMNF-SSDC, OKMNF-$SSDC_1$, and OKMNF-$SSDC_2$ perform much better than KMNF and mostly better than OMNF and MNF. These results imply that the dimensionality reduction results of KMNF are not suitable for image classification. By exploiting both spectral and spatial information for noise estimation, the proposed OKMNF benefits both dimensionality reduction and its post applications (e.g., classification). Compared to linear MNF, the proposed OKMNF not only has good performance in dimensionality reduction for classification but also does better in dealing with nonlinear problems.

To compare the efficiency of feature extraction methods, we took Indian Pines data as an example, and the consumed time (by extracting 30 features) of OKMNF-SSDC, OKMNF-$SSDC_1$, OKMNF-$SSDC_2$, KPCA, KMNF, OMNF, MNF, and PCA are 23.07 s, 25.27 s, 22.80 s, 1.03 s, 1.26 s, 22.87 s, 0.52 s and 0.20 s, respectively. We can find that the proposed OKMNF (OKMNF-SSDC, OKMNF-$SSDC_1$, OKMNF-$SSDC_2$) methods consume comparatively longer time but with better dimensionality reduction performances. However, we can use high performance computing techniques such as graphics processing unit to reduce the processing time of OKMNF. In real applications, the number of features kept for classification should be determined for both classification performance and computing cost. Too few features may not provide adequate class separability. On the other hand, more features might not always bring higher classification accuracy, which can be seen from the results listed in Table 3. It is important to use as few features as possible to avoid overfitting and minimise computational load.

5. Conclusions

This paper proposes an optimized KMNF for dimensionality reduction of hyperspectral imagery. The main reason affecting the original KMNF in dimensionality reduction is the larger error and the instability in estimating noise. Here, we conduct a comparative study for noise estimation algorithms using real images with different land cover types. The experimental results show that the combined spatial and spectral correlation information provides better results than the algorithms only using spatial neighborhood information. OKMNF adopts SSDC, $SSDC_1$, and $SSDC_2$ to stably estimate noise from hyperspectral images. Through this optimization, the overall accuracies of ML classification after OKMNF are much higher than those of KMNF, and the dimensionality reduction results of OKMNF are also better than OMNF, MNF, KPCA, and PCA in most situations. It can be concluded that OKMNF solves the problems existing in original KMNF well and improves the quality of dimensionality reduction. Moreover, OKMNF is valuable to reduce the dimensionality of nonlinear data. We can also expect that OKMNF will enhance the separability among endmember classes and improve the quality of spectral unmixing. Our future work will focus on incorporating more validations on other applications (e.g., target detection).

Acknowledgments: This research was supported by the National Natural Science Foundation of China under Grant No. 41571349, No. 91638201 and No. 41325004.

Author Contributions: Lianru Gao contributed to design the theoretical framework for the proposed methods and to the experimental analysis. Bin Zhao was primarily responsible for mathematical modeling and experimental design. Xiuping Jia improved the mathematical model and revised the paper. Wenzhi Liao provided important suggestions for improving technical quality of the paper. Bing Zhang proposed the original idea of the paper.

References

1. Goetz, A.F.H. Three decades of hyperspectral remote sensing of the Earth: A personal view. *Remote Sens. Environ.* **2009**, *113*, S5–S16. [CrossRef]
2. Plaza, A.; Benediktsson, J.A.; Boardman, J.W.; Brazile, J.; Bruzzone, L.; Camps-Valls, G.; Chanussot, J.; Fauvel, M.; Gamba, P.; Gualtieri, A.; et al. Recent advances in techniques for hyperspectral image processing. *Remote Sens. Environ.* **2009**, *113*, S110–S122. [CrossRef]
3. Hughes, G. On the mean accuracy of statistical pattern recognizers. *IEEE Trans. Inf. Theory* **1968**, *14*, 55–63. [CrossRef]
4. Liu, C.H.; Zhou, J.; Liang, J.; Qian, Y.T.; Li, H.X.; Gao, Y.S. Exploring structural consistency in graph regularized joint spectral-spatial sparse coding for hyperspectral image classification. *IEEE J. Sel. Top. Appl. Earth Obs. Remote Sens.* **2016**, *99*, 1–14. [CrossRef]
5. Jia, X.P.; Kuo, B.; Crawford, M.M. Feature mining for hyperspectral image classification. *Proc. IEEE* **2013**, *101*, 676–697.
6. Benediktsson, J.A.; Palmason, J.A.; Sveinsson, J.R. Classification of hyperspectral data from urban areas based on extended morphological profiles. *IEEE Trans. Geosci. Remote Sens.* **2005**, *43*, 480–491. [CrossRef]
7. Qian, Y.T.; Yao, F.T.; Jia, S. Band selection for hyperspectral imagery using affinity propagation. *IET Comput. Vis.* **2010**, *3*, 213–222. [CrossRef]
8. Falco, N.; Benediktsson, J.A.; Bruzzone, L. A study on the effectiveness of different independent component analysis algorithms for hyperspectral image classification. *IEEE J. Sel. Top. Appl. Earth Obs. Remote Sens.* **2014**, *7*, 2183–2199. [CrossRef]
9. Zabalza, J.; Ren, J.C.; Wang, Z.; Zhao, H.M.; Marshall, S. Fast implementation of singular spectrum analysis for effective feature extraction in hyperspectral imaging. *IEEE J. Sel. Top. Appl. Earth Obs. Remote Sens.* **2015**, *8*, 2845–2853. [CrossRef]
10. Xie, J.Y.; Hone, K.; Xie, W.X.; Gao, X.B.; Shi, Y.; Liu, X.H. Extending twin support vector machine classifier for multi-category classification problems. *Intell. Data Anal.* **2013**, *17*, 649–664.
11. Chen, W.S.; Huang, J.; Zou, J.; Fang, B. Wavelet-face based subspace LDA method to solve small sample size problem in face recognition. *Int. J. Wavelets Multiresolut. Inf. Process.* **2009**, *7*, 199–214. [CrossRef]
12. Gu, Y.F.; Feng, K. Optimized laplacian SVM with distance metric learning for hyperspectral image classification. *IEEE J. Sel. Top. Appl. Earth Obs. Remote Sens.* **2013**, *6*, 1109–1117. [CrossRef]
13. Ma, A.L.; Zhong, Y.F.; Zhao, B.; Jiao, H.Z.; Zhang, L.P. Semisupervised subspace-based DNA encoding and matching classifier for hyperspectral remote sensing imagery. *IEEE Trans. Geosci. Remote Sens.* **2016**, *54*, 4402–4418. [CrossRef]
14. Zhang, L.F.; Zhang, L.P.; Tao, D.C.; Huang, X. On combining multiple features for hyperspectral remote sensing image classification. *IEEE Trans. Geosci. Remote Sens.* **2012**, *50*, 879–893. [CrossRef]
15. Melgani, F.; Bruzzone, L. Classification of hyperspectral remote sensing images with support vector machines. *IEEE Trans. Geosci. Remote Sens.* **2004**, *42*, 1778–1790. [CrossRef]
16. Harsanyi, J.; Chang, C. Hyperspectral image classification and dimensionality reduction: An orthogonal subspace projection approach. *IEEE Trans. Geosci. Remote Sens.* **1994**, *32*, 779–785. [CrossRef]
17. Wang, Q.; Lin, J.; Yuan, Y. Salient band selection for hyperspectral image classification via manifold ranking. *IEEE Trans. Neural Netw. Learn. Syst.* **2016**, *27*, 1279–1289. [CrossRef] [PubMed]
18. Yuan, Y.; Lin, J.; Wang, Q. Dual clustering based hyperspectral band selection by contextual analysis. *IEEE Trans. Geosci. Remote Sens.* **2016**, *54*, 1431–1445. [CrossRef]

19. Yuan, Y.; Lin, J.; Wang, Q. Hyperspectral image classification via multi-task joint sparse representation and stepwise MRF optimization. *IEEE Trans. Cybern.* **2016**, *46*, 2966–2977.
20. Fukunaga, K. *Introduction to Statistical Pattern Recognition*; Academic Press: Cambridge, MA, USA, 1990.
21. Kuo, B.C.; Landgrebe, D.A. Nonparametric weighted feature extractionfor classification. *IEEE Trans. Geosci. Remote Sens.* **2004**, *42*, 1096–1105.
22. Ly, N.; Du, Q.; Fowler, J.E. Sparse graph-based discriminant analysis for hyperspectral imagery. *IEEE Trans. Geosci. Remote Sens.* **2014**, *52*, 3872–3884.
23. Ly, N.; Du, Q.; Fowler, J.E. Collaborative graph-based discriminant analysis for hyperspectral imagery. *IEEE J. Sel. Top. Appl. Earth Obs. Remote Sens.* **2014**, *7*, 2688–2696. [CrossRef]
24. Xue, Z.H.; Du, P.J.; Li, J.; Su, H.J. Simultaneous sparse graph embedding for hyperspectral image classification. *IEEE Trans. Geosci. Remote Sens.* **2015**, *53*, 6114–6133. [CrossRef]
25. Roger, E.R. Principal components transform with simple, automatic noise adjustment. *Int. J. Remote Sens.* **1996**, *17*, 2719–2727. [CrossRef]
26. Green, A.A.; Berman, M.; Switzer, P.; Craig, M.D. A transformation for ordering multispectral data in terms of image quality with implications for noise removal. *IEEE Trans. Geosci. Remote Sens.* **1998**, *26*, 65–74. [CrossRef]
27. Chen, P.H.; Jiao, L.C.; Liu, F.; Gou, S.P.; Zhao, J.Q.; Zhao, Z.Q. Dimensionality reduction of hyperspectral imagery using sparse graph learning. *IEEE J. Sel. Top. Appl. Earth Obs. Remote Sens.* **2016**, *99*, 1–17. [CrossRef]
28. Lee, J.B.; Woodyatt, S.; Berman, M. Enhancement of high spectral resolution remote-sensing data by a noise-adjusted principal components transform. *IEEE Trans. Geosci. Remote Sens.* **1990**, *28*, 295–304. [CrossRef]
29. Bachmann, C.M.; Ainsworth, T.L.; Fusina, R.A. Exploiting manifold geometry in hyperspectral imagery. *IEEE Trans. Geosci. Remote Sens.* **2005**, *43*, 441–454.
30. Mohan, A.; Sapiro, G.; Bosch, E. Spatially coherent nonlinear dimensionality reduction and segmentation of hyperspectral images. *IEEE Geosci. Remote Sens. Lett.* **2007**, *4*, 206–210. [CrossRef]
31. Nielsen, A.A. Kernel maximum autocorrelation factor and minimum noise fraction transformations. *IEEE Trans. Image Process.* **2011**, *20*, 612–624. [CrossRef] [PubMed]
32. Gomez-Chova, L.; Nielsen, A.A.; Camps-Valls, G. Explicit signal to noise ratio in reproducing kernel hilbert spaces. In Proceedings of the IEEE International Geoscience and Remote Sensing Symposium (IGARSS), Vancouver, BC, Canada, 24–29 July 2011; pp. 3570–3573.
33. Nielsen, A.A.; Vestergaard, J.S. Parameter optimization in the regularized kernel minimum noise fraction transformation. In Proceedings of the IEEE International Geoscience and Remote Sensing Symposium (IGARSS), Munich, Germany, 22–27 July 2012; pp. 370–373.
34. Shawe-Taylor, J.; Cristianini, N. *Kernel Methods for Pattern Analysis*; Cambridge University Press: New York, NY, USA, 2004.
35. Li, W.; Prasad, S.; Fowler, J.E. Decision fusion in kernel-induced spaces for hyperspectral image classification. *IEEE Trans. Geosci. Remote Sens.* **2014**, *52*, 3399–3411. [CrossRef]
36. Li, W.; Prasad, S.; Fowler, J.E.; Bruce, L.M. Locality preserving dimensionality reduction and classification for hyperspectral image analysis. *IEEE Trans. Geosci. Remote Sens.* **2012**, *50*, 1185–1198. [CrossRef]
37. Li, W.; Prasad, S.; Fowler, J.E.; Bruce, L.M. Locality-preserving discriminant analysis in kernel-induced feature spaces for hyperspectral image classification. *IEEE Geosci. Remote Sens. Lett.* **2011**, *8*, 894–898. [CrossRef]
38. Zhang, Y.H.; Prasad, S. Locality preserving composite kernel feature extraction for multi-source geospatial image analysis. *IEEE J. Sel. Top. Appl. Earth Obs. Remote Sens.* **2015**, *8*, 1385–1392. [CrossRef]
39. Kuo, B.C.; Ho, H.H.; Li, C.H.; Huang, C.C.; Taur, J.S. A kernel-based feature selection method for SVM with RBF kernel for hyperspectral image classification. *IEEE J. Sel. Top. Appl. Earth Obs. Remote Sens.* **2014**, *7*, 317–326.
40. Schölkopf, B.; Smola, A.; Müller, K.-R. Nonlinear component analysis as a kernel eigenvalue problem. *Neural Comput.* **1998**, *10*, 1299–1319. [CrossRef]
41. Gao, L.R.; Zhang, B.; Chen, Z.C.; Lei, L.P. Study on the issue of noise estimation in dimension reduction of hyperspectral images. In Proceedings of the IEEE Workshop on Hyperspectral Image and Signal Processing: Evolution in Remote Sensing (WHISPERS), Lisbon, Portugal, 6–9 June 2011; pp. 1–4.

42. Zhao, B.; Gao, L.R.; Zhang, B. An optimized method of kernel minimum noise fraction for dimensionality reduction of hyperspectral imagery. In Proceedings of the IEEE International Geoscience and Remote Sensing Symposium (IGARSS), Beijing, China, 10–15 July 2016; pp. 48–51.
43. Zhao, B.; Gao, L.R.; Liao, W.Z.; Zhang, B. A new kernel method for hyperspectral image feature extraction. *Geo-Spat. Inf. Sci.* **2017**, *99*, 1–11.
44. Bioucas-Dias, J.M.; Plaza, A.; Dobigeon, N.; Parente, M.; Du, Q.; Gader, P.; Chanussot, J. Hyperspectral unmixing overview: Geometrical, statistical, and sparse regression-based approaches. *IEEE J. Sel. Top. Appl. Earth Obs. Remote Sens.* **2012**, *5*, 354–379. [CrossRef]
45. Jia, S.; Xie, Y.; Tang, G.H.; Zhu, J.S. Spatial-spectral-combined sparse representation-based classification for hyperspectral imagery. *Soft Comput.* **2014**, 1–10. [CrossRef]
46. Chen, C.; Li, W.; Tramel, E.W.; Cui, M.S.; Prasad, S.; Fowler, J.E. Spectral–spatial preprocessing using multihypothesis prediction for noise-robust hyperspectral image classification. *IEEE J. Sel. Top. Appl. Earth Obs. Remote Sens.* **2014**, *7*, 1047–1059. [CrossRef]
47. Ghamisi, P.; Benediktsson, J.A.; Jon, A.; Cavallaro, G.; Plaza, A. automatic framework for spectral–spatial classification based on supervised feature extraction and morphological attribute profiles. *IEEE J. Sel. Top. Appl. Earth Obs. Remote Sens.* **2014**, *7*, 2147–2160. [CrossRef]
48. Gao, L.R.; Du, Q.; Yang, W.; Zhang, B. A comparative study on noise estimation for hyperspectral imagery. In Proceedings of the IEEE Workshop on Hyperspectral Image and Signal Processing: Evolution in Remote Sensing (WHISPERS), Shanghai, China, 4–7 June2012; pp. 1–4.
49. Roger, E.R.; Arnold, F.J. Reliably estimating the noise in AVIRIS hyperspectral images. *Int. J. Remote Sens.* **1996**, *17*, 1951–1962. [CrossRef]
50. Gao, L.R.; Du, Q.; Zhang, B.; Yang, W.; Wu, Y.F. A comparative study on linear regression-based noise estimation for hyperspectral imagery. *IEEE J. Sel. Top. Appl. Earth Obs. Remote Sens.* **2013**, *6*, 488–498. [CrossRef]
51. Gao, L.R.; Zhang, B.; Sun, X.; Li, S.S.; Du, Q.; Wu, C.S. Optimized maximum noise fraction for dimensionality reduction of Chinese HJ-1A hyperspectral data. *EURASIP J. Adv. Signal Process.* **2013**, *1*, 1–12. [CrossRef]
52. Gao, L.R.; Zhang, B.; Zhang, X.; Zhang, W.J.; Tong, Q.X. A new operational method for estimating noise in hyperspectral images. *IEEE Geosci. Remote Sens. Lett.* **2008**, *5*, 83–87. [CrossRef]
53. Liu, X.; Zhang, B.; Gao, L.R.; Chen, D.M. A maximum noise fraction transform with improved noise estimation for hyperspectral images. *Sci. China Ser. F Inf. Sci.* **2009**, *52*, 1578–1587. [CrossRef]
54. Landgrebe, D.A.; Malaret, E. Noise in remote-sensing systems: The effect on classification error. *IEEE Trans. Geosci. Remote Sens.* **1986**, *24*, 294–299. [CrossRef]
55. Corner, B.R.; Narayanan, R.M.; Reichenbach, S.E. Noise estimation in remote sensing imagery using data masking. *Int. J. Remote Sens.* **2003**, *24*, 689–702. [CrossRef]
56. Gao, B.-C. An operational method for estimating signal to noise ratios from data acquired with imaging spectrometers. *Remote Sens. Environ.* **1993**, *43*, 23–33. [CrossRef]
57. Documentation for Minimum Noise Fraction Transformations. Available online: http://people.compute.dtu.dk/alan/software.html (accessed on 31 March 2017).
58. Wu, Y.F.; Gao, L.R.; Zhang, B.; Zhao, H.N.; Li, J. Real-time implementation of optimized maximum noise fraction transform for feature extraction of hyperspectral images. *J. Appl. Remote Sens.* **2014**, *8*, 1–16. [CrossRef]
59. Camps-Valls, G.; Bruzzone, L. Kernel-based methods for hyperspectral image classification. *IEEE Trans. Geosci. Remote Sens.* **2005**, *43*, 1351–1362. [CrossRef]
60. Acito, N.; Diani, M.; Corsini, G. Signal-dependent noise modeling andmodel parameter estimation in hyperspectral images. *IEEE Trans. Geosci. Remote Sens.* **2011**, *49*, 2957–2971. [CrossRef]
61. Zhang, X.J.; Xu, C.; Li, M.; Sun, X.L. Sparse and low-rank coupling image segmentation model via nonconvex regularization. *Int. J. Pattern Recognit. Artif. Intell.* **2015**, *29*, 1–22. [CrossRef]
62. Zhu, Z.X.; Jia, S.; He, S.; Sun, Y.W.; Ji, Z.; Shen, L.L. Three-dimensional Gabor feature extraction for hyperspectral imagery classification using a memetic framework. *Inf. Sci.* **2015**, *298*, 274–287. [CrossRef]

PERMISSIONS

All chapters in this book were first published in MDPI; hereby published with permission under the Creative Commons Attribution License or equivalent. Every chapter published in this book has been scrutinized by our experts. Their significance has been extensively debated. The topics covered herein carry significant findings which will fuel the growth of the discipline. They may even be implemented as practical applications or may be referred to as a beginning point for another development.

The contributors of this book come from diverse backgrounds, making this book a truly international effort. This book will bring forth new frontiers with its revolutionizing research information and detailed analysis of the nascent developments around the world.

We would like to thank all the contributing authors for lending their expertise to make the book truly unique. They have played a crucial role in the development of this book. Without their invaluable contributions this book wouldn't have been possible. They have made vital efforts to compile up to date information on the varied aspects of this subject to make this book a valuable addition to the collection of many professionals and students.

This book was conceptualized with the vision of imparting up-to-date information and advanced data in this field. To ensure the same, a matchless editorial board was set up. Every individual on the board went through rigorous rounds of assessment to prove their worth. After which they invested a large part of their time researching and compiling the most relevant data for our readers.

The editorial board has been involved in producing this book since its inception. They have spent rigorous hours researching and exploring the diverse topics which have resulted in the successful publishing of this book. They have passed on their knowledge of decades through this book. To expedite this challenging task, the publisher supported the team at every step. A small team of assistant editors was also appointed to further simplify the editing procedure and attain best results for the readers.

Apart from the editorial board, the designing team has also invested a significant amount of their time in understanding the subject and creating the most relevant covers. They scrutinized every image to scout for the most suitable representation of the subject and create an appropriate cover for the book.

The publishing team has been an ardent support to the editorial, designing and production team. Their endless efforts to recruit the best for this project, has resulted in the accomplishment of this book. They are a veteran in the field of academics and their pool of knowledge is as vast as their experience in printing. Their expertise and guidance has proved useful at every step. Their uncompromising quality standards have made this book an exceptional effort. Their encouragement from time to time has been an inspiration for everyone.

The publisher and the editorial board hope that this book will prove to be a valuable piece of knowledge for researchers, students, practitioners and scholars across the globe.

LIST OF CONTRIBUTORS

Wei Cui, Fei Wang, Xin He, Dongyou Zhang, Xuxiang Xu, Meng Yao, Ziwei Wang and Jiejun Huang
School of Resources and Environmental Engineering, Wuhan University of Technology, Wuhan 430070, China

Maofan Zhao, Die Hu and Ying Zhang
Aerospace Information Research Institute, Chinese Academy of Sciences, Beijing 100101, China
University of Chinese Academy of Sciences, Beijing 101400, China
Sanya Institute of Remote Sensing, Sanya 572029, China

Qingyan Meng and Linlin Zhang
Aerospace Information Research Institute, Chinese Academy of Sciences, Beijing 100101, China
Sanya Institute of Remote Sensing, Sanya 572029, China

Mona Allam
Aerospace Information Research Institute, Chinese Academy of Sciences, Beijing 100101, China
Sanya Institute of Remote Sensing, Sanya 572029, China
Environment & Climate Changes Research Institute, National Water Research Center, El Qanater El khairiya 13621/5, Egypt

Yu Meng and Qingqing Huang
Institute of Remote Sensing and Digital Earth, Chinese Academy of Sciences, Beijing 100101, China

Xiaoyi Liu and Lei Lin
Institute of Remote Sensing and Digital Earth, Chinese Academy of Sciences, Beijing 100101, China
University of Chinese Academy of Sciences, Beijing 100049, China

Hichem Sahli
Department of Electronics and Informatics, Vrije Universiteit Brussel, 1050 Brussels, Belgium
Interuniversity Microelectronics Centre (IMEC), 3001 Heverlee, Belgium

Alim Samat and Jilili Abuduwaili
State Key Laboratory of Desert and Oasis Ecology, Xinjiang Institute of Ecology and Geography, Chinese Academy of Sciences, Urumqi 830011, China
CAS Research Center for Ecology and Environment of Central Asia, Chinese Academy of Sciences, Urumqi 830011, China

Claudio Persello
Department of Earth Observation Science, Faculty of Geo-Information Science and Earth Observation (ITC), University of Twente, 7500 AE Enschede, The Netherlands

Paolo Gamba
Department of Electrical, Computer and Biomedical Engineering, University of Pavia, 27100 Pavia, Italy

Sicong Liu
College of Surveying and Geoinformatics, Tongji University, Shanghai 200092, China

Erzhu Li
Department of Geographical Information Science, Nanjing University, Nanjing 210000, China

Yuanyuan Liu, Xuguo Shi and Zhuo Zheng
Department of Information Engineering, China University of Geosciences, Wuhan 430075, China

Xi Gong and Zhong Xie
Department of Information Engineering, China University of Geosciences, Wuhan 430075, China
National Engineering Research Center of Geographic Information System, Wuhan 430075, China

Yifei Xue and Tiejun Wang
Faculty of Geo-Information Science and Earth Observation (ITC), University of Twente, 7500 AE Enschede, The Netherlands

Andrew K. Skidmore
Faculty of Geo-Information Science and Earth Observation (ITC), University of Twente, 7500 AE Enschede, The Netherlands
Department of Environmental Science, Macquarie University, Sydney, NSW 2109, Australia

Lei Wang, K. Andrea Scott and David A. Clausi
Department of Systems Design Engineering, University of Waterloo, Waterloo, ON N2L 3G1, Canada

Gang Fu
Department of Engineering Physics, Tsinghua University, Beijing 100084, China
China Institute of Water Resources and Hydropower Research (IWHR), Beijing 100038, China

Changjun Liu and Tao Sun
China Institute of Water Resources and Hydropower Research (IWHR), Beijing 100038, China

Rong Zhou
Beijing Soil and Water Conservation Center, Beijing 100036, China

Qijian Zhang
Water Resources Information Center of Henan Province, Zhengzhou 450003, China

Chen Ding, Ying Li, Yong Xia, Wei Wei, Lei Zhang and Yanning Zhang
Shaanxi Key Lab of Speech & Image Information Processing (SAIIP), School of Computer Science and Engineering, Northwestern Polytechnical University, Xi'an 710129, China

Haoning Lin, Zhenwei Shi and Zhengxia Zou
Image Processing Center, School of Astronautics, Beihang University, Beijing 100191, China
Beijing Key Laboratory of Digital Media, Beihang University, Beijing 100191 China
State Key Laboratory of Virtual Reality Technology and Systems, School of Astronautics, Beihang University, Beijing 100191, China

Lianru Gao
Key Laboratory of Digital Earth Science, Institute of Remote Sensing and Digital Earth, Chinese Academy of Sciences, Beijing 100094, China

Bin Zhao and Bing Zhang
Key Laboratory of Digital Earth Science, Institute of Remote Sensing and Digital Earth, Chinese Academy of Sciences, Beijing 100094, China
University of Chinese Academy of Sciences, Beijing 100049, China

Xiuping Jia
School of Engineering and Information Technology, The University of New South Wales, Canberra Campus, Bruce ACT 2006, Australia

Wenzhi Liao
Department of Telecommunications and Information Processing, Ghent University, Ghent 9000, Belgium

Index

www.ingramcontent.com/pod-product-compliance
Lightning Source LLC
LaVergne TN
LVHW060523210926
848485LV00005B/1137